Plant Resistance to Insects

Paul A. Hedin, EDITOR
U.S. Department of Agriculture

Based on a symposium
sponsored by the ACS
Division of Pesticide Chemistry
at the 183rd Meeting of the
American Chemical Society,
Las Vegas, Nevada,
March 28–April 2, 1982

ACS SYMPOSIUM SERIES **208**

AMERICAN CHEMICAL SOCIETY
WASHINGTON, D.C.   1983

Library of Congress Cataloging in Publication Data

Plant resistance to insects.

(ACS symposium series, ISSN 0097–6156;208)

"Based on a symposium sponsored by the ACS Division of Pesticide Chemistry at the 183rd meeting of the American Chemical Society, Las Vegas, Nevada, March 28–April 2, 1982."

Includes bibliographies and index.

1. Plants—Disease and pest resistance—Congresses. 2. Insects—Host plants—Congresses. 3. Insects, Injurious and beneficial—Congresses. I. Hedin, Paul A. (Paul Arthur), 1926–   . II. American Chemical Society. Division of Pesticide Chemistry. III. American Chemical Society. National Meeting (183rd: 1982:Las Vegas, Nev.) IV. Series.

SB750.P57   1983   632'.7   82–22622
ISBN 0–8412–0756–9   ACSMC8   208   1–375
                                        1983

ACS Symposium Series

M. Joan Comstock, *Series Editor*

FOREWORD

The ACS SYMPOSIUM SERIES was founded in 1974 to provide a medium for publishing symposia quickly in book form. The format of the Series parallels that of the continuing ADVANCES IN CHEMISTRY SERIES except that in order to save time the papers are not typeset but are reproduced as they are submitted by the authors in camera-ready form. Papers are reviewed under the supervision of the Editors with the assistance of the Series Advisory Board and are selected to maintain the integrity of the symposia; however, verbatim reproductions of previously published papers are not accepted. Both reviews and reports of research are acceptable since symposia may embrace both types of presentation.

CONTENTS

v

PREFACE

B IOLOGISTS HAVE LONG RECOGNIZED the specificity of various insects for plants and the related interactions between plants and herbivorous insects. A degree of understanding has evolved from the realization that different kinds of insects respond differentially to various secondary chemicals occurring in plants. It has also been recognized that plants and insects coevolve, and the continuing adjustments of one to the other reflect the biosynthesis of defensive compounds by the plant and the development of detoxification or avoidance mechanisms by the insect. The dynamic nature of this relationship is illustrated by the ability of insects to induce detoxifying mechanisms within twenty-four hours when challenged by a toxic agent. Plant injury, in turn, can elicit the biosynthesis of additional quantities of resistance agents.

The resources that plants can muster for defense when attacked by insects are limited. Plants that stand exposed to insects over long periods generally develop defense systems that require a relatively large quantity of the resistance agent(s), and these agents often bind dietary protein or otherwise interfere with digestion. In this case, the costs of defense are relatively large and so the yield may be decreased. Plants that are exposed to pests over a short period generally develop defense systems that involve the biosynthesis of a small amount of a highly toxic agent. Although such defense systems require that the plant divert less energy from yield to defense, in evolutionary time the insect may be able to adapt to the plant and defeat the resistance.

Of the expressions of plant resistance that are chemical, the so-called secondary plant compounds appear to be dominant. In most cases they modify or control insect growth, development, and reproduction, but others such as the antifeedants modify behavior. However, antifeedants may also be toxicants and toxicants may also be antifeedants; thus their designation is a function of the bioassay employed. Not all compounds toxic to one insect are toxic to another. In evolutionary time, some insects develop mechanisms by adaption to detoxify compounds in plants on which they must feed whereas others do not. As a result, a compound toxic to one insect may be a feeding stimulant for a second. To the extent that the biosynthesis of the compounds is an expression of genetic information, the elucidation of the compounds and their roles can provide a guide to selections by plant breeders. Additionally, as genetic studies become more sophisticated, the assignments of the role of individual genes in directing biosynthesis of resistance compounds will be expediated. Eventually, improved knowledge about genetic engineering will provide the technology for introducing protective genes into crop plants, thus creating resistant lines.

The book is divided into four sections: Ecological and Histochemical Aspects, Biochemical and Physiological Mechanisms, Insect Feeding Mechanisms, and Roles of Plant Constituents. It is hoped that this volume will help to identify unifying themes by which plants express resistance to insects. With an ever increasing world population exerting greater and greater pressure on food production systems, every potential breakthrough is of critical importance.

As the editor, I am grateful to all of the participants for their contributions to this book. I am also grateful to the U.S. Department of Agriculture for a financial grant.

PAUL A. HEDIN
U.S. Department of Agriculture
Mississippi State, MS 39762

September 30, 1982

ECOLOGICAL AND HISTOCHEMICAL
ASPECTS

Patterns in Defensive Natural Product Chemistry: Douglas Fir and Western Spruce Budworm Interactions

REX G. CATES, RICHARD A. REDAK, and COLIN B. HENDERSON

University of New Mexico, Chemical Ecology Laboratory, Department of Biology, Albuquerque, NM 87131

Studies were undertaken to test various aspects of current plant-herbivore theory. Particular emphasis was given to investigating the effects of Douglas-fir foliage quality on the success of the western spruce budworm. Results indicated that the most important variables influencing budworm success were the concentrations of several specific terpenes and changes in the distribution of foliar terpenes. Additionally, the productivity of trees was inversely correlated with budworm success. The available foliar nitrogen did not appear to be an important factor determining insect success. Finally, escape in time appears to be another factor limiting insect utilization of newly emerged Douglas-fir tissue.

Some of the major objectives in coevolutionary biology are to explain the patterns of interaction between plants and herbivores, the selection pressures maintaining these patterns, the ways in which these patterns differ among plant communities, and how they might change over ecological and evolutionary time (1). The recently developed theory of plant phytophagous insect interactions suggests that the population biology of phytophagous insects is influenced significantly by the diversity of plant species within a community, the intertwining relationships among individual plants, the predictability of the plant resources in space and time to herbivores, the nutritional levels of plant tissues of various growth forms, and the diversity of mechanical and chemical defense mechanisms of plant tissues (1-6). This theory has sparked considerable research effort within several different areas. Appropriate investigations, however, of the ways in which the above factors interact in molding plant and herbivore life histories and community structure are only beginning to surface in the literature (3,4,7-9).

Since we wish to address various aspects of current plant-herbivore theory, a brief discussion of some of the basic premises follows, emphasizing those that pertain to woody

0097-6156/83/0208-0003$06.00/0
© 1983 American Chemical Society

perennials and their associated herbivores. Rhoades and Cates
(6) suggest that plant resources for herbivores vary along a
continuum of increasing predictability. Resources such as
long-lived woody perennials, or mature, evergreen leaf tissues,
are suggested to be predictable, or in Feeny's (5) terminology,
apparent, resources to herbivores. Examples, on the other hand,
of the unpredictable or unapparent resources that may vary
greatly in their occurrence and availability to herbivores are
the ephemeral annuals or the quickly developing young leaf
tissues of perennials (6).

These differences in plant and tissue predictability and
availability as a food resource to herbivores, along with various
tissue developmental constraints, are suggested to be important
determinants of the defensive system evolved by a plant (1,6).
Although plants produce many secondary products, chemical
defensive compounds can be categorized on the basis of function
as toxins or digestibility-reducing substances. Toxins are
treated usually as small molecular weight compounds that affect
the metabolic processes within herbivores and include cyanogenic
glycosides, cardiac glycosides, terpenes, alkaloids, and numerous
others. Digestibility-reducing substances, on the other hand,
are often large molecular weight substances capable of bonding
with proteins and polysaccharides resulting in a complex that may
be difficult to digest. These two classes are not absolutely
distinct but represent a continuum, and in some cases, compounds
may function both as a toxin as well as a protein complexing
substance (6).

The available data suggest that toxins, or qualitative
defenses, are characteristic of ephemeral plants and plant
tissues, and that digestibility-reducing substances, or
quantitative defenses, appear to be characteristic of predictable
plants and plant tissues (5,6). Furthermore, unpredictable plant
resources are thought to escape in space and/or time from their
herbivores to a larger extent than do predictable resources (6).
This escape in space and/or time may be more effective against
specialized or monophagous-oligophagous herbivores than against
the generalized or polyphagous herbivores since the former lack
alternate food resources.

Our objective is to examine some aspects of current plant
herbivore theory using Douglas-fir (Pseudotsuga menziesii) and
western spruce budworm (Choristoneura occidentalis). Both plant
and herbivore are widespread in western North America. Natural
hosts of the budworm include Douglas-fir, species of Abies, and,
on occasion, other conifers (9). Variation in budworm density
occurs on both a geographic and local scale. We have frequently
observed differential defoliation in trees having overlapping
crowns at sites in Montana, Idaho, and New Mexico.
Interestingly, considerable geographic variation is known to
exist in the terpene chemistry of Douglas-fir (10,11). Our

preliminary data show that this variation also exists at a local
level (Table I).

Table I.   Variation in some of the chemical constituents in the
           young needles of Douglas-fir from sites at Boulder,
           Montana (MT) and Taos, New Mexico (NM).

Constituent	Site	x̄	SD	Value Min	Max
Total Nitrogen	MT	1.8	0.5	1.3	3.5
	NM	1.3	0.1	1.1	1.4
α-pinene	MT	176.5	223.6	0.0	1001.0
	NM	105.1	48.4	34.0	221.0
Camphene	MT	62.9	63.3	0.0	211.0
	NM	190.8	84.4	67.0	351.0
β-pinene	MT	134.8	93.7	0.0	429.0
	NM	75.9	52.5	3.0	221.0
Limonene	MT	84.7	69.2	0.0	403.0
	NM	81.5	38.0	24.0	193.0
Bornyl Acetate	MT	63.1	38.4	6.0	189.0
	NM	92.6	42.1	26.0	190.0
Cadinene Isomer	MT	6.9	5.4	0.0	33.2
	NM	4.4	8.1	0.0	32.0
Total Terpenes	MT	713.6	502.1	147.8	2442.5
	NM	616.6	270.2	246.0	1119.0
Polyphenols	MT	0.4	0.2	0.1	1.0
	NM	16.6	9.1	0.0	40.0

Polyphenols = % fresh wt; nitrogen = % dw; terpenes =
counts/20 mg tissue

Objectives

Our focus was upon natural and experimental studies that
were designed to investigate various aspects of the current
plant-herbivore theory using the Douglas-fir/western spruce
budworm system. Specifically, we wished to test whether there
are chemical characteristics in the young needle tissue of
Douglas-fir that reduce the growth, survival, and adult fecundity
of the western spruce budworm. Secondly, we were interested in

elucidating the relationship between budworm success and physical attributes of Douglas-fir. Finally, we desired to determine whether there are any changes in foliage quality due to increased water stress that could account for variability in budworm success.

Resistance is defined as the suite of "heritable characteristics by which a plant species, race, clone or individual may reduce the probability of successful utilization of that plant as a host by an insect species, race, biotype, or individual" (2). The important aspect of this concept is that resistance characteristics are heritable, even though they may be modified by the physical environment. With regard to the above considerations and questions, emphasis was placed upon determining the effects of variation in nutritional (water, nitrogen sources) as well as secondary chemical characteristics (terpenes, resin acids, polyphenols, digestibility-reducing capacity) in the foliage of Douglas-fir trees upon budworm larval growth, survival, density, level of defoliation, adult dry weight, and fecundity. Several physical and phenological measurements of the trees were made at each site including age, height, dbh, crown diameter, crown ratio, bole radius, five year growth increment, average internode length, xylem pressure potential, and time of budburst.

Site Selection and General Methodology

Resistance-Susceptibility Studies

Studies to determine the level of resistance or susceptibility due to foliage quality and tree physical parameters of Douglas-fir to the budworm were conducted at the Boulder, Montana and Barley Canyon, New Mexico sites. The study at the Montana site involved trees exposed to moderate budworm densities, while the study at Barley Canyon was done using budworm that were placed on trees that had very low natural densities of budworm.

Near Boulder, Montana in the Deer Lodge National Forest a site was selected that consisted of a near monoculture of Douglas-fir in which were scattered ponderosa pine, lodgepole pine, and junipers. During May and early June, 1979, 147 trees were located and tagged at the site which was at an elevation of 1620 m. The initial selection of these trees was done such that all were of similar age, height, dbh, and crown diameter. Trees were selected that were growing in similar microenvironments (proximity to drainage, angle and attitude of slope, and distance to neighboring trees). After the initial selection a more exact determination of the above physical parameters was made, and from the initial 147 trees, 80 were selected for this resistance-susceptibility study.

Natural budworm densities were determined by sampling 6 sprays, each 40 cm long, in the same quarter of the tree used to collect tissue for chemical analysis and to collect defoliation data. Densities were expressed as the average number of budworm larvae per 100 buds per tree. A visual estimate of the amount of defoliation also was made in the same area of the crown where the densities and needle tissue were collected. Since budworm may disperse from heavily defoliated trees, (Greenbank, 1963) budworm densities from each tree were weighted by the level of defoliation that each tree sustained. This resulted in an infestation intensity measurement (dependent variable) which was subjected to multiple stepwise correlation analysis using various foliage quality and physical tree parameters as the independent variables. Thirty-one parameters were used as independent variables in this analysis.

Fifteen to 20 g of the young or current year's foliage were collected during the time period that corresponded to the 4th and 5th larval instar. The tissue was frozen, and returned to the laboratory at the University of New Mexico for analysis. Polyphenols and protein complexing capacity were determined by precipitating the polyphenols, weighing this fraction, redissolving in 20% acetone, and finally, measuring the protein complexing capacity of the extract using a 1.5% buffered gelatin solution. This method, with slight modification, is taken from Feeny and Bostock (13) and Feeny (14).

Total nitrogen was determined using standard microkjeldahl digestion, and terpenes were analyzed using a Perkin Elmer 3920 gas chromatograph equipped with a Perkin Elmer 0.10 in x 150 ft capillary column packed with 85% OS-138, 14% CO-880, and 1% V-930. The method for terpene analysis followed Redak (1982) with only minor modifications in column loading time. Dr. D. F. Zinkel (Forest Products Laboratory, USDA, Madison, Wisconsin) kindly analyzed the young foliage for resin acids. The resin acids were not detected and no detailed analysis of the needles for these chemicals was done in any of our studies.

The resistance-susceptibility study, where budworm larvae were placed on trees that naturally had few, if any, budworm, was conducted in 1980 at the Barley Canyon site in the Santa Fe National Forest. This site was approximately 3.2 km long and 0.5 km wide at an average elevation of 2440 m. The dominant vegetation included Douglas-fir, white fir, ponderosa pine, and aspen. Tree selection was identical to that described for the Montana site except that initially 200 trees were selected. From this group 105 trees eventually were used for the study reported here. Thirty-four of the 105 were trenched in 1980 to induce water stress in 1981. One-way multiple analysis of variance showed no significant difference in the measurements of 34 foliage quality and tree physical characteristics between the 34 trees that were trenched in 1980 and the remaining 71 trees (15).

Consequently all 105 trees were used in the analysis reported here.

Larvae were collected from a nearby infested area and transported in the cool of the evening to the University of New Mexico laboratory (15). Approximately 3000, 4th instar larvae were collected from the infested site and placed in vials containing a small amount of young Douglas-fir tissue. Each vial contained 25 larvae. Vials were then returned to the environmental chamber to minimize larval stress. The following day the experimental larvae were placed on the selected trees at the Barley Canyon site. Five larvae were placed on each branch that had at least 10 new, expanding foliage buds, and were contained on the branches using screen sleeves. Seventy percent of full sunlight penetrated these sleeves. Five sleeves, each with five larvae, were placed on each tree to yield a total of 25 larvae per tree. The larvae were allowed to pupate before transferring them to the laboratory where adults emerged and were sexed. Larval survivorship and adult dry weight could then be determined. In addition, for 81 randomly selected females, egg mass dry weight and numbers of eggs per female were determined to examine the relationship between fecundity and female dry weight (15).

During the 5th and 6th instar, approximately 30 g of foliage were collected from the same side and in the same midcrown level where the larvae were placed. The foliage samples were put in zip-loc bags, transported on ice to the laboratory, and frozen until analyzed for nitrogen content and protein complexing capacity. Nitrogen was analyzed as above. Protein complexing capacity was determined using the method described by Bate-Smith (16) with minor modifications (15). For the analysis of foliage terpene content, 30 mg of young foliage were collected and weighed in the field, encapsulated in indium tubing, placed on ice, transported to the laboratory, and frozen until it was analyzed for terpene content. Terpenes were analyzed as described by (15).

A total of 34 variables were used in the multiple stepwise correlation analysis. Twenty-three were used to determine foliage quality, while 11 variables were used to define the physical and phenological attributes of the sample trees. The dependent variables used were average adult female budworm dry weight and average adult male budworm dry weight for each tree. Details are found in Redak (15).

Water Stress Studies

Studies to determine the effect of water stress on foliage quality and budworm success were conducted at 2 sites west of Jemez springs, New Mexico, Santa Fe National Forest. These sites were chosen such that differences in water availability to Douglas-fir trees would be maximized. Thirty trees were selected that were growing on a north facing slope (non-stress), while another 30 were selected from a south facing slope (stress site). Analysis of preliminary data suggested that these trees did not differ initially in foliage quality characteristics. Additionally, at the stress site, about a third of the area beneath each tree was trenched to cut roots in an effort to further maximize water stress. Xylem water potentials were measured using a Scholander pressure bomb when the larvae were in their 5th instar. Budworm larvae were placed on each of the trees at the sites as described above for the Barley Canyon study. Terpenes and nitrogen were analyzed as described above. Polyphenols and protein complexing capacity were measured as described for the Montana site.

Female adult budworm dry weights and the number of survivors of budworm were analyzed by multivariate analysis of variance to test for the effects of site and sex. Stepwise discriminant analysis was used to determine if tree chemical and physical parameters differed between sites (17).

Results

Resistance - Susceptibility Studies Using Natural Budworm Densities and Defoliation

Results of the data gathered at the Montana site indicate that 50% of the variation in the natural budworm infestation intensity variable was explained by 9 variables (Table II). The acetate fraction of the terpenes, myrcene, an unidentified terpene, time of budburst, and bole radius were inversely correlated with budworm infestation intensity, indicating that some aspects of foliage quality may confer resistance against the budworm. The evenness in the quantitative distribution of the terpenes in the foliage among the trees, beta-pinene, total foliar nitrogen, and tree age were all correlated positively with budworm infestation intensity. Examination of the standardized correlation coefficients indicated that the acetate fraction of the terpenes, the quantitative distribution of the terpenes in the foliage among trees, age, and time of budburst were the most important of the included variables in explaining the variation in infestation intensity.

The evenness measurement, calculated from the Shannon-Wiener formula, suggests that trees which have an uneven distribution of terpenes are more resistant to the budworm. It is likely that this imbalance in the terpene distribution is represented by the specific terpenes (acetate fraction, myrcene, and the unidentified terpene) that were found to be important in the analysis. The analysis also indicated that the polyphenol and protein complexing capacity of the extracts from the foliage

were not important in determining the budworm densities and
levels of defoliation.

Studies Using Experimental Female Budworm Levels on Non-Infested Trees

Results of the data gathered at the Barley Canyon site,
where budworm were placed on trees that naturally had few larvae,
indicated that terpenes again were important in determining

Table II. Multiple correlation analysis using infestation
intensity as the dependent variable $R^2 = 0.50$, p <
0.001). Boulder, Montana site, 1979.

Independent Variable	Coefficient	Standard Error	Standardized Regression Coefficient
Acetate Fraction	−0.17828	0.0432	−0.547
Evenness in Terpene Distribution	45.85419	15.7133	0.389
Tree Age	0.38511	0.1327	0.339
Budburst	−3.05608	1.00207	−0.331
Bole Radius	−1.74379	0.6490	−0.302
β-pinene	0.02624	0.0138	0.246
Total Nitrogen	5.17278	2.3350	0.239
Myrcene	−0.03534	0.0162	−0.227
Unidentified Terpene No. 10	−0.22199	0.1247	−0.193

Nitrogen = % dw; Terpenes = area counts/20 mg dw

budworm success (Table III). When the unidentified terpene,
total nitrogen, beta-pinene, and myrcene were in high
concentration in the needles of Douglas-fir trees, the adult
female dry weights were reduced significantly. The five year
growth increment also was inversely related with female dry
weight. When bornyl acetate, terpinolene, and geranyl acetate
were higher in the foliage, budworm success increased. Also,
budburst, tree age, and twig internode length for 1980 were
associated positively with budworm success.
      Examination of the standardized correlation coefficients
indicated that bornyl acetate, the unidentified terpene, total
nitrogen content, and beta-pinene were the most important
variables in determining female adult dry weight. It is
interesting to note that once more protein complexing capacity of
the extract was not important in determining female dry weight.
      Interestingly, several more variables were included in the
model dealing with male budworm success on the trees at the
Barley Canyon site (Table IV). Terpinolene, citronellyl acetate,

alpha-pinene, bornyl acetate, myrcene, an unidentified terpene,
and crown ratio were inversely related with average adult male
dry weight production per tree. Positively correlated with male
dry weight were limonene, young needle water content, gamma-
terpinene, 3 unidentified terpenes, the ratio of soluble to
insoluble nitrogen, and the amount of twig internode growth in
1980. The most important of these 15 variables in influencing
male adult dry weight production were limonene, terpinolene,
citronellyl acetate, alpha-pinene, and bornyl acetate. Once
again, protein complexing capacity explained none of the
variation in average male adult success.

Table III.   Multiple correlation analysis using adult female
              dry weight as the dependent variable. (R^2 = 0.35;
              F = 3.31; p < 0.001). Barley Canyon, New Mexico
              (Redak, 1982).

Independent Variable	Coefficient	Standard Error	Standardized Regression Coefficient
Bornyl Acetate	0.00334	0.00122	0.354
Unidentified Terpene No. 5	-0.04165	0.01862	-0.262
Total Nitrogen	-114.75060	51.62314	-0.232
β-pinene	-0.00541	0.00290	-0.229
Budburst	0.41324	0.22878	0.193
Terpinolene	0.04044	0.02703	0.172
Myrcene	-0.01387	0.000948	-0.165
Tree Age	0.16913	0.12241	0.145
Geranyl Acetate	0.00436	0.00376	0.137
Five Yr Growth Increment	-1.59537	1.31204	-0.133
1980 Internode Length	0.48272	0.37711	0.133

Nitrogen = % dw; Terpenes = Area counts/20 mg dw

Water Stress and Budworm Success

Hypotheses concerning drought stress and its influence on
budworm success and changes in foliage quality were tested using
Douglas-fir trees growing on south facing (stress) and north
facing slopes (non-stress). We first demonstrated that the
degree of water stress was different between the 2 groups of
trees. Xylem pressure potentials averaged 23% higher for the
trees at the stress site when compared to the non-stress site
(p < 0.001).

Next, the male and female adult dry weights and the number
of budworm surviving per tree between the 2 sites were compared.
Table V indicates the means for each of these variables
subdivided by site and sex. These data were used in the
multivariate analysis of variance, and are presented in this
table as a point of reference. Since no dependency was found
between weight and number of survivors (r = 0.158; p = 0.12),
these variables were subjected to multivariate analysis of
variance to test for the effects of site and sex (Table VI).
Budworm success, as measured by adult dry weight and number of
survivors, was found to be significant for both site and sex,
but there was no interaction between these factors. The
relative contributions of weight and numbers of survivors to
the differences between sites and sexes are illustrated by the
characteristic vector coefficients. Not surprisingly, they
indicate that differences between sexes were due to differences
in weight between males and females. The significant result
was that differences between sites were due to differences in
weight and the number of survivors per tree.

Stepwise discriminant analysis was used to determine how
tree chemical, phenological, and physical parameters differed
between sites (Table VII). Only seven of the 18 variables used
were needed to completely differentiate the trees at the 2
sites ($F_{(7,193)}$ = 210.36; p < 0.001). The magnitudes of the
standardized discriminant function coefficients for the
included variables indicated that the differences between sites
were largely due to terpene chemistry (Table VIII). The
discriminant function contrasts primarily the relative
concentration of alpha-pinene versus the concentration of
several terpenes, particularly bornyl acetate and beta-pinene.
Examination of the discriminant scores showed that the stressed
trees loaded negatively on the function ($\bar{x}$ discriminant score =
-2.23), while the non-stressed trees loaded positively ($\bar{x}$
discriminant score = 3.38). In other words, trees from the
stressed site were higher in alpha-pinene while the
non-stressed trees contained more bornyl acetate, beta-pinene,
and other terpenes in their young needles.

Discussion

Douglas-fir Foliage Quality and Resistance to Budworm

Data from all three studies show that, in all cases, the
terpene chemistry of young foliage, or qualitative defenses,
was the most important factor in reducing budworm success. The
protein complexing capacity, or quantitative defenses, of this
tissue was not important in reducing budworm success in any of
the studies.

Table IV. Multiple correlation analysis using adult male dry weight as the dependent variable. (R^2 = 0.35; F = 2.49; p < 0.005). Barley Canyon, New Mexico (Redak, 1982).

Independent Variable	Coefficient	Standard Error	Standardized Regression Coefficient
Limonene	0.00512	0.00200	0.461
Terpinolene	-0.02296	0.00999	-0.300
Citronellyl Acetate	-0.00335	0.00144	-0.256
α-Pinene	-0.00079	0.00055	-0.234
Bornyl Acetate	-0.00062	0.00040	-0.202
Myrcene	-0.00526	0.00315	-0.192
Water Content	6.74118	4.43527	0.177
γ-Terpinene	0.00825	0.00543	0.167
Unidentified Terpene No. 8	-0.02614	0.01863	-0.159
Unidentified Terpene No. 5	0.00770	0.00598	0.149
Crown Ratio	-5.52496	4.97238	-0.124
Unidentified Terpene No. 9	0.00946	0.00831	0.121
Soluble/ Insoluble Nitrogen	0.71473	0.68189	0.119
Unidentified Terpene No. 4	0.00658	0.00686	0.104
1980 Internode Length	0.11978	0.12284	0.102

Nitrogen = % dw; Terpenes = Area Counts/20 mg dw

Table V. Mean adult dry weight and number of budworm surviving per tree on stressed and non-stressed sites. These data were used in the multivariate analysis, the results of which are given in Table VI (Cates et al., 1982).

Site	Sex	N	Weight (mg)	Number Survived
Non-stressed	Male	22	8.95	3.73
	Female	23	18.18	3.04
Stressed	Male	27	10.73	4.63
	Female	26	23.64	5.15

Table VI. Results of multivariate analysis of variance for
           the effects of site and sex on adult dry weight and
           number of survivors. Characteristic vector co-
           efficients indicate the relative contribution of
           the dependent variables to a particular effect
           (Cates et al., 1982).

| Source | $F_{(2,93)}$ | P | Characteristic Vector Coefficients | |
			Number	Weight
Site	10.78	0.0001	0.022	0.018
Sex	73.97	0.0001	−0.007	0.023
Site x Sex	2.42	0.0941	0.018	0.019

Table VII. Chemical, phenological, and physical
            characteristics of Douglas-fir trees that were
            subjected to discriminant analysis.

Nitrogen	Terpenes
Total	13 Individual Compounds
Soluble	Total
Protein Complexing Capacity	Tree Age

Table VIII. Standardized discriminant function coefficients for
             the 7 variables resulting from the analysis of
             chemical and physical parameters among trees
             growing on the stressed and non-stressed sites
             (Cates et al., 1982).

Variable	Standardized Discriminant Function Coefficient
α-pinene	−5.84
Soluble Nitrogen	−0.71
Age	0.34
Unidentified terpene 1	1.22
Unidentified terpene 2	1.66
β-pinene	2.57
Bornyl Acetate	2.81

Nitrogen was found to be of little importance except for
the female model generated from the Barley Canyon Study. In
this case, female success was inversely correlated with foliar
nitrogen levels. However, these levels, based on published
data (18,19), do not appear high enough to be toxic to the
budworm. This observation, in view of the inverse correlation
between total nitrogen and female budworm success, leads us to

suggest that this relationship is more a reflection of tree productivity than its role as a primary nutrient. In other words, higher levels of foliar nitrogen in non-stressed trees that were growing on good quality sites may have been indicative of overall tree productivity and vigor (19,20,21). These trees would be better able, providing they were genetically predisposed, to produce effective defensive chemistry against the budworm. The inverse relationship between female budworm success and the 5 year growth increment, as well as the positive relationship between tree age and female budworm success, are consistent with this hypothesis. Productive trees should be more vigorous, grow better, and hence, possess wider annual rings. The smaller width of annual rings indicates that older trees may be less productive, suggesting a reduced ability to produce effective defensive chemistry.

All of our results indicate that increased productivity is associated with reduced budworm success. In the Montana study this was evidenced by the relationship between budworm infestation intensity and bole radius. In the Barley Canyon study, the relationship between budworm growth and crown ratio, the 5 year growth increment, and total nitrogen support this conclusion as well. Assuming that productivity declines with age, the positive correlation between budworm success and age also implies that a decline in vigor increases tree susceptibility to budworm.

Additionally, our data suggest that Douglas-fir may reduce the risk of damage to new tissue through escape in time. In Montana, trees which burst bud later in the growing season suffered less damage from budworm. This is consistent with theory (1,5,6) which suggests that unpredictability in space and time may be an effective mechanism for reducing the adverse effects of monophagous-oligophagous herbivores on young ephemeral tissues. On the other hand, trees at Barley Canyon show a positive relationship between budburst and budworm size. This appears to be in direct contradiction with the Montana data and theory, until the experimental design is considered. At Barley Canyon, budworm were placed on trees after all had burst bud. Consequently, most budworm were subjected to tissues that had developed for up to 10 days and which possessed more complete defensive systems. Therefore, the correlation we observed with budburst reflected the length of time the foliage was allowed to mature defensively before the budworm feeding occurred. It did not reflect an escape in time component as the Montana study did where natural populations were studied.

Water Stress and Budworm Success

The study involving water stress in the Jemez mountains suggests that changes in foliage quality due to stress positively influences budworm success (17). Female adult budworm from larvae reared on stressed trees were 30% heavier, while the male adults were 20% heavier. In addition, a higher number of both sexes survived on stressed trees (Table VI). The water stress incurred by trees growing on the south site is hypothesized to have modified foliage quality such that trees became highly susceptible to the budworm. This appears to have been primarily due to the decrease in the resistance factors beta-pinene and bornyl acetate accompanied by an increase in alpha-pinene. If alpha-pinene consistently increases in stressed Douglas-fir trees, it is then possible that budworm may use it as an attractant or cue in locating suitable hosts.

The primary emphasis in the literature has been to show the importance of nitrogen in facilitating outbreaks of phytophagous insects (22). More recently, it has been suggested that changes in defensive chemistry are as important as the changes in primary nutrients (23,24). In our studies, nitrogen was not nearly as important as were several terpenes. Budworm success was most strongly associated with changes in terpene defensive chemistry, particularly in the reduction of resistance factors, rather than with changes in the primary nutrition of the herbivore.

Implications for Current Plant-Herbivore Theory

Ephemeral tissues, such as the young needles of Douglas-fir, are suggested to be defended by toxins or qualitative defenses (5,6). Because these tissues are under strong selection for rapid development and growth to a level where they are contributing to the net primary productivity of the plant, selection is postulated to favor a defensive system that does not place a further burden on an already strained energy budget (1,6). Hence, quantitative defenses that require considerable energy to produce, and must be compartmentalized so that they do not interfere with metabolic processes, are postulated to be a minor or missing defensive system in ephemeral tissues. Qualitative, or toxin defensive systems are predicted in these tissues instead. Exceptions to this general prediction may possibly include the young leaves of evergreen trees or shrubs which should show slower growth when compared to short-lived tissues of non-evergreen plants.

While quantitative defenses or digestibility reducing substances are also present in small quantities in the young needles, they do not seem to be effective in reducing female dry weight, larval density, or level of defoliation. This is consistent with the reasoning that young tissue development constraints render difficult the production of a well developed digestibility reducing system.

As discussed in the previous section, we have observed that escape in time, another prediction of current plant-herbivore theory, exists making newly emerged foliage less predictable to the budworm. Some trees may burst bud as late as 10-14 days after the first trees within a stand, and as our data show, this delay is associated with reduced damage by the budworm.

Slansky and Feeny (25) and Feeny (5) suggest that once qualitative defenses are overcome they "affect larval growth to a much lesser extent than do the nutrient characteristics of food plants," and that "once overcome by specific adaptation, they may have little affect on growth or fitness." Our data do not corroborate these predictions. Budworm success was reduced significantly by the presence of higher quantities of terpenes in the young needles of some trees. Additionally, in all 3 studies, nitrogen content was not as important as were the terpene contents of the foliage in influencing budworm success.

In all three studies, but particularly in the 2 dealing with resistance-susceptibility characteristics of foliage quality, specifically the higher quantities of terpenes were correlated with reduced budworm success. At any particular site all of the trees sampled contained the same basic complement of terpenes, albeit, some were present in low quantities. The effect, which we assume is one of toxicity, of the terpenes in reducing budworm success, then, was due to an increase in their quantitative amounts. In other words, the terpenes were acting in a quantitative or dosage dependent fashion.

Qualitative defenses in Douglas-fir also may be successful against the relatively specialized budworm because of the diversity of terpenes that are present. In other words, a large number of qualitative toxic combinations are available for selection to act upon in Douglas-fir. Interestingly, we found that in the Montana study the greater the imbalance in the terpene distribution in the foliage, the less well the budworm performed. This suggests that the more a tree's defensive chemistry deviates from a balanced, or "average" pattern, the less likely it is to be attacked by the chemically well adapted budworm. Additionally, adaptation by budworm to the particular imbalanced terpene pattern of one host tree would confer an advantage to any tree whose terpene pattern was predominated by other terpenes.

Thus we feel that the effect of selection by the budworm for qualitative terpene defenses would be to produce a number of "chemical pattern phenotypes" among individuals at a local or intrapopulational level. Chemical diversity should exist among populations as well, and this has been suggested in our work as well as by others (10,11). Such selection for

diversity in the host plant chemistry should produce races of
budworm that are generally adapted to the average defensive
chemistry of their host plants but that may not be well adapted
to the chemistry of Douglas-fir from other populations. If
this scenario is correct, then some very interesting
silvicultural practices in managing forests could result using
defensive natural product chemistry along with other control
measures in an integrated pest management system.

In a sense, many of the data presented from our work are
preliminary since the above discussion is based on correlation
analysis. Synthetic diets incorporating compounds that are
suggested by these studies to adversely influence budworm
success, as well as other experimental work with natural
populations of Douglas-fir and budworm, are needed to determine
cause and effect. A thorough examination of the natural
product chemistry is needed as some other toxic compounds may
be influencing significantly the patterns being observed.
Considerable effort needs to be given to various aspects of
budworm biology. But in the final analysis, these data suggest
that the Douglas-fir-budworm system may be very useful in
unraveling various aspects of plant-herbivore interactions, and
in understanding ways in which we might better manage our
forests against phytophagous insects.

Acknowledgements

We should like to thank T. I. McMurray, H. J. Alexander,
M. Alexander, J. Horner, M. Freehling, and numerous
undergraduate students for their help and for their constant
questioning and discussion of various aspects of the projects.
We are grateful to Clifford S. Crawford and Fritz Taylor for
providing information and discussion of budworm population
biology. Doug Parker and Mike Chavez (USDA FS, Region 3,
Albuquerque), and Jed Dewey, John Hard, and Larry Stipe (USDA
FS, Region 6, Missoula, MT), were helpful in numerous ways but
particularly in helping us locate appropriate sites. We are
extremely grateful to Linda L. DeVries for typing the
manuscript several times. This research was supported in part
by NSF grants DEB 7619950 and DEB 7927067 to RGC for which we
are grateful. Work leading to this publication was funded by
the Canada/United States Spruce Budworms Program, and
Accelerated Research, Development and Application Program,
sponsored by the USDA Forest Service.

Literature Cited

1. Cates, Rex G.; Rhoades, D. Biochem. Syst. Ecol. 1977, 5, 185-93.
2. Cates, Rex G.; Alexander, H. J. In "Bark Beetles in North American Conifers: Ecology and Evolution": J. Mitton and K. Sturgeon, (Eds.); University of Texas Press; Austin, Texas, in press.
3. Cates, Rex G. Oecologia 1981, 48, 319-26.
4. Cates, Rex G. Oecologia 1980, 46, 22-31.
5. Feeny, P. P. In "Biochemical Interactions Between Plants and Insects"; Wallace, J.; Mansell, R., Eds.; Plenum Press: New York, NY, 1976; pp 1-40.
7. Futuyma, D.; Gould, F. Ecol. Monogr. 1979, 49, 33-50.
8. Rosenthal, G.; Janzen, D., Eds.; "Herbivores: Their Interaction with Secondary Plant Metabolites"; Academic Press: New York, NY, 1979.
9. Johnson, P.; Denton, R. USDA For. Ser. Gen. Tech. Rep. 1975, INT-20, 1-144.
10. von Rudloff, E. Pure Appl. Chem. 1973, 34, 401-10.
11. von Rudloff, E. Can. J. Bot. 1972, 50, 1025-40.
12. Greenbank, D. Entomol. Soc. Can. Mem. 1963, 31, 174-80.
13. Feeny, P.; Bostock, H. Phytochemistry, 1968, 7, 871-80.
14. Feeny, P. Phytochemistry, 1969, 8, 2119-26.
15. Redak, R. "A Determination of the Resistance-Susceptibility Characteristics in Douglas-fir to the Western Spruce Budworm"; MS Thesis, Univ. of New Mexico, Albuquerque, NM, 1982, pp 1-74.
16. Bate-Smith, E. Phytochemistry, 1973, 12, 907-12.
17. Cates, Rex G.; McMurray, T.; Redak, R.; Henderson, C. Nature, (submitted).
18. Mattson, W. Ann. Rev. Ecol. Syst. 1980, 11, 119-61.
19. Gosz, J. Ecol. Bull, 1981, 33, 405-26.
20. Miller, H.; Cooper, J.; Miller, J.; Pauline, O. Can. J. For. Res. 1979, 9, 19-26.
21. Van den Driessche, R.; Dangerfield, J. Plant and Soil 1975, 42, 685-02.
22. White, T. Ecology, 1969, 50, 905-9.
23. Rhoades, D. In (8).
24. Mattson, W.; Addy, N. Science, 1975, 190, 515-22.
25. Slansky, F.; Feeny, P. Ecol. Monogr. 1977, 47, 209-28.

RECEIVED September 13, 1982

Physiological Constraints on Plant Chemical Defenses

H. A. MOONEY, S. L. GULMON, and N. D. JOHNSON

Stanford University, Department of Biological Sciences, Stanford, CA 94305

Plants allocate their carbon and nitrogen
resources to a variety of functions including
constructing and maintaining new tissues (root,
shoot, reproductive), storage, and defense from
herbivores. There is a genetically-determined time
schedule of allocation of resources to various
functions which can be modified to a certain
degree by environmental conditions. The
proposition is examined that this time schedule
represents changing allocation priorities within
the plant which results in time "windows" when it
is more susceptible to herbivore damage than
others. Anti-herbivore "strategies" of plants
must be considered in relation to changing
patterns of resource availability to the whole
plant and in relation to allocation of these
resources to enhance reproductive output.

Plant-insect interaction has been considered extensively
from the view point of the chemistry involved in plant defense
(for example reviews in 1), the growth cycle of the plant as it
affects its "apparency" to herbivores (2), and the seasonal
physical and biotic environment of the herbivores (3). Here we
consider the interaction of the plant's physical environment,
that is, its changing resource availability pattern, with the
utilization of these resources for growth, reproduction and
defense against herbivores.The production of plant defensive
chemicals is, optimally, governed by the cost of the chemicals,
their effectiveness at deterring herbivory, the risk of
herbivory, and the cost of herbivory. The effectiveness of
different chemicals at deterring herbivory has been extensively
studied, and theories concerning the risk of herbivory are well
established. Much less is known about the costs of herbivory
or the costs of defensive chemicals. The costs of defensive

0097-6156/83/0208-0021$06.00/0

chemicals are a function of their biosynthetic maintenance
and turnover costs, but also a function of the value of the
elements used in their construction. The value of these
resources will be determined by their importance to other
activities such as reproduction, photosynthesis, nutrient
uptake, or support. Here we assess the consequences, in terms
of herbivore susceptibility, of simultaneous uses of limiting
resources by plants for a variety of necessary functions. We
earlier discussed costs of defensive compounds in relation to
herbivory (4).

Carbon and nitrogen are the two primary resources which can
be allocated to defensive structures and which may also be
limiting in a plant. They are used directly in the
construction of secondary compounds or protective structures,
and carbon is also used as an energy source to construct these
components. The supply of fixed carbon is an integration of
the availability of all the resources limiting photosynthesis
including light, water, and nutrients. Temperature acts
indirectly to control resource availability. Variation in any
of these results in fluctuations in carbon gain and recurrent
cycles of storage and utilization of carbon in different parts
of the plant (Fig. 1). Limitations on the supply of nitrogen
and carbon could result in "priorities" of use within the plant
in order to maximize fitness. In the first part of this paper
we consider general aspects of both nitrogen and carbon
allocation. Then we examine the way carbon allocation
priorities affect defense from herbivores in a particular
vegetation type, the temperate deciduous forest.

Nitrogen

Nitrogen is a component of several classes of compounds
believed to be important in plant chemical defense, principally
cyanogenic glycosides, glucosinolates, alkaloids, certain
proteins, and non-protein amino acids and peptides. The
nitrogen concentration in leaves is often limiting to insect
larval growth (see review by Mattson, 5), and it is a primary
limiting factor to leaf carbon-gaining capacity (6, 7, 8).
Allocation of nitrogen to chemical defense must be considered
within the context of this complex interdependency.

The quantitative commitment of nitrogen to secondary
compounds varies considerably. At one extreme, the nicotine
concentration of tobacco leaves (Nicotiana tabacum) can range
from 0.17 to 4.93% (9). At the upper level, this is equivalent
to nearly 1% leaf nitrogen concentration, since nicotine is
over 17% nitrogen by weight. The total nitrogen concentration
in leaves rarely exceeds 5%, so nicotine may comprise up to 20%
of the nitrogen in the leaf. Not suprisingly, nicotine
production is correlated with fertilizer application (9).

In another example, the cyanogenic glucoside content in the

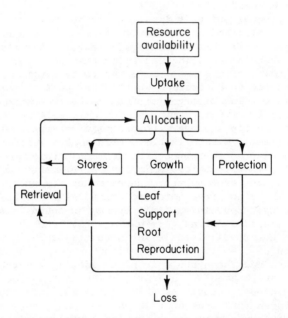

Figure 1.   Flow scheme for a resource within a plant.

leaves of the evergreen chaparral shrub Heteromeles arbutifolia
attains 7% of total dry weight (10). Assuming, as the authors
did, that all of the glucoside was prunasin, this is equivalent
to 0.33% nitrogen by dry weight, or about one-third of the
average total nitrogen in the leaf (11).

On the other hand, most alkaloids contain much less
nitrogen as a fraction of the total weight than nicotine (Table
I), and toxic compounds such as alkaloids are usually found at
very low concentrations in the plant (12). Thus, the
allocation of nitrogen to nitrogenous secondary compounds can
range from substantial to negligible, even with comparable
allocations of carbon.

Another factor which is important in considering the
commitment of nitrogen to chemical defense is the high mobility
of this element within the plant. In forest vegetation, 20-40%
of the nitrogen in senescing leaves is resorbed (13). Nitrogen
is readily moved from senescing leaves to developing fruits
(14), and in deciduous trees a substantial part of spring
growth is made using nitrogen stored in the roots and stem wood
and bark (15, 16).

The nitrogen in secondary compounds also appears to be
readily mobile and metabolizable. When doubly-labeled nicotine
($^{14}C - ^{15}N$) was fed to tobacco (N. rustica), carbon was
recovered in alkaloids, free amino acids, pigments, free
organic acids and free sugars; and nitrogen was recovered in
proteins, alkaloids, free amino acids and pigments (17).
Certain nitrogen-containing toxic compounds found in seeds are
completely degraded during germination (18). Other examples of
alkaloid turnover are given in Robinson (19) and Digenis (20).

Examples of translocation of nitrogen-containing secondary
compounds include the cyanogenic glucoside, linamarin, in the
wild lima bean (Phaseolus lunatus), which is translocated from
seeds to developing seedlings (21). Ricinine in castor beans
is translocated from senescent leaves to young, developing
tissue and to seeds. Caffeine is also translocated among
leaves and into fruits (22). In the case of Heteromeles
arbutifolia, cited earlier, cyanogenic glucosides appear to
move from leaves into developing seeds. McKey (23) gives other
examples of translocation of toxic compounds within the
plant. Thus nitrogen invested in chemical defenses can be
moved throughout the plant to protect vulnerable organs such as
young leaves or fruits, and can later be reused for growth and
development.

Because nitrogenous compounds are readily metabolized or
transported within the plant, it would seem that the major
constraint on use of nitrogen in chemical defense is its
effectiveness and the pattern of its availability in the
habitat. If nitrogen is the major factor limiting to growth,
it is unlikely that large amounts will be diverted to secondary

Table 1. Approximate percentages of nitrogen in some common plant alkaloids

Alkaloids

caffeine	28.9
nicotine	17.3
ricinine	17.0
gramine	16.1
lupinine	8.3
quinine	7.4
mescaline	6.6
colchicine	3.5
tomatine	1 4

Toxic amino acids

βamino propionitrile	40.0
canavanine	31.8
mimosine	14.1

Transport amides

asparagine	21.2
glutamine	19.2

Amino acids

glycine	18.7
alanine	15.7
serine	13.3
proline	12.2
tyrosine	7.7

Cyanogenic glucosides

linamarin	5.7
prunasin	4.7
amygdalin	3.1

compounds, especially since there are many non-nitrogenous
plant products, such as cardiac glycosides, saponins,
flavonoids, quinones, etc., which are also toxic.

Although nitrogen is probably limiting to some extent in
most natural systems, it is not the major limiting nutrient in
all of them, and is not limiting to the same extent at all
times of the year. In large areas of Australia for example,
phosphorus is the principle limiting element (24).

Some plants experience a short-lived abundance of
nitrogen. Most annuals occupy disturbed or semi-disturbed
habitats which are characterized by a sudden release of
nutrients (25). Even many herbaceous perennials occur under
semi-disturbed conditions (eg. old fields), or grow during a
short interval between climatic limitations and biotic
competition when resources are relatively abundant, as is the
case of the forest understory flora (26, 27).

Plant toxins are widely distributed in herbaceous species,
as opposed to protein-complexing polyphenols which are more
limited to leaves of woody perennials (12, 23). This pattern
has been suggested to relate to the low predictability in space
and time of herbaceous foliage in contrast to the foliage of
woody plants (2, 12). However, it should be noted that, at
least in the case of some annuals (9, 28), all of the nitrogen
is taken up during the period of vegetative growth and simply
redistributed during reproduction (Figure 2). Although data
are scarce, this pattern may be widespread among herbaceous
species since, as already stated, they often grow during a
short period of resource availability.

Given a pattern of initial rapid nutrient uptake and
temporal storage in some cases to high concentrations, it is
clearly advantageous to maintain any nitrogen not being used
for photosynthesis or growth in a form that is toxic to
herbivores. Thus, the common occurrence of toxic nitrogenous
compounds in herbaceous plants may simply reflect resource
availability patterns and consequent allocation pathways. This
pattern may also apply to some woody perennials. For example,
chaparral shrubs have periods of active nitrogen uptake in the
fall when growth is not occurring (29). This stored nitrogen
is moved to new growth in the spring and a portion is allocated
to nitrogenous defensive compounds (10).

Carbon

Like nitrogen acquisition, the carbon-gaining capacity of a
plant is environmentally limited. As described earlier, this
limitation results in apparent "priorities" for allocation to
storage, growth of different organs, and elaboration of various
classes of metabolites. These allocation patterns change
during the seasonal growth of an individual. In trees, for
example, root growth generally takes place at times different

from stem elongation which in turn is displaced from stem diameter growth (30, 31) (Figure 3).

Carbon metabolism differs fundamentally from nitrogen metabolism in that virtually all nitrogenous compounds can be recycled within the plant, whereas most of the structural components of the plant, principally cellulose and lignin, are not reusable. There are also putative defensive compounds in this category such as leaf-external resins (32) and possibly some condensed tannins. Thus the constraints on allocation have both an immediate and future time frame.

The interactions between availability of resources, carbon gaining capacity, and the allocation of carbon to plant functions, including chemical defense, will obviously differ among environments. We examine a particular system, the temperate deciduous forest, to illustrate the nature of the constraints on chemical defenses imposed by limitations on carbon gain.

The Deciduous Forest

In spring, 80% of the deciduous forest canopy is produced over a short span of time (31, 33, 34) (Figure 3). Initial leaf expansion, flowering and shoot extension are heavily subsidized from carbohydrate reserves because there is insufficient photosynthetic area to support these processes (35, 36) (Fig. 4). Despite the demand on carbohydrate stores, reserves are not completely depleted and appear to be renewed rapidly following the growth flushes (16). It has been suggested that maintenance of some storage is important in the event the first leaf crop is destroyed due to intense herbivory (37) or a late killing frost (31), and a second crop must be produced.

Deciduous trees vary in the duration of leaf initiation and expansion and new shoot extension (38-41). This variation is typified by two type examples (Figure 5): Quercus species complete all leaf and shoot growth during a short period, usually late April to early June; Betula, by contrast, grows throughout most of the summer. Fagus, Carya, and Fraxinus have growth cycles of the Quercus type, and Liriodendron, Populus, and Liquidamber have extended growth cycles like Betula. Acer species are intermediate. The relationship between duration of stem elongation and dependence on carbohydrate reserves needs to be established for these trees, obviously though oak species which form canopies entirely from pre-formed leaf buds have a heavy drain on carbon reserves for a period.

The young leaves constitute an abundant food resource for phytophagous insects and they are especially vulnerable to herbivores due to a high nitrogen content and lack of sclerophylly. These characteristics appear to be universal in new leaves of deciduous forest species, and this period of

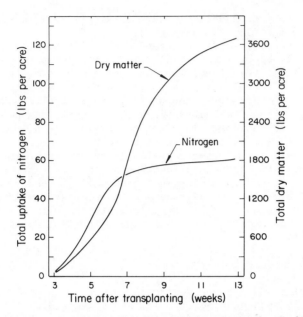

*Figure 2.   Time course of nitrogen and dry weight accumulation in field-grown
tobacco plants. (Reproduced with permission from Ref. 9.)*

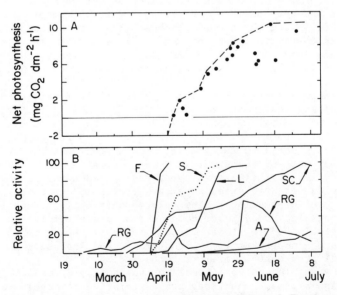

*Figure 3.   Seasonal growth activity of a white oak in Missouri. Development of
canopy photosynthetic capacity is shown in A. Relative seasonal growth activity of
various tree components are given in B. Key: F, flower; S, shoot; L, leaf; SC, stem
circumference; RG, root growth; and A, acorn. (Reproduced with permission from
Ref. 31.)*

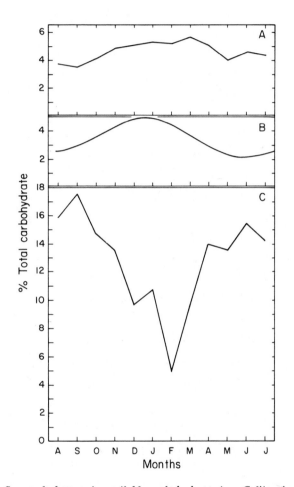

Figure 4. Seasonal changes in available carbohydrates in a Californian evergreen oak (A), a Vermont sugar maple (B), and a Californian buckeye (C). Maple data for sapwood are from Ref. 15. Oak and buckeye data are from Ref. 49, for total branch minus fruits. The evergreen oak has a stable supply of carbon through continuous photosynthesis. The maple draws upon reserves in the spring to build a new canopy. The buckeye draws upon reserves to build fruits in the fall as well as new canopy in late January.

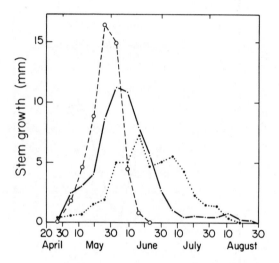

Figure 5.   Seasonal height growth of three deciduous species (39). Key: ○, white ash; − · −, red maple; and ●, gray birch.

vulnerability typically lasts about one month (42, 43, 44) (Figure 6).

Apparently, the potential for reducing this window of vulnerability is fairly limited. The protein content of developing leaves could conceivably be reduced but this would slow the metabolic processes of growth and reduce the gross photosynthetic rate. Both would result in longer development times, prolonging vulnerability, and be a heavier drain on reserves, consequences which would reduce a tree's competitive advantage.

A somewhat similar constraint applies to the early development of sclerophylly, a potential herbivore deterring feature. Construction of rigid cell walls and the production of lignin and other compounds would certainly slow the overall rate of leaf expansion and prolong the period during which the leaf is a net importer of carbon reserves.

We have already mentioned that rapid canopy development is an important competitive advantage and minimizes the time during which young leaves are vulnerable to herbivores. Climate can sometimes be an additional constraint. For example, in the white oak forest in Missouri, leaf development occurs during an interval of optimal growing conditions between the latest killing frost and the midsummer drought (31).

Deployment of defensive chemicals during leaf expansion.

Toxins appear to be effective in small quantities (12), making them an ideal low cost defense against generalist herbivores even though specialists are able to evolve effective detoxification mechanisms. Studies have shown that toxin concentrations are generally higher in new growth (23, 45), but there have been few studies of temperate deciduous trees. It has been shown that generalist herbivores are more common on mature foliage whereas specialist herbivores prefer new foliage (46), but again, there are few studies in the temperate deciduous forest.

Protein-complexing polyphenols are broadly effective against herbivores, especially those not specifically adapted to eat plants containing polyphenols (47). They may constitute 10% or more of the leaf dry weight (12, 37, 44), and this requires allocation of significant amounts of carbon and energy to their construction. The use of limited reserves for such quantitative defenses must be balanced against the requirements of rapid canopy development and new shoot extension.

Patterns of polyphenol accumulation in leaves vary among the few temperate forest species which have been studied. Feeny's (42) classic study on oak leaves (Quercus robur) shows leaf tannin levels rising slowly until August, and then increasing sharply. In the cases of sugar maple (Acer saccharum) and yellow birch (Betula lutea) (44), respectively,

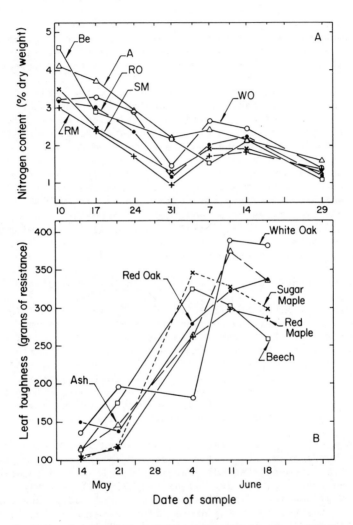

Figure 6.   Seasonal changes in leaf nitrogen (A) and leaf toughness (B) of a series of deciduous tree species. (Reproduced with permission from Ref. 43.)

polyphenol concentrations peak in late June and mid May. In Betula pubescens (48) the concentration of tannins is stable until late summer and then increases.

Feeny (2) has coined the term "apparency" to express the idea that some plants or plant components are more predictable resources in space and time to potential herbivores. He suggests that such predictable plants are more attractive to specialist herbivores (in evolutionary time) because of their predictability, and thus should employ quantitative defensive compounds which are less susceptible to the evolution of detoxification mechanisms.

We propose that the patterns of polyphenol production in leaves are related to the duration of new leaf production. Species which accomplish leaf expansion and shoot extension within a short time period, especially in cold climates, should generally have low polyphenol concentrations in young leaves compared to species with extended periods of active growth.

There are two reasons for the above. First, the new leaves of species with short growth periods are less "apparent" in the sense Feeny (2) has described. They appear early, when predator populations may be inhibited by cold temperatures, and are not present long enough to allow build-up of predator populations. By contrast, new leaves of species with long growth periods are present over an extended period of time, generally much of the summer, and thus constitute a predictable resource for specialized herbivores. Second, new leaf and shoot growth are a greater drain on plant reserves when the entire canopy is produced over a short time period (35). In such species, allocation of 10-25% of new leaf dry weight to polyphenol production would further deplete reserves, slow canopy development, or both. This would decrease the advantage that such species have in terms of escaping herbivory in time, and would also reduce the period of carbon gain which determines reserves and potential growth for the following season. The patterns of polyphenol accumulation in Quercus (42) and Acer and Betula (44) are consistent with this hypothesis, but examination of polyphenol concentrations through time in many more species is necessary. Schultz et al. (44) suggest that the low polyphenol contents of young leaves may be due to the fact that protein synthesis suppresses phenolic synthesis.

To summarize, specific toxins are effective deterrents to generalist herbivores, but polyphenols produced in relatively large quantities are the only effective deterrent to specialists. Since leaf toughness and lower nitrogen content also deter many herbivores, deciduous forest trees are especially vulnerable to attack during the period of canopy development, which occurs yearly at a predictable time. Canopy development is heavily dependent on available reserves of carbohydrate and nitrogen, and the production of quantitative

defensive compounds is expensive in terms of carbon and
energy. It therefore appears that the patterns of production
of polyphenols during canopy development are related to the
duration of new growth. Species which produce the entire
canopy during the shortest period of time, thus incurring the
greatest drain on reserves, may defer maximum polyphenol
production until after the canopy is mature. More information
is needed to examine these ideas.

Summary

     In order to gain an understanding of the options which
plants can utilize to defend themselves against herbivores we
need a more comprehensive view of the carbon and nitrogen
acquisition and allocation pattern of plants of diverse
habitats. For example, there are clear differences in the
allocation patterns between annual versus perennial plants
which relate to the resources available to them through time.
Further, there are differences in the reutilization
possibilities of either carbon or nitrogen compounds within a
plant as it develops. These diverse acquisition and allocation
patterns place constraints on the possible defensive strategies
of plants and hence on the opportunities for herbivore attack.

Acknowledgements. This paper grew out of work supported by NSF
     grant DEB810279. We thank B. Lilley for assistance and
     F.S. Chapin III, J.C. Schultz, G. Puttick and J. Glyphis
     for comments on a version of this paper.

Literature Cited

1. Rosenthal, G.A.; Janzen D.H., Eds.; "Herbivores. Their
     Interaction with Secondary Plant Metabolites". Academic
     Press: N.Y., 1979, p 718.
2. Feeny, P.P. Wallace, J.; Mansell, R. Eds.;"Biochemical
     Interactions between Plants and Insects". Recent Adv.
     Phytochemistry 1979, 10, 1-40.
3. Faeth, S.H.; Mopper, S.; Simberloff D. Oikos 1981, 37, 238-
     251.
4. Mooney, H.A.; Gulmon, S.L. BioScience 1982, 32, 198-206.
5. Mattson, W.J. Ann. Rev. Ecol. Syst. 1980, 11, 119-161.
6. Mooney, H.A.; Gulmon, S.L. "Plant Population Biology".
     Solbrig, O.T.; Jain, S. Johnson G.B.; Raven, P.H. eds.;
     Columbia University Press N.Y. 1979: p 316.
7. Gulmon, S; Chu, C. Oecologia 1981, 49, 207-212.
8. Mooney, H.A.; Field, C.; Gulmon, S.L.; Bazzaz, F.A.
     Oecologia 1981, 50, 109-112.
9. Tso, T.C. "Physiology and Biochemistry of Tobacco
     Plants". Dowden, Hutchinson and Ross, Inc. Pa. p 393.

10. Dement, W.A.; Mooney, H.A. Oecologia 1974, 15, 65-76.
11. Mooney, H.A.; Kummerow, J.; Johnson, A.W.; Parsons, D.J.; Keeley, S.; Hoffmann, A.; Hays, R.I.; Giliberto, J.; Chu, C. "Convergent Evolution in Chile and California. Mooney, H.A. Ed.; Dowden, Hutchinson and Ross Inc. Stroudsberg, Pa. 1977, p 85.
12. Rhoades, D.; Cates, R.G. "Biochemical Interactions Between Plants and Insects" Wallace, J.; Mansell, R.; Eds. Rec. Adv. Phyto Chem. 10, 168-213. Plenum Press N.Y. London.
13. Ryan, D.F.; Bormann, F.H. BioScience 1982, 32, 29-32.
14. Derman, B.D.; Rupp, D.C.; Nooden, L.D. Amer. J. Bot. 1978, 65, 205-213.
15. Taylor, B.K.; May, L.H. Aust. J. Biol. Sci. 1967, 20, 389-411.
16. Kramer, P.J.; Kozlowski, T.T. "Physiology of Woody Plants" Academic Press N.Y., San Francisco, London, 1979
17. Tso, T.C.; Jeffrey, R.N. Arch. Biochem, Biophys. 1961, 92, 253-6.
18. Seigler, D.S. Biochemical Systematics and Ecology 1977, 5, 195-199.
19. Robinson, T. Science 1974, 184, 430-435.
20. Digenis, G.A. J. Pharm. Sci. 1969, 58, 39-42.
21. Clegg, D.O.; Conn, E.; Janzen, D. Nature 1979, 278, 343-344.
22. Waller, G.R.; Nowacki, E.K. "Alkaloid Biology and Metabolism in Plants". Plenum Press. N.Y., 1978, p 194.
23. McKey, D. Am. Nat. 1974, 108, 305-320.
24. Beadle, N.C.W. Ecology 1953, 34, 426-428.
25. Vitousek, P.M.; Reiners, W.A. BioScience 1975, 25, 376-381.
26. Muller, R.N.; Bormann, F.H. Science 1976, 193, 1126-1128.
27. Mahall, B.E.; Bormann, F.H. Bot. Gaz. 1978, 139, 467-481.
28. Herridge, D.F.; Pate, J.S. Plant Physiol. 1977, 60, 759-764.
29. Mooney, H.A.; Rundel, P.W. Botanical Gazette 1979, 140, 109-113.
30. Fritts, H.C. "Tree Rings and Climate". Academic Press, London, 1976, p 567.
31. Dougherty, P.M.; Teskey, R.O.; Phelps, J.E.; Hinckley, T.M. Plant Physiol. 1979, 64, 930-935.
32. Lincoln, D.E. Biochem. Syst. and Ecol. 1980, 8, 397-400.
33. Satoo, T. "Analysis of Temperate Forest Ecosystem". Reichle, D.E. Ed. Springer Verlag, Heidelberg, N. Y., 1970, p. 55
34. Taylor, F.G. Jr. "Phenology and Seasonality Modeling", Helmut Lieth (Ed), Springer-Verlag, N.Y., Heidelberg Berlin, 1974, p.237.
35. Kozlowski, T.T.; Keller, T. Bot. Rev. 1966, 32, 293-382.
36. Hansen, P. Physiol. Plant. 1971, 25, 469-473.
37. Haukioja, E. Holarctic Ecology 1979, 2, 272-274.
38. Farnsworth, C.E. Ecology 1955, 36, 285-292.

39. Kienholz, R. Ecol. 1941, 22, 249-258.
40. Kramer, P.J. Plant Physiol. 1943, 18, 239-251.
41. Ahlgren, C.E. Ecology 1957, 38, 622-682.
42. Feeny, P.P. Ecology 1970, 51, 565-581.
43. Hough, J.A.; Pimental, D. Envir. Entomol. 1978, 7, 97-102.
44. Schultz, J.C.; Nothnagle, P.J.; Baldwin, I.T. Am. J. Bot.
    1982, 69, 753-759.
45. Scriber, J.M.; Stansky, F. Jr. Ann. Rev. Entomol. 1981, 26,
    183-211.
46. Cates, R.G. Oecologia 1980, 46, 22-31.
47. Bernays, E.A. Ecol. Ent. 1981, 6, 353-360.
48. Haukioya, E.; Niemelä, L.; Iso-Iivari; Ojala, H.; Aro, E.
    Rep Kevo Subarctic Res. Stat. 1978, 14, 5-12.
49. Mooney, H.A.; Hays, R.L. Flora 1973, 162, 295-304.

RECEIVED September 28, 1982

Impact of Variable Plant Defensive Chemistry on Susceptibility of Insects to Natural Enemies

JACK C. SCHULTZ

Dartmouth College, Department of Biological Sciences, Hanover, NH 03755

The major role of chemical defenses in plants is hypothesized to be increasing the impact of insect diseases, parasites, and predators. None of these factors alone provides an explanation of why evolutionarily labile insects rarely defoliate their long-lived hosts. However, interactions among all of them could increase the useful evolutionary lifetime of each and the effectiveness of all. In particular, chemical variability is observed to place insects in compromise situations which increase their exposure and susceptibility to natural enemies. Forest trees are shown to be highly variable in space and time, and the impact of this variability on caterpillars is explored in several examples.

Despite the impression made by occasional widespread pest outbreaks such as those of the gypsy moth, severe defoliation of forested ecosystems is quite unusual. Fewer than 10% of the species listed in the Canadian Forest Survey of Lepidoptera (1, 2) exhibit periodic or occasional outbreaks. Generally, defoliation in forests is less than 7% of primary production per year (3, but see 4). The vast majority of forest Lepidoptera are quite rare almost all of the time, and their numbers do not fluctuate to a noticeable degree. These observations suggest that some factor or factors normally regulate forest insect populations and keep defoliation at low levels.

A number of regulatory factors have been proposed, and various potential regulators have been shown to operate in certain systems (5,6). However, none of these factors can be shown to be generally effective in most or all forests. For example, a given parasitoid species may be the most important influence on its host at one site but not at another (7). It is difficult to identify emergent generalizations about the relative importance of various potential controls.

0097-6156/83/0208-0037$06.00/0
© 1983 American Chemical Society

    Indeed, empirical observation, evolutionary theory, and
common sense all suggest that single-factor approaches are not
likely to identify underlying causal relationships. Consider the
impact of plant defensive chemistry. In recent years it has
become clear that secondary compounds and the relative concentrations of primary nutrients in plant tissues may restrict feeding
and growth of herbivorous insects (see 8). Clearly the availability of plant tissues is a function of tissue quality as well
as quantity, helping to explain why herbivorous insects may
appear food-limited before their host tissues are exhausted (9).
However, to maintain forest insect populations at the stable,
extremely low levels we normally observe, plant chemistry would
have to deter feeding, poison insects, or reduce digestion so
strongly that population dynamics of the insects are profoundly
depressed. This necessitates drastically reducing the fitness of
organisms having enormous potential fecundities, population
growth rates, and very short generation times. Many forest
Lepidoptera have potential fecundities of from 200-1000 eggs/
female, yet year-to-year population levels indicate s survivorship of only 1 or 2 per female (10).
    Such a strong impact on survivorship or fecundity, and on the
fitness of individuals, means exerting strong natural selection on
herbivorous insects. This should favor the rapid evolution of
insect adaptations which overcome it. This is, of course, a
common occurrence in the application of pesticides or the development of resistant crop plant cultivars (11). The supposition that
plant defenses select for detoxication adaptations in insects is
the foundation of the concept of coevolution (12).
    Forest trees represent a particularly vulnerable paradox. An
individual tree may live for 300 or more years. During this time
it does not move, and presumably cannot adapt to environmental
change. If tree defenses were responsible for the strong regulation of insects and herbivory, the lifetime of a single tree
ought to provide enough time for the evolution of highly virulent
insects which could defoliate their host trees repeatedly. This
does not appear to happen.
    Feeny (13) attempted to resolve this dilemma by proposing
that forest trees may have developed a particularly recalcitrant
defense, one which even insects could not overcome in hundreds of
generations. His suggestion was that protein-complexing polyphenols, or tannins, could provide such protection. However,
there are many insects which feed preferentially on high-tannin
content tissues (14,15), and specific adaptations exist which can
nullify or reduce the digestion inhibition effects of tannins
(16).
    One must conclude that no uniform physical or chemical
defense should be regarded as insurmountable by evolving insects.
Any uniform chemical plant defense should select for pests
capable of defeating it. Obviously, however, there is a solution

to this dilemma, since, as I have pointed out, forests are not
stripped year after year.

The Evolutionary Importance of Chemical Variability

In fact, there are at least 3 possible solutions. All three
have one thing in common: they focus on the rapidly-growing body
of evidence that trees are not uniform in defensive chemistry.
Instead, most plants are highly complex, dynamic mosaics of
variable chemistry and nutrient value. This observation suggests
ways in which defensive chemistry may remain effective over many
insect generations:

Complex resistance. Qualitative and/or quantitative
chemical variation in plants may expose insects to more than one
deterrent (or poison) concurrently. Several authors have proposed (12,18,19) and Pimentel and Belotti (20) have shown in the
laboratory, that insects may be slower to adapt to such complex
chemical mixtures. As a result, even sublethal doses of toxins
may remain effective over long periods of time.

Resource restriction. If chemical defenses vary quantita-
tively within or between individual plants, then some tissues may
be defended while others are not. As a result, insects have
available to them the evolutionary option of avoidance; they may
develop the ability to recognize poor quality food and avoid it,
rather than evolving detoxication mechanisms (12,18). This should
result in feeding activity concentrated on a restricted set of
tissues or plant individuals. There are two important consequences of this. First, contact rates with defenses can be
lowered by avoiding them. Hence, the evolution of detoxication is
less likely or less rapid (18). Second, and perhaps more
important, the effectiveness of natural enemies may be enhanced
(below).

Multiple-factor interactions. Each potential regulatory
factor may interact synergistically with the others and enhance
their effectiveness. For example, plant chemistry can influence
the effectiveness of predators, parasitoids and diseases in a
variety of ways (21,22,23). However, the selective pressure
exerted by uniform chemical defenses should be strengthened by
interactions with natural enemies, and their useful life will be
shortened.
Consequently, although the ways in which plant chemistry can
influence the effectiveness of natural enemies are dizerse, they
can remain effective through evolutionary time only if variability
is part of the picture as well. In fact, although reviews have
tended to focus on chemical enhancement of natural enemy regulation (23), there are probably as many ways in which uniform plant
chemistry can interfere with the actions of these enemies as there

are positive effects (24,25,26). It is not even clear that the
impact of uniform plant defenses will always be positive from the
plant's point of view.

I suggest that variable plant chemistry, by restricting
resource availability and focusing the activities of herbivores
on a few tissues, promotes compromises between food-finding and
risks from natural enemies which are not readily countered by
most insects. The spatial and temporal heterogeneity which
appears to be common in forest trees is the most important part
of the tree's defensive system, and is the only way a plant's
chemical defenses can remain effective over evolutionary time.
This variable impact on natural enemies may be more important in
regulating consumption than any single factor can be.

Variability in Tree Defenses

There are many possible causes of chemical and nutrient
variability in tree tissues (27,28) which result in a wide range
of spatial arrays of suitable and unsuitable food for insects
(29). Although large-scale spatial variation may influence
insect host race formation and have interesting consequences for
insect biogeography and host race formation (30), the scale of
variation with which the individual insect deals most often is
more local, on the individual tree or tissue basis.

Most antiherbivore traits have been found to be highly
variable on this smaller scale; significant variation in nutrient
content and secondary chemistry is commonly observed within tree
canopies, on a branch-to-branch basis (28,29).

On an even finer scale, we have found that adjacent leaves on
single sugar maple (Acer saccharum Marsh) and yellow birch
(Betula alleghaniensis Britt.) may differ greatly in several
traits important to herbivorous insects (31; Figure 1). Some of
these differences, e.g., in tanning coefficient (32), vary by
factors of 2 or more from leaf to leaf (Figure 1). The pattern of
such variation appears random in sugar maple, but may be age-
related and hence spatially predictable (young leaves occur only
at certain growing points) in yellow birch (31). Insects such as
caterpillars foraging along sugar maple branches may have little
information available to them about the spatial distribution of
leaf quality, while those foraging in yellow birch may be able to
locate leaves with particular traits by searching in certain
places (e.g., ends of branches).

The significance of such spatial arrays lies in the
behavioral responses of insects foraging in these trees. If
certain leaf types are unavailable while others are preferred,
then such spatial arrays force insects to move about in search of
good feeding sites (29). For insects which spend much time (or
all of their lives) feeding in one place (sessile species, such as
aphids), this search is performed once; after a suitable site is
located, these insects are restricted to one portion of their

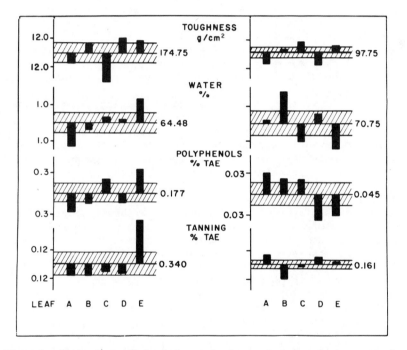

Figure 1.   Leaf-to-leaf variation in four traits along a single branch of sugar maple (left) and yellow birch (right) on 6/23/81. Horizontal axis is mean of each measure for that branch; hatched area is one standard deviation. Each black bar represents the actual value for one leaf, plotted as deviation from the mean. Branch terminus is to right; yellow birch leaves D and E are at least 10 days younger than the others.

hosts. More mobile species which require more food to complete
development (e.g., caterpillars) must repeatedly search for new
feeding sites throughout their lives.

Many insects do indeed recognize and respond to tissue
quality variation. There are numerous examples of leaf-age-
specific preferences among folivorous insects, including some
which make rigid feeding decisions based on tissue ages which
differ by less than a few weeks (29). The apparent resistance of
individual trees in otherwise susceptible stands has been recognized for some time (e.g., 33). The few species of Lepidoptera
larvae whose foraging behavior has been studied travel considerable distances (sometimes several meters) and spend large
proportions of their time sampling, rejecting and judging the
acceptibility of leaves on individual host trees (29). Gall
forming aphids select leaves of a certain size and may engage in
territorial disputes to protect their choices (34). It appears
clear that individual insects respond to spatial variability in
forest tree leaves.

Temporal variability in tree tissue quality is well known
(see 29,35 for review). Year-to-year, seasonal, day-to-day, and
diurnal shifts in nutrient contents and secondary chemistry have
all been observed. Particularly intriguing is the increasing
body of evidence that damage by insects and pathogens may result
in short-term or year-to-year changes in secondary chemistry (36,
32). We have found that ongoing defoliation by gypsy moth
(Lymantria dispar L.) larvae is associated with profound changes
in phenolic chemistry of red oak (Quercus rubrum) leaves (38).
Over a period of a month, tanning coefficients increased
dramatically, and seasonal (2 month) increases in hydrolyzable
tannins were observed in trees undergoing defoliation. Preliminary studies of yellow birch and sugar maple suggest that day-
to-day responses in phenolic production may be generated by
damage to leaves (39).

The importance of seasonal changes in secondary chemistry
and nutrients to the feeding success and life history patterns of
some forest Lepidoptera is well established (40,41). Year to
year changes in chemical phenology may influence tree susceptibility and insect population dynamics (42,43). Hence, a given
tree may not present the same distribution of leaf quality in
every year. Insects attempting to assess host quality for offspring which will feed during the next year or even later in the
same season may not have very complete information available for
selecting oviposition sites.

Shorter-term temporal variation in leaf quality should act to
complicate the spatial arrays described above. Thus, not only may
a foraging insect have difficulty locating suitable feeding sites
in space, but their locations may shift from time to time or
continuously, as seasonal changes, induction effects, or even
plant pathogen attack (44) alter tissue quality. A suitable
tissue at one time may not be suitable later in the day, or later
in the insect's life.

The picture of a forest tree that I wish to portray, then, is one of great spatial heterogeneity, complicated by ongoing change. For an insect capable of dealing with some subset of the great number of tissue quality factors which could influence feeding, there may be only a limited array of suitable tissues in a canopy. These suitable sites may be widely scattered, forcing long searches and much traveling. The pattern is complicated further by constantly changing tissue qualities. The situation can be said to resemble a "shell game", in which a valuable resource (suitable leaves) is "hidden" among many other similar-appearing but unsuitable resources. The insect must sample many tissues to identify a good one. The location of good tissues may be spatially unpredictable, and may even change with time. For a "choosy" or discriminating insect, finding suitable food in an apparently uniform canopy could be highly complex.

Impact on Natural Enemies

Although chemical variability may not alter all of the potential effects of plant chemistry on the effectiveness of natural enemies, there are a number of important qualitative differences in the kinds of interactions possible. In some cases the impact variable chemistry may have on an insect's susceptibility to risks is simply greater than it would be were plant chemistry uniform. In other cases wholly different relationships are possible.

Toxic substances acquired from the host plant may provide resistance to parasitoids (24), pathogens (25), and predators (45). By avoiding some toxins in plant material and selecting superior food tissues, insects feeding on variable hosts may become more susceptible to some enemies. Of course, other substances in preferred tissues may still be toxic to certain of these enemies, but this is less likely than it would be were plant compounds uniformly encountered by the host insect.

An insect host's exposure to parasites and predators may be increased by variable plant defenses in three ways. First, by restricting feeding activity to certain tissue types or portions of the host plant, the position of insect hosts becomes more predictable. Parasites (24,46,42) or predators (48) able to recognize physical plant traits such as tissue color or form, or those capable of employing the unique chemistry of the preferred tissues as cues (47,49) would be able to locate their hosts more readily by focusing their search on these traits.

Second, the increased movement necessary for locating widely dispersed feeding sites should increase contact rates with enemies. Movement makes insects more conspicuous to parasitoids or predators sensitive to it (50,51). Random encounters with arthropod predators or parasites should increase with searching activity, as would risk of dislodgement and fallout.

Moving long distances and tasting many surfaces (29) may also greatly increase contact rates with pathogens. Pathogens are often distributed on plant tissue surfaces by other host insects (52), and are transmitted by subsequent contact with the same surfaces. Increased movement should disperse pathogens more widely and increase the probability of encountering them. Moreover, the chemistry of plant tissues may influence the composition of their surface faunas/floras (53). Hence, by focusing feeding activities on tissues with certain chemical traits, insects may simultaneously find themselves feeding on tissues which promote the growth of pathogens. This may be particularly important when leaf age is a criterion for choice. We have observed consistently elevated viral mortality (70% vs 30%, N = 80) among individuals of the noctuid, Orthosia hibisci Guenee, when fed older (45 days) yellow birch leaves as compared with those fed young (less than 10 days) leaves from the same tree. Although this effect could be due to differential chemistry in the two leaf age classes (31), a more parsimonious hypothesis is that older leaves have had more time to collect more pathogens. Hence, travelling on older leaves may be quite risky.

Some insect species may avoid the risks of cuing visual predators (e.g., birds) while moving by being active only at night (29,54). In many habitats this would mean reducing the time available for feeding by one-half or two-thirds, resulting in a decrease in growth rate of as much as 40 or 50% (55,56). Slowing the growth rate by this much adds to the length of time an insect is available to all risks (57,58); there is a tradeoff between restricted feeding and risks over the lifetime of the insect as well as during each feeding bout. This is the third means by which the insect's exposure to risks is increased by variable host quality.

These tradeoffs can be depicted graphically (Figure 2). I have suggested (29) that the form of expected food yield during an insect's foraging among variable resources should be represented by a rising, asymptotic curve if the insect selects some subset of tissues from those it encounters. We can plot the survivorship probability of an individual as a negative exponential function of the capture rate; such a function for a constant capture rate of 30% is shown in Figure 2. A capture rate of 30% represents the median rate of removal of caterpillars by birds in a north temperate forest as determined by Holmes et al (59), and is a conservative estimate for parasitism rates at moderate host densities (e.g. 60). The more variable the leaves from which an insect must select a meal, the lower its expected yield over a given time interval (the lower yield curve in Figure 2 represents 1/2 the available leaves of the upper curve, or twice the variability). As a result of the increased time spent searching, the contact rate with and probability of capture by a predator or parasite increases greatly for a given food yield as tissues become more variable. There is a direct influence of variability on risk.

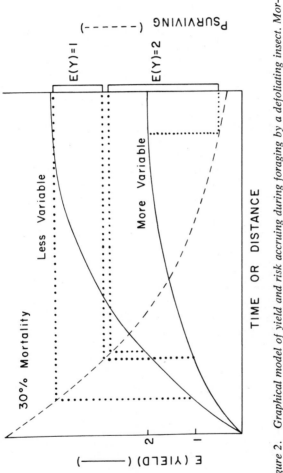

Figure 2. Graphical model of yield and risk accruing during foraging by a defoliating insect. Mortality from natural enemies is assumed to be constant at 30%, resulting in exponentially decreasing survivorship curve (dashed). Yield (solid lines) accumulates curvilinearly (see 29) and more rapidly when leaves are less variable (because most leaves can be consumed) than when they are highly variable (many do not contribute to the diet). For an expected yield of 1 hypothetical unit, foraging longer (because leaves are variable) reduces the probability of surviving by an amount labeled E(Y)=1 on right axis. The reduction in survivorship is much greater by the time E(Y)=2. Leaf variability decreases the probability of surviving.

An example employing data for gypsy moth larvae and their tachinid fly parasite, Blepharina pratensis Meigen, is depicted in Figure 3. The parasitoid infection rate is derived from studies done in Centre County, PA, under moderate gypsy moth densities (60). The flies begin to oviposit on foliage when gypsy moth larvae are in the 3d instar, and the microtype eggs are consumed by larvae. There is differential parasitoid survivorship in caterpillars of different instars, and the "survival" curve in Figure 3 represents successful caterpillar kills corrected for parasitoid mortality in the host. Caterpillar dry weights are taken from a study (61) of the effects of gypsy moth defoliation on host plant food quality and larval growth. The "normal foliage" growth curve approximates the growth rate of gypsy moth larvae on normal oak foliage through the last 4 instars. The "induced foliage" curve approximates the growth of larvae on foliage from defoliated trees (61). Development time for these larvae is about 4 days (3-4%) longer than it is for "normal foliage" larvae (61). Most of the retardation occurs in the first 3 instars; by the 5th and 6th instars reduced food quality no longer depresses growth rates below control larvae (M. Montgomery, pers. comm.).

As a consequence of an apparently induced change in food quality (probably due to increased tannin contents; 38), development time is lengthened. This in turn results in an increase in parasitism rates. The depiction in Figure 3, although somewhat schematic, shows a decrease in survivorship of almost 20% resulting from a growth rate reduction of 3%. Interestingly, were growth rates slowed enough, the caterpillars could escape parasitism by this fly. B. pratensis eggs last about 2 weeks on foliage. Were development of some caterpillars delayed enough, they might enter the 3d instar late enough to avoid viable parasitoid eggs. On the other hand, the adult flies apparently track caterpillar population development and time oviposition to coincide with entry into the 3d caterpillar instar (60).

Thus a constant mortality or susceptibility, from a complex of enemies or from generalized predators or parasites, results in a steep increase in risk with time (Figure 2). The time necessary to accumulate materials for growth and the level of risk while doing so may be increased greatly when food plant quality is variable. Both spatial variability and temporal variability (e.g. induction) can have this effect. Even when the risk accumulation is slower and growth is slowed a very small amount (as in the fly-gypsy moth case), host plant variation can have a major impact on exposure to enemies (Figure 3).

Finally, density-dependent mortality from various enemies may be enhanced by host plant variation. Again, focusing feeding activities on a restricted set of suitable tissues should also focus the activities and abundance of pathogens, parasitoids, and predators. Sessile insects, such as gall-forming aphids (55,62),

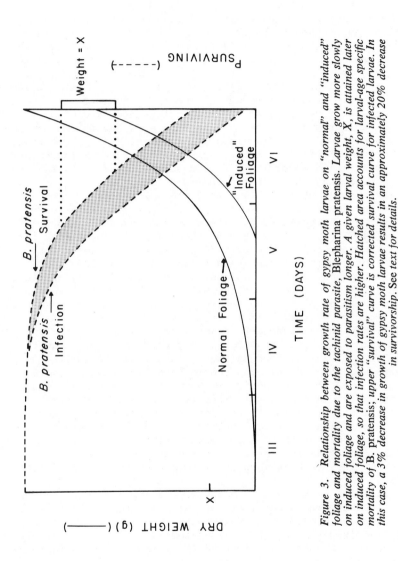

Figure 3. ʻRelationship between growth rate of gypsy moth larvae on "normal" and "induced" foliage and mortality due to the tachinid parasite, Blepharina pratensis. Larvae grow more slowly on induced foliage and are exposed to parasitism longer. A given larval weight, X, is attained later on induced foliage, so that infection rates are higher. Hatched area accounts for larval-age specific mortality of B. pratensis; upper "survival" curve is corrected survival curve for infected larvae. In this case, a 3% decrease in growth of gypsy moth larvae results in an approximately 20% decrease in survivorship. See text for details.*

experience increased contagion in terms of mortality from enemies. By occurring predictably on particular plant surfaces or in particular locations, insects may focus the searching activities of predators and parasites in a density-dependent fashion (e.g., 55,62,63).

More complex, second order interactions may be imagined, involving more than one natural enemy. For example, consider insects to which tannins are important deterrents and digestion inhibitors. As mentioned above, elevated gut pH appears to be a way of dealing with tannins, since tannin-protein complexes are dissociated or inhibited at alkaline pH (16,32). Indeed, using a model in vitro system in which hemoglobin is employed as a protein substrate, we found that several natural tannins and phenolic extracts do not precipitate this protein when the pH exceeds about 8.5 (Figure 3; 32); binding is quite complete from pH 4 through 8. Although hemoglobin is not a plant protein, it resembles several plant proteins in molecular size and solubility (unlike casein, for example) and is a useful comparison (32).

It is interesting to note that the solubility of the crystalline toxin of a common, important caterpillar pathogen, Bacillus thuringiensis (Bt), runs from just over pH 8 to about pH 9.5 (64,65,66). Above pH 9.5, there is some doubt that the protein toxin remains effective (66). Hence, a caterpillar adapted for feeding on high-tannin foods is in a precarious situation, caught between increasing the digestibility of its food and the risk of pathogen susceptibility. The solubility of the protein coats of several nuclear polyhedrosis viruses (NPV)- and hence their virulence in the insect gut- ranges from pH 4.5 to pH 8.5 (67,68). Hence tannin-tolerant insects with elevated gut pH's may be relatively resistant to these pathogens. According to theory (69,70), early successional plants should have low tannin contents and their herbivores should have lower gut pH values (16). An emerging hypothesis would be that caterpillar species feeding on late successional trees would be more susceptible to Bt and less susceptible to NPV than are their relatives on earlier-successional plants. This hypothesis is as yet untested. It could have great practical importance, since these pathogens are currently being developed and promoted as biological control agents for forest pests on both high-tannin and low-tannin tree species.

Microbial chitinase has been proposed as a synergist for Bt (71). Its role would be to digest holes in the insect gut wall and facilitate penetration of Bt toxin. However, unless the caterpillar's gut pH can be manipulated (71), this is unlikely to be effective with Bt, but might be feasible with NPV (Figure 4).

How does chemical variability enter into this pH scenario? First, by concentrating on low-tannin tissues, an insect may be able to feed on a tree species with high average tannin values while maintaining a lower gut pH. However, this could increase pathogen risk (Figure 4). Second, gut pH may fluctuate with the

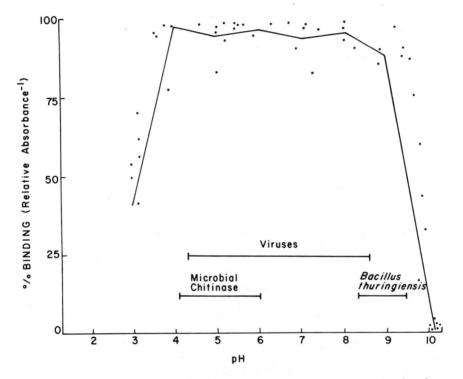

Figure 4.   Binding of a protein (hemoglobin) to several tannin extracts (tannic acid, sugar maple tannins, yellow birch tannins, quebracho tannins; see 29) at various pH values. Ranges of microbial chitinase activity, NPV activity, and Bt toxicity are given. See text for discussion and references.

food actually ingested, and may decrease in animals starved for 12 hours (72). Thus any insect experiencing long periods between meals may experience lowered gut pH and possible digestibility problems when feeding begins. More interesting, a high gut pH may decline into a region of maximum pathogenicity for organisms like Bt.

But why would a caterpillar not feed for up to 12 hours? If suitable food is widely scattered and risks of movement among feeding sites are high (above), many insects may be forced to feed only at night (29,54). In north temperate forests, such an insect will "starve" for from 8 to 14 hours. One consequence of this tactic may be that the first meal of the evening may be very risky.

Some aspects of plant variation could interfere with the impact of natural enemies. Some enemies may be unable to associate microhabitat cues (e.g., chemical, physical, color, position) with prey or host location. For these enemies, prey or host feeding on restricted tissues will tend to appear widely spaced and they may not be readily encountered. It appears to me that many, if not most, parasitoids and predators can be found to use one or more cues. This negative effect could be counteracted by increased encounter rates during herbivore searching movements.

The metabolic costs of travelling among feeding sites and resting long periods without feeding could be translated directly into reduced insect fecundity (29). Were this effect strong enough, it is conceivable that insect densities might be reduced directly. A possible consequence of this would be reduced density-dependent mortality. There are no data available for the metabolic costs of walking for insects such as caterpillars.

Conclusions and Management Prospects

I have argued that uniform chemical defenses cannot be evolutionarily stable. For trees, this means they should not remain effective even for a single tree generation. But trees are not uniform; they are dynamic, highly diverse habitats and food sources for insects. Although plant chemistry can influence natural enemies directly, it may do so in either positive or negative fashion, and this influence should not remain effective over evolutionary time, either. However, chemical variability influences susceptibility of herbivores to natural enemies by forcing costly tradeoffs upon insects which involve unavoidable risks and metabolic costs. These difficulties are less easily overcome by adaptation. In addition to these synergistic effects, chemical variability could maintain the effectiveness of plant defenses through a general slowing of adaptation due to lowered contact rates with specific defenses or the difficulty of dealing with multiple factors.

These observations are of practical importance. The kinds of variability I have described appear to exert control on insect

populations and consumption in nature. For example, induction
responses may be critical to limiting occasional pest outbreaks
(35,36,38). The timing and type of human intervention in such
events could disrupt such delicate controls. Early reduction in
insect densities during an outbreak through the use of pesticides
or a biological control agent acting early in the life history of
a pest like the gypsy moth could reduce the impact of caterpillar
feeding on host trees and slow or halt tree induction responses.
Just such a situation, although with little consideration of the
biology of the participants, has been modelled by several authors
(73,74). As a result of untimely intervention, interactions
which may naturally limit outbreaks could be frustrated and
artificial controls may become necessary over extended periods.

     Some biological control efforts may be faulty or may be
improved when plant chemistry and variability are taken into
account. For example, it seems reasonable to hypothesize that Bt
may work well on high-tannin adapted pests (with elevated gut pH),
such as those feeding on late successional or slow-growing tree
species, but it may be less effective on early successional
species or early in the growth season on high-tannin trees. NPV
may be more effective in early successional situations or any
situation where tannins are not important plant defenses. In
addition, some plant chemicals may make certain biological control
agents less effective. Examples include plant chemicals which are
toxic to parasitoids (24) and those which are antibacterial (e.g.,
monoterpenes in confers; 26). Knowledge of natural variation in
plant chemistry could greatly aid in improving such control
methods.

     Finally, I would suggest that plant variability, genotypic
and/or phenotypic, is as important to trees as it is to
herbaceous species such as crop plants. It should thus be as
important in tree plantations and forest management as it has
become in agriculture. As forest management takes on more
characteristics of large scale agriculture, perhaps we should take
a lesson from the mad scramble for old "new" genes in corn and
other crops and avoid the mistakes inherent in large, uniform
plantations (11,75,76). Tree defense variability may be as
important or more important than uniform resistance per se. It
seems reasonable to suggest maintaining it or mimicking it under
intense management conditions. Certainly, there are substantial
grounds for concentrating research efforts on studies of variance
as well as means.

Acknowledgements

     Ideas were developed in conversation with I.T. Baldwin, R.T.
Holmes, and P.J. Nothnagle. I.T. Baldwin carried out chemical
analyses of birch and maple leaves, and M.J. Richards drew the
figures. I thank Michael Montgomery, USDA Forest Service, for
permission to use unpublished data. Supported by NSF grant

DEB-8022174 to JCS and RTH as part of continuing studies of herbivory at the Hubbard Brook Experimental Forest, W. Thornton, NH.

Literature Cited

1. Prentice, R. M. "Forest Lepidoptera of Canada, Recorded by the Forest Insect Survey 2"; Canada Dept. Forestry Bull. 128.
2. Prentice, R. M. "Forest Lepidoptera of Canada, Recorded by the Forest Insect Survey 3"; Canada Dept. Forestry, Publ. 1013.
3. Mattson, W. J.; Addy, N.D. Science 1975, 190, 515-522.
4. Journet, A. R. P. Austral. J. Ecol. 1981, 6, 135-138.
5. Andrewartha, H. G.; Birch, L. C. "The Distribution and Abundance of Animals"; Univ. Chicago Press, Chicago, 1954.
6. DeBach, P. "Biological Control by Natural Enemies"; Cambridge Univ. Press, NY, 1974.
7. Hassell, M. P. J. Animal Ecol. 1980, 49, 603-628.
8. Rosenthal, G. A.; Janzen, D. H. "Herbivores: Their Interactions with Plant Secondary Compounds"; Academic Press, NY.
9. Ehrlich, P. R.; Birch, L. C. Amer. Natur. 1967, 101, 97-107.
10. Lawton, J. H.; McNeill, S. Symp. British Ecol. Soc. 1979, 20, 223-244.
11. Lupton, F. G. H. In "Origins of Pest, Parasite, Disease and Weed Problems"; J. M. Cherritt, G. R. Sagar, eds., Blackwell, London, 1977; pp. 71-83.
12. Ehrlich, P. R.; Raven, P. H. Evolution 1964, 18, 586-608.
13. Feeny, P. Rec. Adv. Phytochem. 1976, 10, 1-40.
14. Feeny, P. Ecology 1970, 51, 562-581.
15. Schultz, J. C.; Holmes, R. T. in prep.
16. Berenbaum, M. Amer. Nat. 1980, 115, 138-146.
17. Dolinger, P. M.; Ehrlich, P. R.; Fitch, W. L.; Breedlove, D.E. Oecologia 1973, 13, 191-204.
18. Maiorana, V. Biol. J. Linn. Soc. 1979, 11, 387-396.
19. Atsatt, P. R.; O'Dowd, D. J. Science 1976, 193, 24-29.
20. Pimentel, D.; Bellotti, A. C. Amer. Nat. 1976, 110, 877-888.
21. Greenblatt, J. A. J. Appl. Ecol. 1981, 18, 1-10.
22. Greenblatt, J. A.; Barbosa, P. J. Appl. Ecol. 1981, 18, 1-10.
23. Price, P. W.; Bouton, C. E.; Gross, P.; McPheron, B. A.; Thompson, J. N.; Weis, A. E. Ann. Rev. Ecol. Syst. 1980, 11, 41-65.
24. Campbell, B. C.; Duffey, S. S. Science 1979, 205, 700-702.
25. Maksymiuk, B. J. Invert. Path. 1970, 15, 356-371.
26. Smirnoff, W. A.; Hutchison, P, M. J. Invert. Path. 1965, 7, 273-280.
27. Whitham, T. G.; Slobodchikoff, C. N. Oecologia 1981, 49, 287-292.

28. Whitham, T. G. In "Insect Life History Patterns", R. F. Denno, H. Dingle eds., Springer-Verlag, NY, 1981; pp. 9-27.
29. Schultz, J. C. In "Impact of Variable Host Quality on Herbivorous Insects", R. F. Denno, M. S. McClure, eds., Academic Press, NY, in press.
30. Fox, L. R.; Morrow, P. A. Science 1981, 211, 887-893.
31. Schultz, J. C.; Nothnagle, P. J.; Baldwin, I. T. Amer. J. Bot. 1982, 69, 753-759.
32. Schultz, J. C.; Baldwin, I. T.; Nothnagle, P. J. J. Agric. Food Chem. 1981, 29, 823-826.
33. Painter, R. H. In "Breeding Pest-Resistant Trees", H. D. Gerhold, E. J. Schreiner, R. E. McDermott, J. A. Winieski, eds., Pergamon Press, NY, 1966; pp. 349-355.
34. Whitham, T. G. Amer. Nat. 1980, 115, 449-466.
35. Rhoades, D. F. In "Herbivores: Their Interaction with Plant Secondary Compounds", G. A. Rosenthal, D. H. Janzen, eds., Academic Press, NY, 1979; pp. 4-54.
36. Haukioja, E.; Niemelä, P. Ann. Zool. Fenn. 1977, 14, 48-52.
37. Benz, G. In "Eucarpia/IDBC Working Group: Breeding for Resistance to Insects and Mites", Bull. SROP, 1977; pp. 155-159.
38. Schultz, J. C.: Baldwin, I. T. Science 1982, 217, 149-151.
39. Baldwin, I. T.; Schultz, J. C. in prep.
40. Schweitzer, D. F. Oikos 1979, 32, 403-408.
41. Hough, J. A.; Pimentel, D. Envir. Entomol. 1978, 7, 97-102.
42. Campbell, I. T. In "Breeding Pest Resistant Trees", M. D. Gerhold, E. J. Schreiner, R. E. McDermott, J. A. Winieski, eds., Pergamon Press, NY, 1966; pp. 129-134.
43. Mattson, W. J.; Lorimer, N.; Leary, R. A. Proc. IUFRO Conf. Genetics of Host/Parasite Interactions, Centre for Agric. Publ. and Document., Wageningen, Netherlands, 1982; in press.
44. Matta, A. Plant Disease 1980, 5, 345-361.
45. Brower, L. P.; Van Zandt Brower, J. Zoologica 1964, 49, 137-159.
46. Arthur, A. P., Can. Entomol. 1962, 94, 337-347; IBID 1966, 98, 213-223; IBID 1967, 99, 877-886.
47. Read, D. P.; Feeny, P. P.; Root, R. B. Can. Entomol. 1970, 102, 1567-1578.
48. Heinrich, B. Oecologia 1979, 42, 325-337.
49. Ryan, R. B. Can. Entomol. 1979, 111, 477-480.
50. deRuiter, L. Behaviour 1952, 4, 222-232.
51. Richerson, J. V.; DeLoach, C. J. Ann. Entomol. Soc. Amer. 1972, 65, 834-839.
52. Aizawa, K. In "Insect Pathology, an Advanced Treatise Vol. 1, E. A. Steinhaus, ed., Academic Press, NY, 1963; pp. 381-412.
53. Lovett, J. V.; Duffield, A. M. J. Appl. Ecol. 1981, 18, 283-290.

54.   Windsor, D. M.   In "Ecology of Arboreal Folivores", G. G.
      Montgomery, ed., Smithsonian Inst. Press, Washington, DC,
      1978; pp. 101-113.
55.   McGinnis, A. J.; Kasting, R.   Can. J. Zool. 1959, 37, 259-
      266.
56.   Muthukrishna, J.; Delvi, M. R.   Oecologia 1974, 16, 227-236.
57.   Feeny, P. P.   In "Coevolution of Animals and Plants", L.
      Gilbert and P. R. Raven, eds., Univ. Texas Press, Austin,
      1975; p. 246.
58.   Schultz, J. C.   Evolution 1981, 35, 171-179.
59.   Holmes, R. T.; Schultz, J. C.; Nothnagle, P. J.   Science
      1979, 206, 462-463.
60.   Doane, C. C.; McManus, M. L.   USDA Tech. Bull. 1584, USDA-FS,
      Washington, DC, 1981; p. 389.
61.   Wallner, W. E.; Walton, G. S.   Ann. Entomol. Sco. Amer. 1979,
      72, 62-67.
62.   Whitham, T. G.   In "Impact of Variable Host Quality on
      Herbivorous Insects", R. F. Denno, M. S. McClure, eds.,
      Academic Press, NY, in press.
63.   Solomon, B. P.; McNaughton, S. J.   Oecologia 1979, 42, 47-56.
64.   Sharp, E. S.; Detroy, R. W.   J. Invert. Path. 1979, 34,
      90-91.
65.   Raun, E. S.; Sutter, G. R.; Revelo, M. A., J. Invert. Path.
      1966, 8, 365-375.
66.   Fast, P. G.; Milne, R.   J. Invert. Path. 1979, 34, 319.
67.   Gudauskas, R. T.; Canerday, D.   J. Invert. Path. 1968, 12,
      405-411.
68.   Ignoffo, C. M.; Garcia, C.   J. Invert. Path. 1966, 8, 426-
      427.
69.   Rhoades, D. F.; Cates, R. G.   Rec. Adv. Phytochem. 1976, 10,
      168-213.
70.   Cates, R. G.; Orians, G. H.   Ecology 1975, 56, 410-418.
71.   Daoust, R. A.; Gunner, H. B.   J. Invert. Path. 1979, 33,
      368-377.
72.   Heimpel, A. M.   Can. J. Zool. 1955, 33, 94-106.
73.   Anderson, R. M.; May, R. M.   Science 1980, 210, 658-661.
74.   Carpenter, S. R.   J.Theor. Biol. 1981, 92, 181-184.
75.   Walsh, J.   Science 1981, 214, 161-164.
76.   Marshall, D. R.   Ann. NY Acad. Sci. 1977, 287, 1-20.

RECEIVED September 28, 1982

Responses of Alder and Willow to Attack by Tent Caterpillars and Webworms: Evidence for Pheromonal Sensitivity of Willows

DAVID F. RHOADES

University of Washington, Department of Zoology, Seattle, WA 98195

Red alder (Alnus rubra) and Sitka willow (Salix sitchensis) trees subjected to attack by tent caterpillars
(Malacosoma californicum pluviale) or webworms (Hyphantria cunea), respectively, exhibited a change in foliage
quality such that bioassay insects fed leaves from the
attacked trees grew more slowly than those fed leaves
from unattacked control trees. In contrast, bioassay
of leaf quality of S. sitchensis, subjected to attack
by tent caterpillars, indicated that altered leaf quality
had been induced not only in the attacked trees but also
in nearby unattacked control trees. This suggests that
S. sitchensis is sensitive to and can respond to signals generated by attacked trees or the caterpillars.
Since no evidence was found for root connections between attacked and control willows, the message may be
transferred through airborne pheromonal substances.

During the last several years there has been an increasing
interest in, and demonstration of, the fact that plants subjected
to attack by insects or other herbivores can decrease the quality
of their tissues as food by increasing their content of defensive
substances, decreasing their content of nutrients, or both (1-5).
There is evidence for both short and long term plant responses to
attack. Short term responses occurring during the period of
attack can be expected to influence the fitness of the attacking
herbivores, whereas long term responses may influence the fitness
of subsequent herbivores. These induced plant responses could
have profound effects on the population dynamics of herbivores.
The experiments described here were designed to detect short term
changes in leaf quality of red alder (Alnus rubra Bong; Betula-
ceae) and Sitka willow (Salix sitchensis Sanson; Salicaceae) in
response to attack by two species of polyphagous, univoltine,
colonial, defoliating Lepidoptera, western tent caterpillars

0097-6156/83/0208-0055$06.00/0
© 1983 American Chemical Society

(Malacosoma californicum pluviale Dyar; Lasiocampidae) and fall
webworms (Hyphantria cunea Drury; Arctiidae).

If herbivore attack can lead to reduced food quality of
plants, then it seems reasonable that naturally attacked plants
should exhibit decreased food quality compared to naturally
unattacked ones. On the other hand, there is considerable evidence that herbivores preferentially attack plants or tissues of
high food quality (3). In other words, plant food quality can
probably act both as a dependent and an independent variable as
far as degree of herbivory is concerned. Therefore, conclusions
drawn from comparisons of foliage properties of naturally
attacked versus naturally unattacked plants are likely to be
confounded by the interaction of these two effects. To avoid
this problem the following experiments were conducted by
subjecting plants to attack by insects placed on them by the
investigators.

Red Alder Attacked by Tent Caterpillars

In western Washington State tent caterpillars were abundant
during the springs of 1975, 1976, and 1977, attacking alders,
willows, and other species of broad-leafed trees. Alders and
willows at our study site near Kent, King Co., Washington, were
heavily attacked in 1977, some trees suffering almost complete
defoliation. In 1978 the local populations crashed, producing
few viable egg masses. Since the spring of 1979, no natural
colonies have been observed within a radius of 10 km of the
Kent site.

In spring of 1979 seven pairs of 3-year-old alder trees of
average height 3.1 $\pm$ 0.1 (S.E.) m and volume 3.5 $\pm$ 0.5 m^3 were
selected. Pairs were chosen on the basis of proximity and similarity in size and exposure to the sun. For each pair one member
was randomly assigned as test tree, the other as control. Distance between each control tree and the nearest test tree averaged
6.1 $\pm$ 1.1 m. Two tent caterpillar egg masses, collected the
previous winter from alder trees, were attached to each of the test
trees. A sample of the egg masses (n = 55) contained 214 eggs
per egg mass of which 3.2% were parasitized and 6.6% did not hatch
due to unknown causes. The egg masses had all hatched by April
26. Migration of larvae from the test trees and onto control
trees was prevented by tanglefoot bands over aluminum foil on the
trunks of all trees. Netting bags were placed over assay branches
on both test and control trees to provide equivalent leaves,
protected from the tent caterpillars, for later bioassay and
chemical analysis. This ensured that any differences found between leaves from test and control trees were due to changes in
leaf quality rather than within-plant heterogeneity combined with
preferential consumption of high quality leaves by the insects.
It also ensured that any plant responses observed were systemic
rather than localized wound responses. In mid-May the colonies

were observed to be growing more slowly than expected, constructing
very small tents and causing little leaf damage to the trees, so
on May 20 an average of 3.4 $\pm$ 0.4 (S.E.) additional colonies
collected from alder were placed on each of the test trees.
Commencing May 23, third and fourth instar tent caterpillars
collected from alder were raised in the laboratory (one replicate
per tree, 25 larvae per replicate) on the assay branches detached
from the test and control trees. The assay larvae were weighed,
survivors counted, and dead larvae removed periodically (Figure
1A). Assay branches were replaced each time that the larvae were
weighed and censused. Larvae feeding on leaves from the test
trees grew more slowly and had lower survivorship than those fed
leaves from the control trees (Figure 1A). Biomass per replicate
of larvae, until the initiation of pupation, was significantly
higher for larvae fed control leaves than for those fed leaves
from attacked trees on each occasion measured ($p < .025$, one-
tailed paired t test). Survivorship was poor in both groups but
significantly higher for the controls. For pupae survivorship
averaged 16.0 $\pm$ 6.3% (S.E.) for controls versus 6.9 $\pm$ 3.5% for
those produced from larvae feeding on leaves from attacked plants
($p < 0.05$, one-tailed paired t test). Death of larvae in both
laboratory groups and the insects in the field was associated with
a condition in which, shortly before death, the larvae produced
liquid feces instead of the normal fecal pellet. Pupal weights
were not significantly different between test and control groups
for either male or female pupae, but adults from pooled test and
pooled control groups produced only one egg mass in the former
case and eight egg masses in the latter ($\chi^2 = 5.59$, $p < 0.025$).
By June 11 all of the field load insects had died or left the
trees. Leaf damage to the test trees when measured on June 3 was
relatively light. Leaves exhibiting noticeable damage averaged
27.6 $\pm$ 2.1% (S.E.) for the control trees and 49.0 $\pm$ 4.7% for test
trees ($p < 0.01$, one-tailed paired t test). Estimated leaf area
loss averaged 2.5 $\pm$ 0.2% for controls and 11.3 $\pm$ 2.1% for test
trees ($p < 0.005$, one-tailed paired t test). Damage to control
trees was due to unidentified insects other than tent caterpillars.
     Samples of fresh leaves from the assay branches of all trees
were extracted with 85% aqueous methanol on June 3. These
extracts were assayed for total phenolic content by the Folin-
Denis Method (6) and proanthocyanidins (7). On a dry weight basis,
the proanthocyanidin specific extinction coefficients [$E_{1\,cm}^{1\%}(550\,nm)$]
indicated 24% higher levels, on average, in leaves from attacked
versus control trees ($p < 0.05$, one-tailed paired t test). Folin-
Denis specific extinction coefficients [$E_{1cm}^{1\%}(725\,nm)$] averaged 12%
higher, on a dry weight basis, for leaves from attacked trees, but
this difference was not significant ($p < 0.1$, one-tailed paired t
test). The chemical nature of alder proanthocyanidins has not
been investigated, but proanthocyanidin $E_{1cm}^{1\%}$ values have been
commonly used to indicate levels of condensed tannins in plants
(8, 9, 10).

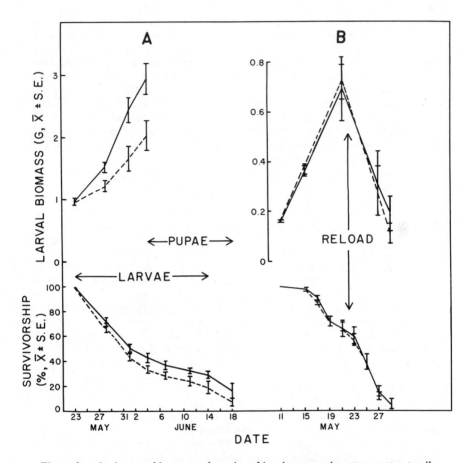

Figure 1. A: Average biomass and survivorship of groups of western tent caterpil-lar larvae fed leaves in the laboratory from red alder trees under attack by tent caterpillars in the field (– – – –) compared to those fed leaves from unattacked con-trol alders (—). B: Average biomass and survivorship of groups of tent caterpillar larvae fed leaves in the laboratory from Sitka willow trees under attack by tent cater-pillars in the field (—) compared to those fed leaves from unattacked control willows (– – – –). The test trees were reloaded with additional colonies of tent caterpillars on the indicated date.

Since test and control alders had been randomly assigned, no significant pre-treatment differences in foliage quality were to be expected. Thus attack of alder by tent caterpillars produced a change in foliage quality which caused decreased growth, survival, and egg mass production in tent caterpillars raised in the laboratory on detached branches. Furthermore, this change was induced by relatively light grazing levels and occurred within a time period of 27 days, or shorter, of the initiation of attack. The change in foliage quality was systemic since it occurred in leaves on branches protected from attack, and it was associated with an increase in proanthocyanidin content of the foliage. Even though significant differences in biomass accumulation and survivorship were observed between test and control assay insects, the survivorships of both groups were low, and this was reflected in the poor performance of the attacking insects in the field. This may have been due to low foliage quality of all the trees caused by repeated attack from tent caterpillars during the 1975–1978 outbreak (3).

Sitka Willow Attacked by Tent Caterpillars

During the same spring (1979) that the above experiment was performed, a similar experiment in which Sitka willow was subjected to attack by tent caterpillars was carried out. Seven pairs of four-year-old willow trees of average height 4.5 ± 0.2 (S.E.) m and volume of 6.1 ± 0.7 m^3 were selected at the Kent site. Test and control trees were randomly assigned for each pair. Distance between each control and the nearest test tree averaged 3.5 ± 0.4 m. An average of 7.4 ± 0.3 egg masses collected from alder and from the same bulk collection used in the previous experiment were placed on each of the test trees, all hatching by April 24. Migration of larvae from the test trees, onto control trees and onto assay branches, was prevented by tanglefoot bands over aluminum foil. All test and control trees were treated similarly. On May 11 the laboratory feeding experiment was initiated. Second and third instar tent caterpillar larvae, raised on willow in the field from egg masses collected from alder, were fed detached leaves in the laboratory (25 larvae per replicate, one replicate per tree). Feed leaves, detached from the assay branches, were replaced every second day. In contrast to the experiment with alder, no difference in growth rates or survivorship between insects raised on leaves from attacked or control trees was observed (Figure 1B).

Due to the very poor performance of the insects in the field, as had been noted in the alder experiment above, the test trees were reloaded with an average of 8.0 ± 1.0 (S.E.) additional colonies of tent caterpillars, obtained from alder, since none were available from willow, on May 22. These additional colonies were mainly in the 4th and early 5th instar, whereas the original larvae hatched on the trees were still mainly in the third instar. Thus, at the time of reload, the trees experienced a large

increase in biomass of attacking insects. Coincident with the
reload operation, a large drop in average biomass was observed for
both test and control groups of assay insects. This was due mainly
to an increased rate of mortality from May 25 onward (Figure 1B).
It should be noted that biomass and survivorship data (Figure 1B)
represent performance of assay insects fed leaves obtained two
days previously so the first post-load survivorship is that
obtained on May 25.

Death of the assay insects was associated with the production
of liquid feces. Very little feeding by the reload insects was
observed. They spent most of their time resting on the tents or
wandering in apparent attempts to leave the trees. By the end of
May all insects had either died or dropped from the trees. This
behavior strongly contrasts with that observed when tent caterpillars are transferred to willow, previously unattacked by tent
caterpillars during that season, from other plants within their
host range. Under these conditions the larvae readily feed,
survive, grow and remain on the trees for extended periods (for
example see following experiment). So, in spite of the large
numbers of egg masses and mature colonies placed on the test trees,
leaf damage was low when measured on June 1. An average of
37.7 ± 5.4% (S.E.) of leaves on the test trees were attacked.
Estimated leaf area lost from the test trees averaged 6.2 ± 2.4%.
The corresponding figures for the controls were 11.5 ± 1.4%
(p < 0.0025, one-tailed paired t test) leaves attacked and
1.7 ± 0.5% (p < 0.0025) leaf area lost, due to unidentified
insects other than tent caterpillars.

Thus, in contrast to the results obtained the same season
with alder, no differences were found in tent caterpillar growth
or survival when fed leaves from attacked versus control willows.
It is possible that the rapid drop in biomass of the assay insects
feeding on leaves from attacked and control trees coincided with
reload of the test trees by chance, possibly due to the rapid
spread of a pathogen through the laboratory insect population (11).
On the other hand, the possibility that reload of the test trees
caused the biomass drop in both test and control assay insects
cannot be discounted. Both attacked and nearby unattacked control
willows may have rapidly decreased the food quality of their
leaves in response to a sudden increase in biomass of insects on
the attacked trees. If so, this suggests that unattacked willows
are sensitive to signals from the insects or nearby attacked
willows. This possibility was investigated in the following
experiment.

In the spring of 1981, 20 six-year-old willows of average
height 5.6 ± 0.1 (S.E.) m and volume 7.5 ± 1.0 m^3 were selected
from the same stand used in the preliminary experiment in 1979.
None of these trees had been used in the previous experiment.
Stems and assay branches were banded with tanglefoot as before,
and 10 test and 10 nearby control trees were randomly assigned
in pairwise fashion. Distance between each control and the

nearest test tree averaged 3.3 ± 0.6 (S.E.) m. A far control group of 20 six-year-old willows 1.6 km from the test site was also selected and banded with tanglefoot. These trees averaged 4.8 ± 0.2 m in height and 8.9 ± 1.7 m³ in volume.

In the preliminary experiment of 1979, egg masses had been placed on the test willows. Approximately one month after the eggs had hatched the numbers of larvae attacking the trees had been reinforced with additional colonies (reload). In 1981, insufficient egg masses were available to repeat this procedure, but as the season progressed we were able to locate tent cater-pillar colonies in the field. These colonies, containing larvae from third to early fifth instar, were collected from wild rose (Rosa nutkana), bitter cherry (Prunus emarginata), domestic apple (Pyrus malus), hawthorn (Crataegus monogyna), and Scouler willow (Salix scouleriana). On May 28 the colonies were attached to the test trees, unattacked until that time, at an average density of 3.9 ± 0.4 (S.E.) colonies per tree. This corresponds to approxi-mately 600 larvae per tree. The larvae were censused on the trees periodically until the end of June (Figure 2). Leaf quality before and after adding insects (load) was bioassayed in the laboratory using tent caterpillar larvae raised from the egg stage on willow. In contrast to the bioassay used in previous experi-ments, in which larvae had been grown continuously on periodically replaced leaves over the entire course of the experiment, a dif-ferent assay method was used. In this procedure batches of larvae (4–14 larvae per batch, depending on availability at the time of assay, one replicate per tree) were raised on leaves from assay branches of test and control trees for an average time period of 22.1 hours (range 21.1 to 23.0) after which time the larvae were weighed and discarded. Fresh larvae were then used in subsequent assays. An advantage of this method is that it greatly facili-tates the detection of rapid changes in foliage quality. On the other hand, it produces no information concerning the cumulative effects of foliage quality, and changes thereof, on survival or egg production. Since the amounts of foliage consumed or frass produced by the insects were not measured, it is not known whether differences in growth rates were due to differences in amounts eaten or to differences in amounts assimilated or both. Whatever the mechanism may be, differences in larval growth imply differ-ences in leaf quality.

Relative growth rates of larvae fed leaves from test, near control, and far control groups were calculated as % increase in fresh weight mass per unit time, over and above that of starved in-sects, normalized with respect to the far control group (Figure 2). On May 11, prior to load, there were no significant differences in growth rates between the test and control groups, nor were there any significant differences in growth for the first three assays subsequent to load ($0.1 < p < 0.25$, ANOVA, for the most significant case). However, on June 9, 11.5 days after load, larvae fed leaves from the test trees grew significantly more slowly than those fed

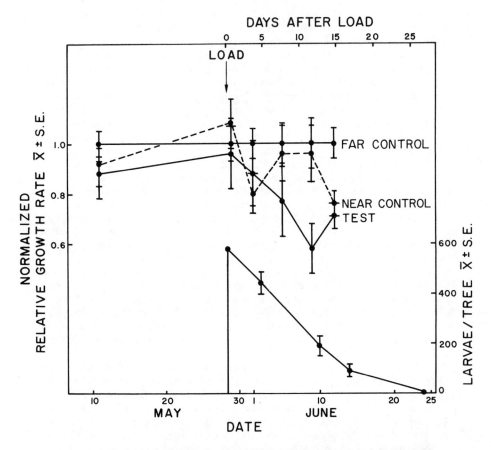

Figure 2. Top: normalized average relative growth rates of groups of western tent caterpillar larvae raised in the laboratory of leaves from test, nearby control, and far control Sitka willow trees. The test trees were loaded with tent caterpillar colonies on the indicated date. Bottom: density of tent caterpillar larvae attacking the trees on various dates.

leaves from the near or far controls (Figure 2, p < 0.05, Newman-Keuls multiple range test). In the subsequent assay on June 12th, 14.5 days after load, larvae fed leaves from both test and nearby controls grew more slowly than those fed leaves from the far controls (Figure 2, p < 0.01, Newman-Keuls multiple range test).

Leaves damaged averaged 44.8 ± 1.2% (S.E.) for the test trees, 6.4 ± 0.6% for the near controls, and 10.4 ± 0.8% for the far controls, when measured at the beginning of July. All these differences were significant (p < 0.01 for the least significant case, Newman-Keuls multiple range test). Estimated leaf area losses were 15.8 ± 1.0% for the test trees, 1.1 ± 0.1% for the near controls, and 1.6 ± 0.2% for the far controls. Leaf area lost by the test trees was significantly greater than that lost by each of the control groups (p < 0.001), but leaf areas lost by each of the two control groups were not significantly different (p < 0.5). Damage to leaves of control trees was due to insects other than tent caterpillars. Leaf area losses due to sampling were estimated at 2.6 ± 0.4 (S.E.)% for the test trees versus 3.3 ± 0.5% for the near controls and 2.4 ± 0.3% for the far controls. The differences in sampling pressure were not significant (p > 0.5, ANOVA).

These results supported the hypothesis that altered leaf food quality can be induced in both willows attacked by tent caterpillars and nearby unattacked willows. The chemical basis of the changes in leaf food quality are unknown. Leaf Folin-Denis total phenolic and proanthocyanidin specific extinction coefficients showed little difference between test and control trees throughout the experiment. Willows commonly form clonal stands and, conceivably, subterranean root connections could link each nearby control tree to a test tree. If so, altered leaf quality of the nearby controls could be due to a long-range systemic response. However, Salix sitchensis is not known to form clonal stands (12), and excavation of 20 S. sitchensis trees (1-3 m in height) at the same site and of the same age as the study trees gave no evidence of cloning or root grafting. Therefore, airborne pheromonal substances emitted by the tent caterpillars or the attacked trees may be involved.

Sitka Willows Attacked by Fall Webworm

Fall webworms, like tent caterpillars, are polyphagous, colonial Lepidoptera that attack a wide variety of broad-leafed trees and shrubs. Unlike tent caterpillars, of which the larval stage occurs in the spring, webworm larvae occur from mid-summer to early fall in western Washington. In July 1980 an experiment to investigate possible changes in willow leaf quality induced by fall webworm attack was initiated. The experiment was designed to take account of possible changes in leaf quality of unattacked willows near attacked ones. Two groups of willows, 10 willows per group and approximately 60 m between groups, were selected at the Kent site, a test group A to be loaded with webworms and a control

group B (Figure 3). These trees were members of five-year-old
even-aged stands, and their average heights and volumes were
5.4 $\pm$ 0.2 m (S.E.), 7.5 $\pm$ 1.9 m^3 for group A and 4.7 $\pm$ 0.2 m,
4.1 $\pm$ 0.5 m^3 for group B. Two additional groups of willows (C and
D), 10 willows per group and approximately 69 m between groups,
were selected at a site near Seattle-Tacoma International Airport,
approximately 8 km from the Kent site. Groups C and D were
members of a five- and four-year-old stand and measured 4.6 $\pm$ 0.3
m, 12.0 $\pm$ 2.1 m^3, and 4.3 $\pm$ 0.3 m, 11.5 $\pm$ 1.8 m^3 respectively.
These trees served as far control groups.

Trunks and assay branches of all trees were banded with
tanglefoot as previously described. On August 13 an average of
1.9 $\pm$ 0.2 (S.E.) colonies of webworms, obtained from alder, were
loaded onto each of the test trees (Group A, Figure 3). Webworms
transferred from alder to willow readily feed, and a sample of
the larvae were raised to the adult stage in the laboratory on
willow. Difficulties were experienced in counting the larvae on
the trees because they remain concealed within their tent when
not feeding, so on September 1 all colonies were removed and
the larvae counted. At this time leaves attacked (%) for the test
group A and groups B, C, and D averaged: A, 56.6 $\pm$ 2.7 (S.E.);
B, 8.6 $\pm$ 0.7; C, 10.4 $\pm$ 1.8; D, 14.4 $\pm$ 1.6. Estimated leaf area
loss (%) averaged: A, 20.7 $\pm$ 2.8; B, 1.1 $\pm$ 0.4; C, 1.6 $\pm$ 0.3;
D, 2.6 $\pm$ 0.4. For both leaf damage measures Newman-Keuls multiple
range tests showed that the test group had suffered greater leaf
area damage than the other three groups ($p < .001$) among which
damage was not significantly different.

On September 9 fresh colonies obtained from alder were reloaded onto the test trees at approximately the same density as
in the original load (Figure 3). A census was obtained on September 15, and the remaining larvae were removed and counted on September 21. At this time leaves attacked (%) averaged: A (test
group), 61.4 $\pm$ 3.1; B, 8.5 $\pm$ 1.1; C, 9.4 $\pm$ 1.7; D, 14.7 $\pm$ 2.1.
Estimated leaf area loss (%) averaged: A, 30.6 $\pm$ 2.7; B,
1.3 $\pm$ 0.2; C, 1.7 $\pm$ 0.4; D, 2.9 $\pm$ 0.3. Again, both measures
showed significantly greater leaf damage ($p < .001$) for the test
group A than the other three groups which were not significantly
different from each other. Leaves removed in sampling (%)
averaged : A, 8.9 $\pm$ 1.0; B, 11.2 $\pm$ 1.3; C, 7.9 $\pm$ 1.5; D,
9.7 $\pm$ 1.4. Sampling pressure was not significantly different
among the four groups ($p > 0.5$, Newman-Keuls multiple range test).
Damage to leaves of control trees was due to unidentified insects
other than webworms.

Leaf quality of the four groups of trees before and after
load was compared by measuring the relative growth rates of webworm larvae (14-20 larvae per replicate, one replicate per tree)
fed leaves from the assay branches over a one to three day period
in the laboratory (Figure 3). These assay larvae were collected
from alder and fed willow for two days prior to the feeding experiments. Since mass changes of starved insects were not measured,

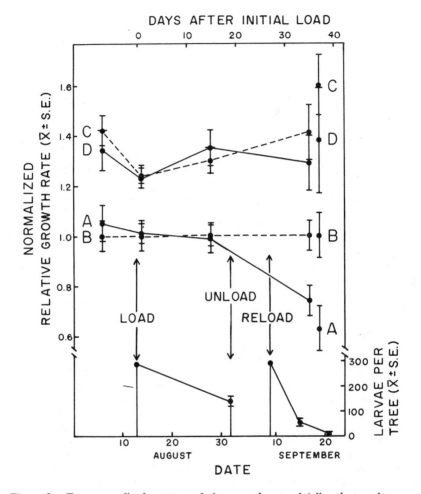

Figure 3. Top: normalized average relative growth rate of fall webworm larvae raised on leaves from test (A), nearby control (B), and far control (C and D) groups of Sitka willow trees. The test trees were loaded with webworms and unloaded on the indicated dates. Connected points represent growth rates of larvae fed detached leaves in the laboratory. Unconnected points at the right represent growth rates in the field of larvae on the various groups of willow trees. Bottom: density of webworm larvae attacking the test trees on various dates.

the results are expressed as averages of the % increases in mass
per unit time over the initial values, normalized with respect to
the B group (Figure 3). The webworms grew faster on the distant
controls at the airport site (groups C and D) than on leaves from
the test and nearby controls at the Kent site (groups A and B) in
all four laboratory assays. There were no significant differences
in growth on leaves from groups C and D throughout the experiment.
Growth rates of the laboratory assay insects on leaves from the
test group A and nearby control group B were not significantly
different from each other before load and during the first two
assays subsequent to load. However, following the reload operation, 35 days after the original load, growth rates of the laboratory insects were lower on leaves from the test trees (A) than on
those of the far and nearby controls (Figure 3). This was confirmed by measuring the growth rates in the field of groups of
larvae (14 larvae per group, one replicate per tree) confined in
netting bags on the assay branches of all trees for a three-day
period on September 17-20 (Figure 3). Relative growth rates in
the field feeding experiment showed a pattern very similar to that
of the final laboratory experiment (Figure 3). Absolute growth
rates in the laboratory were much higher than in the field assays,
however. They ranged between 5.5 times higher (group A) and 4.1
times higher (group C).

Probabilities that the observed differences in larval growth
between groups C and D, D and B, and B and A in the final laboratory feed are merely by chance are > 0.5, < 0.05, and < 0.1
respectively (Newman-Keuls multiple range test). Corresponding
probabilities for the field feeding experiment are > 0.5, < 0.1,
and < 0.1, respectively. Since these two feeding assays constitute independent tests, we can conclude that during the period of
final laboratory and field growth assays the test trees exhibited
an altered leaf quality such that the assay larvae grew more
slowly on these leaves than on leaves from the nearby control
trees ($p = 0.1 \times 0.1 = 0.01$). The plant response occurred within
a period of 35 days from the time of initial load, and it was
systemic.

Higher growth rates of the larvae fed leaves from willows at
the airport site than those fed leaves from willows at the Kent
site, throughout the experiment, show that significant and fairly
constant differences in leaf quality can exist between trees at
different sites. Normalized relative growth rates of webworms fed
detached leaves in the laboratory were similar to those obtained
by growing webworms on the trees.

The chemical basis for the observed differences in leaf food
quality is unknown. Little differences were found in Folin-Denis
total phenolic or proanthocyanidin extinction coefficients of 85%
aqueous methanol leaf extracts from the various plant groups
throughout the experiment. There were no indications of changes
in leaf quality of control willows 60 m distant from willows
attacked by fall webworm.

Summary and Discussion

These results lend support to previous suggestions (1, 2, 13, 14) that plants may commonly decrease the quality of their tissues as food in response to herbivore attack. Red alder trees subjected to attack by western tent caterpillars exhibit a change in foliage quality such that tent caterpillars fed unattacked leaves from attacked trees grew more slowly, died at a faster rate, and produced fewer egg masses than those fed leaves from unattacked trees (Figure 1A). This change in foliage quality was associated with an increase in proanthocyanidin extinction coefficients of leaf extracts. Similarly, fall webworms fed unattacked leaves from Sitka willows subjected to attack by fall webworms grew more slowly than those fed leaves from unattacked willows (Figure 3).

On the other hand, no differences were found in growth or mortality of tent caterpillars fed leaves from unattacked willows compared to those fed leaves from willows attacked by tent caterpillars (Figure 1B). Reload of the attacked trees with additional tent caterpillars coincided with a rapid decrease in biomass and an increased mortality of insects fed leaves from both attacked and unattacked control trees. This could have been by chance, but it suggested the possibility that both attacked and control trees exhibited a rapid decrease in food quality in response to the addition of more insects to the attacked trees. If so, this suggested that unattacked willows were sensitive to signals from nearby attacked willows or the attacking insects. An experiment to test this hypothesis produced positive results (Figure 2). Tent caterpillars fed leaves from willows attacked by tent caterpillars grew more slowly than larvae fed leaves from nearby and far controls 11.5 days after the initiation of attack. Three days later, larvae fed leaves from both attacked willows and nearby controls grew more slowly than larvae fed leaves from the far controls. No evidence was found for root connections between willows of the same age as the study trees at the same site. This suggests that the results may be due to airborne pheromonal substances!

The burden of proof for such an unprecedented effect should be high, and the foregoing experiments with willows and tent caterpillars cannot be considered to constitute such proof. However, at the very least, they show that the results of experiments designed to test for changes in leaf quality of attacked plants should be interpreted with caution, particularly if control plants are near attacked ones.

Acknowledgements

I thank G. H. Orians for discussion and A. F. G. Dixon, G. Batzli, and unknown reviewers for commenting on the manuscript. Special thanks are due to L. Erckmann who participated in most of the work. N. E. Beckage, C. J. Baron, A. B. Adams, R. Hagen, and J. C. Bergdahl have all contributed to some aspect. I thank Mr.

R. H. DeBoer, Stoneway, Inc., Mr. W. D. Robertson, Mr. R. A. Marr
and the Seattle-Tacoma International Airport Authorities for use
of their facilities, and Dr. J. P. Donahue for identifying the
insects. This research was supported by National Science Foundation grants DEB 77-03258 and DEB 80-05528 to G. H. Orians and D.
F. Rhoades.

Literature Cited

1. Benz, G. _Eucarpia/IOBC Working Group Breeding for Resistance
    to Insects and Mites._ Bull. SROP, 1977/78, pp 155-159.
2. Rhoades, D. F. "Herbivores: Their Interaction with Secondary
    Plant Metabolites"; Rosenthal, G. A.; Janzen, D. H.; Eds;
    Academic Press: New York, 1979, Chapter 1.
3. Rhoades, D. F. "Impact of Variable Host Quality on Herbi-
    vorous Insects"; Denno, R.; McClure, M.; Eds; Academic Press,
    in press.
4. Ryan, C. A. "Herbivores: Their Interaction with Secondary
    Plant Metabolites"; Rosenthal, G. A.; Janzen, D. H.; Eds.;
    Academic Press: New York, 1979, Chapter 17.
5. Haukioja, E. _Oikos_, 1980, _35_, 202-213.
6. Swain, T. S.; Goldstein, J. L. "Methods in Polyphenol
    Chemistry"; Pridham, J. B., Ed.; Macmillan Press: New York,
    1964, pp 131-146.
7. Swain, T.; Hillis, W. E. _J. Sci. Food Agric._, 1959, _10_,
    63-68.
8. Bate-Smith, E. C. _Phytochemistry_, 1975, _14_, 1107-1113.
9. Bate-Smith, E. C. _Phytochemistry_, 1977, _16_, 1421-1426.
10. McKey, D.; Waterman, P. G.; Mbi, C. N.; Gartlan, J. S.;
    Struhsaker, T. T. _Science_, 1978, _202_, 61-64.
11. Riddiford, L. M. _Science_, 1967, _22_, 1451-1452.
12. Hitchcock, C. L.; Cronquist, A. "Flora of the Pacific North-
    west"; University of Washington Press: Seattle, 1973, p 66.
13. Benz, G. _Z. ang. Ent._, 1974, _76_, 196-228.
14. Haukioja, E.; Hakala, T. _Rep. Kevo Subarctic Res. Stat._,
    1975, _12_, 1-9.

RECEIVED September 27, 1982

Function and Chemistry of Plant Trichomes and Glands in Insect Resistance

Protective Chemicals in Plant Epidermal Glands and Appendages

ROBERT D. STIPANOVIC

U.S. Department of Agriculture, Agricultural Research Service, National Cotton Pathology Research Laboratory, P.O. Drawer JF, College Station, TX 77841

Plants have developed various mechanisms of defense against phytophagus insects. Two defensive morphological features are trichomes and glands. Trichomes may be hair-like or glandular. Plant hairs act as physical barriers keeping smaller insects away from the leaf surface. Glandular trichomes and plant glands may exude a sticky substance that entraps and immobilizes small insects, or they may contain toxic constitutents which spill into the surrounding tissue when the gland is ruptured, making it unpalatable or toxic. These toxins are generally weak and do not kill the insect directly, rather they retard insect growth and delay pupation. As a result, the insects are more vulnerable to disease, predation, and the environment. The balance which has evolved between plants and insects could be seriously disrupted if secondary plant toxin analogs are synthesized and used as insecticides. Insects developing resistance to the analogs might develop resistance to the natural toxin.

Plants have served as a food source for fish, insects, and mammals since early biotic times. In response plants have developed intricate physical as well as chemical protective mechanisms. The two defensive structures that are the primary subject of this chapter are trichomes and glands. Trichomes are epidermal appendages of diverse form and structure, such as protective and glandular hairs and scales or peltate hair. Hairs, whether they are unicellular, multicellular, or peltate, may be glandular. Glandular trichomes elaborate various substances, such as volatile oils, resins, mucilages, and gums. Rupture of the cuticle allows the glandular contents to escape. Other types of glands are also found on the surface of the plant. These include: 1) the glandular epidermis covering the leaf teeth as found in <u>Prunus</u>; 2) nectary glands which produce a

sugary secretion associated with flowers and other plant parts;
3) glandular excretory structures which discharge a variety of
oils and resins; 4) oil glands which contain various terpenoid
oils.

The utility of secondary plant chemicals as defensive agents
against pest attack has only recently been widely accepted.
However, this view is still not universally accepted as
illustrated by Luckner's (1) recent comment:

"....secondary metabolism is characterized by a high degree
of order, i.e. by the precise regulation of enzyme level and
activity in metabolic pathways, the compartmentation and
channeling of enzymes, precursors, intermediates, and
products as well as the integration of secondary metabolism
in the programs of cell specialization of the producer
organism. However, characteristics of secondary metabolism
are also the bizarre chemical structures of the formed
products, the restricted occurrence of the different
secondary compounds within groups of living beings and their
usefulness which usually is small or absent, i.e.,
properties which give secondary metabolism its erratic
features."

However, I believe this and other chapters will more than
adequately show that secondary plant chemicals were and still are
important weapons in the plants arsenal of defense.

The following pages review the function and chemistry of
trichomes and glands. The concluding sections address the
insect's response to the plant's defensive chemicals, and end
with an appeal to the chemical pesticide industry.

Extrafloral Nectaries

It is generally conceded that floral nectaries evolved as an
attractant to pollinating insects. The function of extrafloral
nectaries is not so obvious. Some suggest that the secretion of
sugars is associated with a shift from a "sink" to a "source" of
carbohydrates during development (2, 3). Others propose that
sugars are excreted incidental to the excretion of water and
salts (4, 5, 6). However, B. L. Bentley convincingly argues in
her review that extrafloral nectaries are attractants for ants
which act as "pugnacious body guards" in defense of the plant
(7). She bases this contention on four observations. 1) Extrafloral nectaries follow no recognizable diurnal pattern and
produce nectar throughout the day and night. 2) Active nectaries
are found on younger portions of the plant and are often
associated with developing reproductive organs. The nectaries
are fully developed before the reproductive organ, and secretion
ceases soon after the organ matures. 3) Secretory activity is
usually greatest at the height of the growing season and may
continue throughout the year in moist tropical regions. 4)

Attack by sucking insects often increases the rate of secretion
in a positively correlated manner with increased infestation
levels. Bentley argues that nectaries frequently remain a carbohydrate source until maturity, while developing buds, flowers,
and fruit never become a carbohydrate source; sucking insects
reduce the carbohydrate source pressure of an organ and yet
nectar secretions increase with increasing infestations. Thus,
the source/sink hypothesis does not seem to apply.

In Catalpa, Elias and Newcombe report that glandular
trichomes on leaves are not only very similar to nectaries but
are their precursors (8). These authors propose these multiple
nectaries in the lower leaf surface vein axies act as attractants
for beneficial insects to control or minimize the effects of
herbivorous insects. Predatory ants are generally considered to
be beneficial insects. Their presence on plants containing
extrafloral nectaries is a widely observed phenomena. However,
these mercenaries are hired at a cost. For example, Curtis and
Lersten report that one of the most devastating insects observed
to be feeding on cottonwood (Populus deltoides) was an Aphis
species herded by an unidentified species of large black ant (9).
Thus, this "body guard" may "pugnaciously" defend the plant from
some pests only to usher in a plant pest of its choosing.

Trichomes as a Form of Physical Resistance

In medieval times, cities commonly erected walls as a
protective barrier to invasion. Plants have evolved similar
protective mechanisms. Thorns and nettles which repel most
herbivores are two examples of protective barriers. To ward off
insect attack, plants have evolved non-glandular trichomes which
act in a similar fashion. Thus, in some plants, the density,
length, or branching of trichomes have been negatively correlated
with insect survival (10).

The degree of leaf pubescence greatly affects the behavior
of cereal leaf beetle gravid females, Oulema melanopus (11).
Densely pubescent wheat had only one-third as many eggs as the
glabrous control. On pubescent leaves it was often found that
eggs were laid in areas where adults had disrupted the trichome
cover by feeding. Field cage tests showed that the viability of
eggs decreased with increasing pubescence. Eggs laid or
mechanically placed on pubescent wheats were more susceptible to
desiccation. Larval survival and weight gain were also minimized
on densely pubescent leaves (12). This was especially true in
first-instar larvae, while third-instar and fourth-instar larvae
were little affected by pubescence. However, long or dense
trichomes did not correlate in a greenhouse test with resistance
of wheat to greenbug, Schizaphis graminum (13).

A hairy cotton variety that had the most trichomes, the
longest trichomes, and the most branched trichomes was the most
resistant to spider mites and leafhoppers (14). In a no-choice

cage test with three cotton isolines, Lygus herperus oviposition
response was highest, but growth rate slowest on densely
pubescent cotton as compared to normal and smooth cottons (15).
Nymphal survival was essentially equal on all lines, but the
effect of predators was not a factor in these cage tests.
However, pubescence is not always an advantage to plants, as
illustrated by the greater resistance of smooth leaf cotton to
fleahoppers (16).

Hooked trichomes on the French bean, Phaseolus vulgaris have
been reported to capture a diverse group of insect species
including the potato leafhopper, Empoasca fabae (17), and the
aphids, Myzus persicae (18), Aphis fabae (19), and Aphis
craccivora (20). A recent study has shown that the major factor
affecting potato leafhopper damage on the French bean was the
density of hooked trichomes (21). Leafhopper nymphs were impaled
by the trichomes, leading to wounding and eventual death. This
same positive correlation between capture mortality and trichome
density also has been reported for adult leafhoppers on field
bean cultivars, P. vulgaris and P. lunatus (22). Hooked
trichomes growing at angles less than 30° are reportedly
ineffective in capturing leafhoppers.

The inheritance of pubescence type (23) and the morphology
of pubescence in soybean, Glycine max, have been studied, and
densely pubescent plants were found to be the most vigorous (24).
Growth differences were associated with E. fabae infestations.
Glabrous soybeans supported a higher population of E. fabae and
had a higher incidence of oviposition than pubescent varieties
(25, 26). This same observation has been made on other pubescent
host plants (17). The differences in populations of E. fabae
were related to orientation, length, and the erectness of the
leaf trichomes (27, 28). When populations of E. fabae, which is
1.0 - 4.0 mm in length, were compared to that of a springtail,
Deuterosmiathurus yumanensis, which is 0.2 - 0.4 mm in length, on
near isogenic lines of soybean, it was found that both species
had the highest populations on the glabrous genotype. On a
deciduous variety, with only misshapen and severely appressed
trichomes, the population of the small phytophagus species was
depressed, while that of the larger species was not affected.
Populations of E. fabae decreased with increasing trichome
length (0.9-1.6 mm) regardless of trichome density, whereas the
springtail's population decreased with increasing trichome
density. The populations of thrips, Sericothrips variabilis, and
bandedwing whitefly, Trialeurodes abutilonea, showed no
consistent response to trichome variations (29).

Trichomes and Glands as a Form of Chemical Resistance

In addition to physical forms of resistance, plants also
rely on chemicals to immobilize, repel and poison phytophagus
insects. These chemicals may be located in various types of oil

or secretory glands or in glandular trichomes. Glandular hairs or trichomes are widely distributed in vascular plants. A careful histochemical study showed that 39 of 43 plant species from 26 different plants families had stem hairs (30). Some of these plants, such as tomato, had several morphologically distinct hairs. Plant hairs are also chemically distinct as evidenced by their reaction with various histochemical reagents.

Reviews on the intracellular compartmentation of flavonoids and other secondary metabolites have been published (31, 32). A review on the chemical constituents of these glands is complicated by a number of factors. The diversity and the complexity of chemicals produced in a gland or trichome are two such factors. However, one of the most formidable obstacles to a review on glandular constituents is the details reported on the extraction procedure. It is common for investigators to simply divide the plant into its obvious components, e.g. roots, foliar parts, flowers and buds, stems, bark, etc. When examining leaves, for example, there is often no attempt to differentiate between various morphological entities. The leaf is ground or extracted whole. All evidence indicating the occurrence of a particular chemical in a trichome or gland is obliterated.

In the study of resistance mechanisms, it is recognized that the location of a particular toxic chemical may be as important as the presence or absence of that chemical. For example, a high concentration of a toxic chemical in a plant part that the insect does not eat or eats only in the later stages of larval growth, will probably have minimal affect on the resistance of that plant to its host. Conversely, a low concentration of a toxic chemical in a specific site at which the insect feeds in an early larval stage may significantly affect the plant's resistance.

In host-plant resistance studies, it is therefore imperative for investigators to report not only what chemicals are present, but also to report as accurately as possible the principal site(s) at which the chemical is concentrated. This information is generally not reported, therefore this review will undoubtedly omit reported chemicals that are located in glands and glandular trichomes and improperly indicate glandular components that are actually located elsewhere in the plant. However, information on the location of chemicals will be given wherever possible. This section will review the major types of chemicals that are present in glands and glandular trichomes.

Immobilizing Chemicals. Some plants produce a sticky, gummy exudate from glandular trichomes. These exudates effectively immobilize small insects.

A number of plants of the Solanum and Nicotiana genera are particularly adept at producing sticky leaf exudates. In some wild potato species, an exudate is discharged from glandular hairs when aphids mechanically rupture the cell walls (33). The clear, water soluble exudate is stable in the absence of O_2,

but rapidly darkens and forms a precipitate on the aphids limbs
when exposed to air. Eventually the aphid becomes immobilized
and dies. Investigators found several morphologically different
types of glands; however, only one was an apparent source of the
exudate. Once this gland was ruptured, the material was not
replaced. The clear material gave a positive test with the
Folin-Denis phenol reagent. The formation of the dark
precipitate was inhibited by the copper chelating agent, sodium
diethyldithiocarbamate. Polyphenol oxidase enzymes may be
involved in this reaction since copper is essential for their
reactivity. In other research on Empoasca fabae, the percentage
mortality of nymphs, females, and males confined to the glanded
species, S. polyadenium, was 78, 64, and 94 respectively,
compared to less than 20% on 2 nonglandular species (34).
Trichome exudates were found on the mouthparts of E. fabae in 75,
67, and 29% of dead nymphs, females, and males, respectively.
Scanning electron microscopy showed that the tip of the labium
was totally occluded by the trichome exudate. The surviving
leafhoppers had no observable exudate on their bodies. The
viscous exudate, which rapidly darkens and hardens, could be
dissolved in 95% ethanol.

Aphids, attempting to feed on the tomato, Solanum pennellii,
are entangled in a sticky leaf exudate and soon die (35). This
exudate could be removed by washing with cotton saturated with
95% ethanol, but the washed leaves rapidly secreted a new exudate
that also was fatal. Trichomes were a deterrent to oviposition
by the tobacco whitefly, Bemisia tabaci, on both tomato and
tobacco leaves (36). The sticky exudate may partially account
for this since whiteflies were found glued to the glandular
hairs.

Glandular hairs also have been implicated in the resistance
of certain annual Medicago species to the alfalfa weevil, Hypera
postica (37). The annual species, M. disciformis, and M.
scutellata, possess erect glandular hairs, and the exudate from
these glands acts as a glue to immobilize larvae. The perennial
species, M. sativa, which is susceptible to the alfalfa weevil,
possesses only procumbent glands. The exudate from these glands
does not prevent larval movement (38).

Stylosanthes hamata and S. scabra, which are highly
productive and nutritious species of tropical pasture legumes,
are covered with glandular trichomes. The trichomes secrete a
viscous secretion that immediately immobilize larvae of the
cattle tick, Boophilus microplus (39). Ticks have a natural
tendency to climb plants, and wait for a host animal. The
secretion has no repellant properties, so that ticks do not
attempt to seek alternative plants. In addition to immobilizing
the larvae, the plants produce an unidentified volatile
compound(s) that poison the larvae within 24 hours.

As with many plant defensive systems, glandular trichomes
may also be detrimental to the plant. It has been proposed that

the tobacco introduction 1112 (T.I. 1112) may be resistant to
some insects because its leaves, although hairy, lack glandular
trichomes (40). The glandular trichomes that cover the aerial
portions of most tobacco plants are a conspicuous source of
odors, as well as a source of the sticky exudate. It has been
proposed that the trichome exudate serves as an attractant or
oviposition stimulant for tobacco budworm moths and as an
arrestant for alate green peach aphids. The absence of glandular
trichomes on the leaves of T.I. 1112 may account for its partial
resistance to these insects. However, T.I. 1112 has glandular
trichomes on the calyces which may explain the readiness of
tobacco budworm moths to oviposition on its flowers and flower
buds. The absence of glandular trichomes also increased
Trichogramma parasitism on Heliothis eggs on T.I. 1112.
Heliothis egg mortality from this parasite on other tobacco is
negligible because the tiny parasitic wasp becomes stuck in the
exudate (41).

Toxic and Repellent Allelochemics

Levin has proposed to classify toxic plant chemicals
involved in resistance into two broad categories, those that
confer specific resistance and those that confer general
resistance (42). Compounds exhibiting specific resistance are
characterized as extremely toxic to a small group of specialized
pathogens or herbivores. Each compound is present only in a few
species, and is often tissue specific. They tend to reach their
highest concentration in young leaves and fruits and decrease as
they mature. Examples would be sinigrin, tomatine, solanine, and
gossypol. Compounds classified as showing general resistance
deter, repel, or are weakly toxic to most microorganisms and/or
herbivores. These compounds are present in several plant species
and sometimes in families of different orders. They are
generally not tissue specific, and their concentration increases
as the tissue matures. Examples include chlorogenic acid,
quercetin, and tannins.

Although a toxic plant chemical may not fit either category
perfectly, those chemicals discussed below that are tissue
specific would generally be considered to show specific
resistance. It is interesting to note that those phytochemicals
that are especially toxic to one group of insects are quite often
essential dietary ingredients or feeding stimulants to other
insects that feed primarily on that plant.

The phytochemicals that are discussed below may be divided
into three classical categories: alkaloids, flavonoids, and
terpenes. Straight chain carbon compounds, such as hydrocarbons,
waxes, fatty acid and alcohols also occur in glands. The
alkaloids tend to react as feeding deterents and as poisons
(43). They are probably the major class found in any lists of
plant toxins (44, 45). Flavonoids are well known antibacterial,

antimicrobial and antiviral agents (46-49). Flavonoids also
inhibit the growth of Heliothis zea larvae (50). They may also
act as either feeding deterrents or stimulants for different
insects (22, 43, 51, 52, 53). It was found that vicinal
hydroxylation was necessary, but not sufficient for growth
inhibition (50). Flavanones, flavones, and flavonoid glycosides
inhibit Na-independent passive transport of sugars by intestinal
epithelial cells (54). The inhibitory action is influenced by
the extent and position of hydroxylation of the flavonoid
nucleus, the degree of planarity of the molecules, the presence
of a carbonyl group, and glycosylation. The occurrance of
flavonoid aglycones throughout the plant kingdom has been
carefully cataloged (55).

Terpenoids are a diverse group of compounds showing a wide
spectrum of biological activities. Among the most biologically
active terpenoids are the sesquiterpene lactones. Their
biological activity has been reviewed (56, 57, 58).
Sesquiterpene lactones show cytotoxicity, mammalian toxicity,
allergic contact dermatitis, allelopathy, antibiotic activity to
bacteria and fungi, schistosomicidal activity, antifeedant
activity toward mammals and insects, insect larval growth
inhibition, and deter insect oviposition. Their activity can be
explained in many cases by Michael addition reactions between a
nucleophile and the exocyclic α,β-unsaturated γ-lactone function.
Other types of terpenoids are antifeedants to the African
armyworm (59), and the obscure root weevil (60). General
terpenoid antifeedants from east African tropical plants have
been reviewed (61, 62). Terpenes may act as feeding stimulants
or inhibitors depending on concentration (63). Terpenoids also
exhibit insect growth regulation (64), antiecdysone activity
(65), and are active against a wide range of bacteria and fungi
(66-71).

Alkaloids and Phenols from Tobacco, Tomato, and Potato.
Tobacco, tomato, and potato plants contain a number of toxic
alkaloids. Probably the most widely studied is nicotine. The
insecticidal properties of this and other tobacco alkaloids have
been reviewed (72). A study of Nicotiana showed that alkaloids
are secreted by trichomes in the seven species tested (73).
Nicotine was the major alkaloid identified in the trichome
secretion (74). Anabasine and nornicotine were also identified
in two species   (75). Other ether soluable constitutents have
also been identified (76). Aphids were killed by contact with
these secretions. Trichome secretions from Nicotiana also were
toxic to the tobacco hornworm and the two spotted spider mite
(77, 78). In a physiological study of insect's response to
nicotine, it was found that the penetration rate of nicotine into
the isolated nerve cord was pH dependent (79). At the pH of
insect haemolymph the penetration of nicotine into the nerve cord
was higher on a unit weight basis in the nicotine-susceptible

silkworm (Bombyx mori) than in the nicotine-resistant hornworm,
Manduca sexta. This was interpreted as indicating a less
efficient ion-impermeable barrier of the silkworm nervous
system.

The effects of the Solanum alkaloids on the Colorado potato
beetle, Leptinotarsa decemlineata, have been carefully studied.
Solanine, chaconine, leptine I, leptinine I, leptinine II,
demissine, and tomatine were all found to be feeding deterrents
(74, 80). Nicotine, even in quite small concentrations, was very
toxic (74). The Colorado potato beetle was found to feed more on
young tomato plants which had a low concentration of tomatine
than on mature or flowering plants that had a higher
concentration of tomatine (81). Feeding was also inhibited
(20-80%) when tomatine was infiltrated into tomato leaf disks at
concentrations between 65 and 165 mg/100 g fresh weight.
Nornicotine, solanine, and tomatine were all toxic to nymphs of
the two-striped grasshopper, Melanoplus bivittatus, which is a
general feeder (82). Trichome secretions of tomato are also
toxic to the green peach aphid (83). An extensive list of plant
alkaloids and other phytochemicals that act as insect feeding
deterrents and repellents as well as feeding stimulants and
attractants has been published (45). In the case of the tobacco
hornworm, M. sexta, norvalatine and nonpolar or weakly polar
fractions of tomato leaves stimulate feeding, while feeding
deterrents were present in water extracts of tomato leaves (84).
Tomato trichome exudates are topically toxic to the spider mite,
Tetranychus urticae (85). The flavonal glycoside, rutin, has
been isolated from the tetracellular glandular trichomes of the
tomato plant, Lycopersicon esculentum (86). This and other
trichome exudates reduce larval growth of Heliothis zea. A
morphologically different type of glandular trichome on the wild
tomato, L. hirsutum, contains 2-tridecanone, which is toxic to H.
zea and M. sexta when applied topically (87). Total
glycoalkaloids in foliage of wild potatoes were significantly
correlated with resistance to the potato leafhopper (88).

Flavonoids and Terpenoids in Bud Exudate. The occurrence of
flavonoid aglycones in buds has been reviewed by Wollenweber and
Dietz (55). They note that the flavonoid aglycones with a low
number of hydroxyl groups and/or a high number of methoxyl
groups, because of their lipophilic nature, do not accumulate in
the cell sap. Thus, these types of compounds are found in plants
with glandular trichomes, excretion cells, or cavities. A
variety of plant families produce bud exudates containing widely
varying flavonoids. It was previously mentioned that flavonoids
have a wide range of biological activities. However, the
biological activity of only a small fraction of the flavonoids
has been investigated against an even smaller fraction of plant
pests. For this reason, specific compounds will not be listed,
but it is clear that many, if not most, of the flavonoids confer
some protection on their parent plants.

Table I. Flavonoid Aglycones in Bud Exudates

Family	Species	Flavones	Flavonols	Flavanones	Dihydroflavonols	Chalcones	Dihydrochalcones
Salicaeae	Populus [a,b]	4	12	5	1	2	1
Betulaceae	Alnus [c,d,e]	6	14	1		1	
	Betula [a,c,f,g,h]	9	20	2			
	Ostrya [c]	2	12				
Rubiaceae	Elaegia [i]	1					
Asteraceae	Acanthospermum [f]		1				
Hippocastumaceae	Aesculus [k,l]		11				
Rosaceae	Prunus [a,m]		9	1			
Rhamnaceae	Rhamnus [k,m]		4				
Didieraceae	Decrya [n]		1				

[a]Ref. (55)   [h]Ref. (91)
[b]Ref. (110)   [i]Ref. (94)
[c]Ref. (111)   [j]Ref. (116)
[d]Ref. (112)   [k]Ref. (92)
[e]Ref. (113)   [l]Ref. (117)
[f]Ref. (114)   [m]Ref. (118)
[g]Ref. (115)   [n]Ref. (119)

   The flavonoid aglycones reported in buds of Alnus, Betula, Ostrya, Elaegia, Acanthospermum, Aesculus, Prunus, Rhamnus, and Decrya are listed in Table 1. The biosynthesis of flavonoids in subcellular glands has been investigated in Populus nigra (89), and the flavonoids in the lipophilic coating in Populus buds have been thoroughly analyzed (90). In addition to this wide variety of flavonoids, mixtures of terpenoids have also been reported in bud exudates. In Betula nigra the bud exudate is reported to be mainly unidentified terpenoids (91). Unidentified terpenoids are found in bud exudates from Aesculus sp. (92), Alnus sp., Betula sp., and Ostrya sp. (93). Some specific terpenoids have been identified. The triterpenes isofouquierol, dammarenediol-11, and (20S)-dammar-24-ene-3 , 20, 26-triol have been isolated from the bud exudate of Elaegia utilis (94), and δ-amyrenone is reported to be a major constituent of the bud exudate from alder trees

(<u>95</u>). Several melampolide type sesquiterpene lactones have been
isolated from the Tanzanian plant <u>Acanthospermum</u> <u>glabratum</u> (<u>96</u>,
<u>97</u>); however, their occurence in the bud exudate is uncertain.
Although the constituents of bud exudates are usually
ascribed to flavonoids and terpenes, other constituents may also
be present. The flower buds of <u>Alnus</u> <u>pendula</u> produce a viscous
material which includes a wide variety of compounds including
acids (cinnamic, β-phenylpropionic, and benzoic), an ester (β-
phenylethyl cinnamate), ketones (<u>trans</u>, <u>trans</u>-1,7-diphenyl-1,3-
heptadiene-5-one and benzylacetone), an alcohol (β-phenylethyl
alcohol), an aldehyde (cinnamaldehyde), stilbenes (pinosylvin and
pinosylvin monomethyl ether), and phenols (eugenol and chavicol),
as well as paraffins, flavonoids (pinocembrin, pinostrobin,
alpinetin, and galangin), and triterpenes (δ-amyrenone and
taraxerone) (<u>98</u>). In <u>A</u>. <u>sieboldiana</u> two ketones were also
isolated (yashabushiketol and dihydroyashabushitol).

Farinose Exudates of the Polypodiaceae and Primulaceae.
Members of the <u>Polypodiaceae</u> produce a yellow or white powdery
deposit on the lower surface of their fronds. These deposits are
usually referred to as farinose exudates. Farina from species of
<u>Pityrogramma</u>, <u>Cheilanthes</u>, <u>Adiantum</u>, and <u>Notholaena</u> have been
carefully studied. The farinose coating of these plants is
formed by the terminal cell of small hairs usually found on the
lower surface of the frond.
Wollenweber (<u>99</u>) has reviewed the morphology and chemistry
of these ferns which are prodigious producers of farina,
representing 0.9%-5.0% of the dry weight of the fronds.
Wollenweber reports that flavonoids are the major constituents of
the farina. Chalcones (<u>100</u>-<u>106</u>) dihydrochalcones (<u>102</u>, <u>103</u>, <u>107</u>,
<u>108</u>, <u>109</u>), flavonols (<u>102</u>, <u>106</u>, <u>108</u>, <u>120</u>-<u>128</u>), flavones (<u>102</u>,
<u>108</u>, <u>121</u>, <u>123</u>, <u>125</u>, <u>128</u>, <u>129</u>), flavanones (<u>130</u>), and the
chalcone-like compound, ceroptene (<u>124</u>, <u>131</u>), have been reported
in the farina from ferns. The flavonoid pattern in the farina of
220 samples from 14 species of the goldenback and silverback
ferns has been reviewed (<u>132</u>). Flavanones with methyl
substituents (<u>133</u>) and flavonols esterified at the 8 position
with butyric and acetic acid (<u>134</u>) have also been identified in
farina. The triterpenes diplopterol, fernene, adiantone,
isoadiantone (<u>135</u>, <u>136</u>), and 6 β, 22-dihydroxyhopane (<u>136</u>) and
various phytosterols (<u>137</u>-<u>140</u>) have been isolated from fern.
However, the morphological source of these compounds is not
clear. This is also true for the sesquiterpenes (<u>141</u>, <u>142</u>, <u>143</u>),
the lactone, calomenanolactone (<u>144</u>), the ecdysone analogues
(<u>145</u>, <u>146</u>), and p-hydroxystyryl-β-D-glucoside (<u>147</u>). However,
the novel dihydrostilbene, 5-hydroxy-3,4'-dimethoxy-6-carboxylic
acid bibenzyl, is a farinose constitutent of <u>Notholaena</u> <u>dealbata</u>
and N. <u>limitanea</u> (<u>148</u>). Glycosides of flavonols have also been
isolated from farina (<u>121</u>). A group of hydrocarbons, terpenoids,
and fatty acids have been detected in the glandular lipids of

Dryopteris assimilis ferns by gas chromatography (149). The
dihydrochalcone, 2',6'-dihydroxy- 4,4'-dimethoxydihydrochalcone,
from Pityrogramma calomelanos has shown marked antifungal
properties (150).Other flavonoids from Pityrogramma are reported
to be allelopathic agents. 2',6'-Dihydroxy-4'-methoxydihydro-
chalcone inhibited spore germination and gametophyte development
by P. calmelanos at all concentrations tested, but 2,6-dihydroxy-
4'-methoxychalcone inhibited germination at 5 X 10^{-6} M and
stimulated germination at 5 X 10^{-7} M. Izalpinin showed
similar effects, inhibiting germination at 5 X 10^{-3} M and
stimulating germination at 5 X 10^{-4} M. These flavonoids
appear to act as germination inhibitors around the parent plant
(151).
    Many primroses (Primula) also produce a farinose exudate on
their stems and leaves. Eighteen species have been investigated
(152). The major components were found to be flavone,
5-hydroxy-and 5,8-dihydroxyflavone and the three 2'-hydroxy
derivatives.

    Sesquiterpene Lactones and Other Glandular Trichome
Components from Veronia, Parthenium, Helianthus, and Artemesia.
Sesquiterpene lactones are major constituents of Veronia,
Parthenium, Helianthus and Artemesia. As indicated previously,
the sesquiterpene lactones as a group are active in a wide range
of biological systems. Mabry, et al. found that the
sesquiterpene lactone, glaucolide A, found in the glandular
trichomes of some species of Veronia, protects these plants
against some insects (153). Leaf diets were prepared with and
without glaucolide A for six insect larvae: 1) the yellow
woolybear, Diacrisia virginica; 2) cabbage looper, Trichoplusia
ni; 3) yellowstriped armyworm, Spodoptera ornithogalli; 4)
saddleback caterpillar, Sibine stimulea; 5) fall armyworm,
Spodoptera frugiperda; 6) southern armyworm, S. eridania (57).
In a free choice situation, all insects preferred the diet
without the lactone, and the Veronia species that did not contain
glaucolide A. Larvae of 3,5, and 6 were significantly smaller
when fed diets containing glaucolide A compared with a control
diet, while larvae of 1 and 2 were unaffected. Furthermore, the
days to pupation increased for all larvae except 1. Other work
has shown that plants containing glaucolide A deterred ovipostion
in some insects (154). Other sesquiterpene lactones have been
isolated from Veronia species (155, 156, 157) as well as
triterpenes (158, 159), long chain alkanes (159), and an iridoid
glucoside (160). The morphological origin of these latter
compounds is uncertain.
    Parthenin is a major component found in the trichomes of
Parthenium hysterophorus, making up about 8% of the dry weight of
the plant (161). It causes dermatitis in humans and cattle (162,
163) and acts as an allelochemic (164). Parthenin acted as an
antifeedant and was toxic orally at 0.01% to Dysdereus koenigi,

Aedes aegypti, Tribolium castaneum, Periplaneta americana, and
Phthormea operculella (165). Two other sesquiterpene lactones,
coronopilin and tetraneurin-A (166), and 34 flavones (19
glycosides and 15 aglycones) have been isolated from Parthenium
species (167).

  Sesquiterpene lactones may offer some insect resistance in
sunflower. Glandular trichomes occurring on the leaves,
phyllandries, and anthers of Helianthus maximiliani contain a
sesquiterpene lactone, maximilin-C. The first instar larvae of
the sunflower moth, Homeosoma electellum, suffered a high
mortality rate when fed on a wheat germ diet containing
concentrations of 1.0 and 10.0% of maximilin-C (168, 169). A
number of other sequiterpene lactones have also been isolated
from Helianthus species (170-175). The diterpenoid acids,
trachyloban-19-oic acid and (-)-kaur-16-en-19-oic acid, have also
been implicated in resistance to the sunflower moth (176).
However, as with maximilin-C, relatively large dosages (0.5-1.0%)
were required to reduce growth on synthetic diets by one-half, as
compared to a control. Other diterpenoids (177, 178, 179),
triterpenoids (180, 181, 182), two acetylinic compounds (172,
183), a flavone (175), and volatile constituents (184) have also
been reported from Helianthus species. A preliminary report
indicates that brittle brush contains a glandular trichome
sesquiterpene that is a feeding deterrent to moth larvae (185).

  The sesquiterpene lactones in Artemesia species have been
reviewed (186). Histochemical tests have shown that A. nova has,
in addition to nonglandular trichomes, glandular trichomes
covering 21-35% of its leaf surface (187). These glands hold a
clear fluid which contains some of the monoterpenes and all of
the sesquiterpene lactones present in the leaves.

  Components of Glandular Trichomes in Populus, Prunus,
Newcastelia, and Salvia. In Populus deltoides, the marginal
teeth of the first leaves to emerge are covered with
non-glandular trichomes. In successive leaves, the teeth have
glands that secrete a resin as the lamina unrolls. Extrafloral
nectaries occur proximal to each glandular tip. Field
observations and a laboratory feeding experiment indicate that
the resin acts as an insect repellant (9). In P. deltoides,
insect feeding was greater on newly expanded leaves in which the
resin had dried than on buds covered with fresh liquid resin
(188). Larvae of the cottonwood leaf beetle, when given leaves
of P. deltoides in which only the margin was resinous, fed
normally, pupated, and emerged as normal adults. Larvae given
resin-covered leaves did not eat them. The few larvae that
pupated failed to complete their life cycle. Crude extracts of
poplar (Populus) tree leaves exhibited antibacterial and
antifungal activity (189); responsible agents have not been
identified.

  The leaf teeth in Prunus are covered by glandular trichomes.

A coumarin glucoside, tomenin, has been isolated from Prunus
tomentosa, but its morphological origin is not indicated (190).
     Five different types of terpenoid secreting trichomes have
been described in the Western Australian shrub Newcastelia
viscida (191). A resin is released beneath the cuticle of the
glandular hair, which expands and eventually breaks, releasing
the resin. Once terpenoid production has ceased, the gland is
closed off by leaf cutinization of the walls. As the gland
matures, a drop of resin forms and runs down onto the adaxial
leaf surface. After rupture, the glands apparently become
functionless but new glands are formed during leaf expansion, and
resin is continuously produced until the leaf is fully expanded.
Terpenes identified in N. viscida include the triterpenoic acids,
oleanolic and betulic acids (192), and the tricyclic diterpene,
isoprimara-9(11),15-diene-3,19-diol (193).
     A quick dip of the aerial parts in ether has been used to
extract the glandular trichome contents of Salvia glutinosa
(194). The main component was the triterpenol, α-amyrin. Small
amounts of flavones and flavonols were also isolated.

Leaf Exudates from Didymocarpus, Larrea, Hymenaea, and
Beyeria. Didymocarpus pedicellata is a small herbaceous plant
found in the western Himalayas. It produces a reddish, dusty
leaf exudate from which 7-hydroxy-5,6,8-trimethoxyflavone (195),
two chalcones, pedicin, and pedicellin, and flavanone, and
isopedicin, (196) have been isolated. Two other chalcones,
pashonone and methylpedicin, have been isolated from the leaves
of D. pedicellata, and the latter is reported to be one of the
major components (197). Leaf components are reported to be toxic
to fish (198, 199). Diterpenoid acids have been isolated from
the leaves of D. oblonga (200, 201).
     Larrea tridentata and L. divaricata are arid and semiarid
plants found in North and South America. Larrea produces a leaf
resin that accounts for 10-15% of the dry weight of the leaves.
The resin is composed of about half nordihydroquaiaretic acid,
which is one of the most powerful antioxidants known (202). The
other half is primarily composed of flavonoids (202, 203).
Volatile constituents of Larrea have been reported (204) and
their antiherbivore chemistry reviewed (205).
     The tropical legume, Hymenaea courbaril, produces a leaf
resin that was tested as a defense against the generalist
herbivore, beet armyworm (Spodoptera exigua) (206). Larvae
showed a dose-response in the decrease of pupal weight and delay
in pupation. In a palatability test, S. exigua strongly
preferred untreated to treated bean leaf disks. The primary
components in leaf resin were found to be the sesquiterpene
hydrocarbons, caryophyllene, α- and β-selinene, and β-copaene
(207).
     The leaves of Beyeria brevifolia, from western Australia,
have a hard resin coating. Diterpenols and diterpenoic acids

have been identified in this resin (208). Other diterpenoids
have been isolated from B. calycina (209, 210).

   Cotton Pigment Glands. Plants belonging to the genera
Gossypium (cotton), Cienfuegosia, Thespesia, and Kokia contain
subepidermal pigment glands from which the phenol, gossypol, has
been isolated (211). Structure elucidation studies, chemical
reactivity (212, 213, 214), and biosynthesis (215) of gossypol
have been reviewed. Because gossypol is present in raw
cottonseed meal, its toxicity, especially to monogastric animals,
has been carefully studied. The toxicology, physiological
effects and metabolism have recently been reviewed (216, 217).
The Chinese reports that gossypol may act as an antifertility
agent in men (218) have renewed interest in this and related
compounds.
   Gossypol has been isolated in an optically active form from
Thespesia populnea ([α]$_D^{19}$ +445° ± 10°, CHCl$_3$) (219) and in a (+)
and (±) form from cottonseed (219, 220, 221). Glands in the
foliar parts of G. hirsutum produce, in addition to gossypol, the
terpenoids, p-hemigossypolone (222), heliocides H$_1$, (223), H$_2$
(224), H$_3$ (225), and H$_4$ (223). In addition to these
compounds, G. barbadense also produces the methyl ether
derivatives, p- hemigossypolone methyl ether and the heliocides
B$_1$, B$_2$ (226), B$_3$, and B$_4$ (227). G. raimondii produces
raimondal (228). The Schiff base, gossyrubilone, has been
detected in the glands of G. hirsutum (226). Anthocyanins (229,
230) and flavone (230) have been reported in glands (229) and
cyanidin-3-glucoside has been isolated from G. barbadense glands
(231). Hemigossypol, hemigossypol-6-methyl ether,
6-deoxyhemigossypol (232), gossypol-6-methyl and 6,6'-dimethyl
ether (233), desoxyhemigossypol and desoxyhemigossypol-6-methyl
ether (234) have been isolated from glands in roots, stems, or
seed.
   Several of these compounds have exhibited antibiotic
activity. Zaki, et al. found that hemigossypol and
desoxyhemigossypol were more active than gossypol against the
fungus, Verticillium dahliae, (235). Hemigossypolone was also
active against V. dahliae (236). Terpenoids were found to be
exuded into the rhizosphere by cotton roots (237) and may be
responsible for resistance to root rot (238). Gossypol has been
shown to have antibacterial (239), antiviral (240, 241, 242) and
antitumor   (243) activities. Apogossypol, which has a lower
mammalian toxicity, retains this antiviral activity (244).
   The chemical constituents in the cotton plant have been
extensively studied. Volatile or steam distillable compounds
present in cotton buds (245-253) and leaves (254, 255, 256) have
been extensively cataloged. A number of the surface lipids have
been identified (257). Two anthocyanins, cyanidin-3-β-glucoside
(231, 258) and pelargonidin (259), and the flavonols and flavonol
glycosides, quercetin, kaempferol, isoquercitrin, quercetin-7-

glycoside, and quercetin-3'-glycoside (260) have also been
isolated. Isoquercitrin, quercitrin, and quercetin are toxic and
inhibit growth of the bollworm, H. zea, tobacco budworm, H.
virescens, and the pink bollworm, Pectinophora gossypiella, in
laboratory tests on artificial diets (261). These compounds were
found to be more toxic to the tobacco budworm than to the
bollworm. Some of these compounds, such as cyanidin-3-β-
glucoside and the terpene essential oils, are undoubtedly in, but
not necessarily confined to, the pigment glands. Others may be
located in glandular trichomes.

Variations in the concentration of gossypol and other
constituents, such as tannins and flavonoids, have been measured
with respect to such variables as cultivar (244, 262, 263), plant
age (264), and plant part (265). In each study, insect growth
was negatively correlated with gossypol or terpenoid aldehyde
content. Hanny, et al. found cabbage looper (Trichoplusia ni)
damage correlations of -0.46, -0.60, and -0.31 for flowerbud,
terminal leaf, and leaf terpenoids, respectively (262). Yellow
cotton anthers were found to have a higher gossypol content than
cream-colored anthers. This is believed to be responsible for
suppressing Heliothis virescens larvae growth (265).

Cook was among the first to propose that the pigment glands
of cotton might act as a repellant to the bollworm (266).
Lukefahr and Houghtaling found that cultivars of cotton with a
flowerbud gossypol content of 1.7% significantly reduced the
populations of tobacco budworm in large, replicated field tests
(267). It also was found that this experimental Upland cotton
was utilized less efficiently by the bollworm than a standard
line containing 0.5% gossypol (268). Food consumption by the
tobacco budworm larvae decreased with increasing gossypol
content. Seaman, et al. observed a strong correlative
coefficient between a plant's resistance level and the
concentration of gossypol and the heliocides (269). This same
study reports that in lines differing greatly in resistance,
hemigossypolone and the four heliocides occur in roughly the same
proportion in buds, with gossypol and the heliocides each
comprising about 44% of the terpenoids. This is in contrast to a
subsequent study which reported gossypol as the primary terpenoid
(270) in buds. These differences may be accounted for by the age
of the tissue examined. Seaman also reported that the
concentration of each terpenoid aldehyde in resistant individual
plants is approximately twice the amount found in susceptible
plants (269).

The toxicity of the cotton terpenoid aldehydes to different
larvae has been reported by several groups (227, 259, 270-275).
These results are summarized in Table 2. Although there is
variation in the test results among the different laboratories,
it is obvious that gossypol and probably the other terpenoid
aldehydes have a wide range of toxicity to many moth larvae.

Waiss, et al. report that the first instar larvae of

Table II.   Concentration of Cotton Terpenoids Required to Reduce Laval Growth by 50% Expressed as Percent Diet

Compound	*Heliothis virescens*			*H. zea*	*Pectinophora gossypiella*		*Spodoptera littoralis*	*Earias insulana*
	Stip.[a]	Chan[b]	Hedin[c]	Chan[b]	Chan[b]	Stip.[a]	Abou Dona[d]	Meisner[e]
Gossypol	0.04	0.12	0.12	0.07	0.07	0.93	0.31	0.43[h]
Heliocide H_1	0.10	0.12	–	–	0.09	0.03	–	–
Heliocide H_2	0.46	0.13	–	–	0.09	0.10	–	–
Heliocide H_3	0.16	–	–	–	–	–	–	–
Hemigossypolone	0.29	0.08	–	<0.07	0.07	0.04	–	–
Heliocide B_1	0.20	–	–	–	–	–	–	–
Heliocide B_2 & B_3	N.E.	–	–	–	–	–	–	–
Hemigossypolone methyl ether	N.E.	–	–	–	–	–	–	–

[a]Ref. (227)
[b]Ref. (284)
[c]Ref. (259)
[d]Ref. (272)
[e]Ref. (273)
[f]Added to the diet as a mixture of heliocides B_2 (67%) and B_3 (33%).
[g]No effect.
[h]As calculated.

Heliothis zea generally avoided consuming the pigment glands
(276) indicating that gossypol may act as a feeding deterrent.
Gossypol apparently acts as a feeding deterrent to Spodoptera
littoralis, Egyptian cotton leafworm. When polystyrene lamellae
were painted on one side with sucrose and on the other side with
gossypol, the 1% gossypol lamellae strongly suppressed feeding
over that of a control (277). The feeding rate was negatively
correlated with the gossypol concentration. Larvae fed about
half as much on lamellae painted with a leaf extract of a high-
gossypol cotton as on lamellae painted with an extract of a
glandless cotton.

Larvae (90-110 and 170-190 mg) of S. littoralis, raised on
cotton plants with a mean content of 1.23% gossypol in dry leaf
powder, weighed less, had delayed pupation, and a lower pupal
weight compared to larvae raised on a cotton having an
intermediate gossypol content (278). In neonates the effect was
even more pronounced. When S. littoralis larvae were offered
diets containing 0.5% gossypol acetate throughout their life
span, larval mortality was nearly 70% after the first 10 days and
only 0.3% of the larvae pupated. At a concentration of 0.25%
gossypol acetate, 44% of the larvae died within 10 days and 26%
of the larvae pupated (279).

Studies on the spiny bollworm, Earias insulana, have given
similar results (273). Survival and average weights of larvae
raised on diets containing gossypol were lower than the control.
The same was true with regard to percent pupation and adult
emergence. Seventy percent of larvae raised on a cotton cultivar
with a gossypol content of 0.10% in the bolls pupated, while only
9% of those raised on a cultivar with 2.34% gossypol in the bolls
pupated.

The effect of gossypol on three consecutive generations of
S. littoralis has been studied by El-Sebae, et al. (280).
Larvae of S. littoralis were treated topically with different
concentrations of gossypol. After emergence, the moths were
treated topically at the same concentrations of gossypol and
paired. The number of eggs and percent hatchability were
calculated. First instar larvae were reared on an artificial
diet containing different gossypol concentrations. The
treatments were repeated in the next generation. At the lowest
concentration of gossypol in the diet (0.25%), the number of eggs
laid decreased to about 52% of the control in the first
generation. In the second and third generations, the number of
eggs laid decreased to about 10% and 2%, respectively, of the
control. Hatchability dropped in the first, second, and third
generations to 50%, 43%, and 19%, respectively, of the control.
At a concentration of 0.50% gossypol in the diet, the effect was
even more dramatic, with the hatchability dropping below 1% in
the third generation.

Meisner, et al. reported inhibition of growth and protease
and amylase activity in S. littoralis larvae after feeding on

high gossypol cultivars of cotton (281). El-Sebae, et al. also
studied the midguts of S. littoralis (280). They found that
gossypol: 1) inhibited protease activity and lipid peroxidation,
2) increased microsomal N-demethylation activity, 3) stimulated
mitochondrial ATPase activity at low concentrations (10 μM), and
4) inhibited ATPase activity at higher concentrations. The
inhibition of protease and amylase activity and lipid
peroxidation agree with gossypol's ability to retard growth by
reducing protein biosynthesis. The inhibition of ATPase, which
provides energy for biosynthesis, is also compatible with the
observed results. The implications of increased microsomal
N-demethylation activity is discussed below.

     Larvae of S. littoralis treated topically with gossypol (100
μ g/g body weight) 24 hours before being treated with various
insecticides had a higher LD_{50} than control insects (272).
The increase in LD_{50} was highest for the chlorinated
insecticide, endrin (200%), followed by the organophosphate
insecticide, Cyolane (154%), the phosphorothioate insecticide
leptophos (123%), and the carbamate insecticide, Zectran (87%).

     S. littoralis larvae had a lower mortality rate when allowed
to feed on leaves treated with the insecticide, phosfolan, from a
cotton cultivar with a high gossypol content as compared to a
cultivar with little or no gossypol (282). The ability of S.
littoralis larvae to rapidly detoxify insecticides after
treatment with gossypol, agrees with the results of El-Sebae
(280) that microsomal N-demethylation reaches its maximum
activity after 24 hours. An increase in microsomal oxidases has
also been observed in the liver of rats which were fed gossypol
(283).

     As opposed to the results indicated above, the effect of
gossypol on the boll weevil is quite different. Gossypol is a
feeding stimulant to the boll weevil (260). Boll weevils feeding
on an artificial diet were healthier and had improved egg hatch
when the gossypol fraction from cotton seed was used as the
principle protein source.

Insect Resistance to Secondary Plant Allelochemics

     The introduction of synthetic pesticides heralded a new era
in which agricultural production flourished. However, the use of
pesticides introduced unexpected problems. In the case of
insecticides, beneficial as well as harmful insects were killed.
Insects rapidly developed resistance to the insecticides,
requiring the introduction of new and more potent chemicals.
This same scenario has not been observed with plant
allelochemics. Phytophagous insects and plants have coexisted
for eons. In general, plants, through their allelochemics have
exacted their toll on insects, while falling prey to these same
insects. The reasons for these differences deserve comment.
     Companies manufacturing pesticides, in addition to
considerations such as cost and safety, select only those

chemicals that kill the majority of the target organisms with
which they come in contact. Indeed, the agricultural producer
demands such products. Even federal licensing agencies might
balk at permitting the sale of a pesticide that was only
marginally effective. Yet this is the path plants commonly
follow. Experience has shown that given adequate pressure,
nearly every species of insect is capable of developing some
tolerance to a particular insecticide (285). The development of
resistance depends on the presence of resistance genes and
adequate selection pressure by which these genes are concentrated
and integrated into the genome of the population. If the genetic
potential for resistance is present, the rate of this development
will depend on factors such as the frequency of the resistance
genes, their dominance, the selection pressure and previous
exposure to the insecticide. Assuming a sufficient survival
rate, resistance will develop more rapidly the more intense the
selection pressure (285).

     With plant allelochemics, the selection pressure is
generally not high enough to concentrate the resistance genes
in the genome of the population. The plant allelochemics slow
insect growth and extend the time of pupation. The larvae thus
become more susceptible to disease, predators and environmental
stress. However, some insects with susceptible genes do survive
and their genes are continually included in the pool.

     Plants that produce "specific" toxins may be plagued by
insects that develop a tolerance to these toxins in much the same
way as insects develop tolerance to synthetic insecticides. Two
examples from this chapter are the tobacco hornworm and the boll
weevil which have developed a high tolerance to nicotine and
gossypol, respectively. Some occurrence in the distant past may
have placed sufficiently high selection pressure on these
insects that they developed tolerance to these compounds.
Alternatively, the same effect could have occurred by a low
selection pressure applied over a very long period time. Other
plants protect themselves by employing "general" toxins.
Selection pressure from this type of toxin would be even less.

     Thus plants have evolved which produce chemicals which are
only marginally toxic to insects. Production of chemicals that
are extremely toxic to insects would give a plant only a
temporary adaptive advantage.

Host-Plant Resistance and the Pesticide Industry

     The recent advances in identifying and utilizing
allelochemics involved in host-plant resistance has drawn the
attention of the pesticide industry. A potential problem that
may not be recognized, is the effect on insects if analogs of
plant protective chemicals are sprayed on agricultural crops.
Insects treated with such analogs, could rapidly become tolerant
not only to the analog, but also to the natural allelochemic.

Insects treated with such analogs, could rapidly become tolerant not only to the analog, but also to the natural allelochemic. There are examples of cross-resistance in which an insect treated with one chemical was found to have resistance to a chemical with which it had never been treated. A strain of house-fly which had developed resistance to the synthetic insecticide, DDT, was found to be resistant to the natural pyrethrin insecticides (286). These two insecticides are chemically very different, and yet they may react at the same site inducing cross-resistance.

  In cotton, for example, it has been observed that some insects that are not normally cotton pests, become a pest on glandless cottons which do not contain gossypol. Such insects, which are minor pests because of their avoidance of gossypol, if sprayed continuously with the proper insecticide, could develop, through cross-resistance, a tolerance for gossypol and appear as a new major pest on this crop. Thus, the potential for developing such resistance is already present even with synthetic insecticides that are quite distinct from the plant's allelochemic. The potential becomes infinitely greater if the allelochemic and synthetic insecticide are chemically similar. Some may believe that an insect cannot develop a tolerance fo natural insecticides. On the contrary, the Mexican bean beetle has developed resistance to the natural rotenoid insecticides (287). There are already scattered reports of Heliothis developing resistance to the pyrethroids after use of only a few years.

  With the one exception of selection pressure, all other prerequisites for the development of resistance are present in the plant-insect relationship. Therefore, careful studies are needed before such chemicals are introduced as a control measure.

Acknowledgements

  I thank Jan Cornish, Terri Mutchler, Susan Chiles, and Glenda Ward for their assistance in the preparation of this manuscript and Howard Williams for helpful discussions.

Literature Cited

1. Luckner, M. in "Secondary Plant Products"; Bell, E. A.; Charlwood, B. V., Eds.; Springer-Verlag: New York, 1980; p. 24.
2. Frey-Wyssling, A. Ber. Schweiz. Bot. Ges. 1933, 42, 109.
3. Rhyne, C. L. Advan. Front. Plant Sci. 1965, 13, 121.
4. Dahlgren, K. V. O. Impatiens, Sv. Bot. Tidskr. 1940, 34, 53.
5. Dahlgren, K. V. O. Impatiens. Ark. Bot. 1945, 32, 53.
6. Stahl, E. Flora 1920, 113, 1.

7.   Bentley, B. L. in "Annual Review of Ecology and
     Stystematics", Vol. 8; Johnston, R. F., Ed.; Annual Reviews
     Inc.: Palo Alto, 1977, p. 407.
8.   Elias, T. S.; Newcombe, L. F. Acta Bot Sin 1979, 21, 215.
9.   Curtis, J. D.; Lersten, N. R. Amer. J. Bot. 1974, 61,
     835.
10. Levin, D. A. Quart. Rev. Biol. 1973, 48, 3.
11. Schillinger, J. A.; Gallun, R. L. Ann. Entomol. Soc. Am.
     1968, 61, 900.
12. Everson, E. H.; Ringlund, K. Crop Sci. 1968, 8, 705.
13. Starks, K. J.; Merkle, O. G. J. Econ. Entomol. 1977, 70,
     305.
14. Kamel, S. A.; Elkassaby, F. Y. J. Econ. Entomol. 1965, 58,
     209.
15. Benedict, J. H.; Leigh, T. F.; Hyer, A. N. Envir. Entomol.
     in press.
16. Lukefahr, M. J.; Cowan, C. B., Jr.; Bariola, L. A.;
     Houghtaling, J. E. J. Econ. Entomol. 1968, 61, 661.
17. Poos, F. W.; Smith, F. F. J. Econ. Entomol. 1931, 24,
     361.
18. McKinney, K. B. J. Econ. Entomol. 1938, 31, 630.
19. Fluiter, H. J. de; Ankersmit, G. W. Tijdschr.
     Plantenziekten 1948, 54, 1.
20. Johnson, B. Bull. Entomol. Res. 1953, 44, 779.
21. Pillemer, E. A.; Tingey, W. M. Entomol. Exp. Appl. 1978,
     24, 83.
22. Pillemer, E. A.; Tingey, W. M. Science 1976, 193, 482.
23. Bernard, R. L.; Singh, B. B. Crop Sci. 1969, 9, 192.
24. Singh, B. B.; Hadley, H. H.; Bernard, R. L. Crop Sci. 1971,
     11, 13.
25. Wolfenburger, D. A.; Sleesman, J. P. J. Econ. Entomol.
     1963, 56, 895.
26. Robbins, J. C.; Daugherty, D. M. Proc. North Cent. Branch
     Entomol. Soc. Am. 1969, 24, 35.
27. Broersma, D. B.; Bernard, R. L.; Luckman, W. H. J. Econ.
     Entomol. 1972, 65, 78.
28. Turnipseed, S. G.; Sullivan, M. J. in "World Soybean
     Research"; Hill, L. D., Ed; Interstate Printers: Donville,
     1976, p. 549.
29. Turnipseed, S. G. Environ. Entomol. 1977, 6, 815.
30. Beckman, C. H.; Mueller, W. C.; McHardy, W. E. Physiol.
     Plant Pathol. 1972, 2, 69.
31. Charriere-Ladreix, Y. Proc. FEBS Meetings 1979, 55, 101.
32. Luckner, M.; Diettrich, B.; Lerbs, W. in "Progress in
     Phytochemistry", Vol. 6; Reinhold, L.; Harborne, J. B.;
     Swain, T., Eds.; Pergamon Press: New York, 1979, p. 103.
33. Gibson, R. W. Ann. Appl. Biol. 1971, 68, 113.
34. Tingey, W. M.; Gibson, R. W. J. Econ. Entomol. 1978, 71,
     856.
35. Gentile, A. G.; Stoner, A. K. J. Econ. Entomol. 1968, 61,
     1152.

36. Ohnesorge, B.; Sharaf, N.; Allawi, T. Z. Angerv. Entomol. 1980, 90, 226.
37. Shade, R. E.; Thompson, T. E.; Campbell, W. R. J. Econ. Entomol. 1975, 68, 399.
38. Kreitner, G. L.; Sorenson, E. L. Crop Sci. 1979, 19, 380.
39. Sutherst, R. W.; Jones, R. J.; Schnitzerling, H. G. Nature 1982, 295, 320.
40. Elsey, K. D.; Chaplin, J. F. J. Econ. Entomol. 1978, 71, 723.
41. Rabb, R. L.; Bradley, J. R. J. Econ. Entomol. 1968, 61, 1249.
42. Levin, D. A. in "Annual Review of Ecology and Systematics", Vol. 7; Johnston, R. F., Ed.; Annual Reviews Inc.: Palo Alto, 1976, p. 121.
43. Norris, D. M. ACS Symp. Ser. 1977, 62, 215.
44. Schoonhoven, L. M. in "Structural and Functional Aspects of Phytochemistry", Vol. 5; Runeckles, V. C.; Tso, T. C., Eds.; Academic Press: New York, 1972, p. 197.
45. Hedin, P. A.; Maxwell, F. G.; Jenkins, J. N. in "Proceedings of the Summer Institute on Biological Control of Plant Insects and Diseases"; Maxwell, F. G.; Harris, F. A. Eds.; University Press of Mississippi: Jackson, 1974, p. 494.
46. Vibhute, Y. B.; Wadje, S. S. Indian J. Exp. Biol. 1976, 14, 739.
47. Hufford, C. D.; Lasswell, W. L., Jr. Lloydia 1978, 41, 156.
48. Wacker, A.; Eilmes, H. G. Arzneim-Forsch 1978, 28, 347.
49. Mitscher, L. A.; Park, Y. H.; Omoto, S.; Clark, G. W., III; Clark, D. Heterocycles 1978, 9, 1533.
50. Elliger, C. A.; Chan, B. C.; Waiss, A. C., Jr. Naturwissenschaften 1980, 67, 358.
51. Matsuda, K. Tohuko J. Agric. Res. 1976, 27, 115.
52. Nielsen, J. K. Entomol. Exp. Appl. 1978, 24, 562.
53. Matsudo, K. Appl. Entomol. Zool. 1978, 13, 228.
54. Kimmich, G. A.; Randles, J. Membr. Biochem. 1978, 1, 221.
55. Wollenweber, E.; Dietz, V. H. Phytochem. 1981, 20, 869.
56. Rodriguez, E.; Towers, G. H. N.; Mitchell, J. C. Phytochem. 1976, 15, 1573.
57. Burnett, W. C., Jr.; Jones, S. B., Jr.; Mabry, T. J. in "Biochemical Aspects of Plant and Animal Coevolution", Harborne, J. B., Ed.; Academic Press: New York, 1978, p. 233.
58. Ivie, G. W.; Witzel, D. A. "Sesquiterpene Lactones: Structure, Biological Action, and Toxicological Significance", in press.
59. Kubo, I.; Lee, Y.; Pettei, M.; Nakanishi, K.; Pilkiewicz, F. J. Chem. Soc., Chem. Commun. 1976, 1013.

60. Doss, R.; Luthi, R.; Hrutfiord, B. F. Phytochem. 1980, 19, 2379.
61. Kubo, I.; Nakanishi, K. Adv. Pestic. Sci., Plenary Lect. Symp. Pap. Int. Congr. Pestic. Chem., 4th 1979, 2, 284.
62. Kubo, I. in "Konchu no Seiri to Kagaku";Hidaka, T.; Takahashi, S.; Isoe, Y., Eds.; Kitami Shobo: Tokyo, 1979, p. 153.
63. Alfaro, R. I.; Pierce, H. D., Jr.; Borden, J. H.; Oehtschlager, A. C. Can. J. Zool. 1980, 58, 626.
64. Nakajima, S.; Kawazu, K. Heterocycles 1978, 10, 117.
65. Slama, K. Acta Entomol. Bohemoslov. 1978, 75, 65.
66. Makarenko, N. G.; Schmidt, E. N.; Raldugin, V. O.; Dubovenko, Zh. V. Mikrobiol. Zh. (Kiev) 1980, 42, 98.
67. Andrews, R. E.; Parks, L. W.; Spence, K. D. Appl. Environ. Microbiol. 1980, 40, 301.
68. Agarwal, I.; Mathela, C. S. Indian Drugs Pharm. Ind. 1979, 14, 19.
69. Morris, J. A.; Khettry, A.; Seitz, E. W. J. Am. Oil Chem. Soc. 1979, 56, 595.
70. McEnro, F. J.; Fenical, W. Tetrahedron 1978, 34, 1661.
71. Szabentai, M.; Verzar-Petri, G.; Florian, E. Parfuem. Kosmet. 1977, 58, 121.
72. Schmeltz, I. in "Naturally Occurring Insecticides"; Jacobson, M.; Crosby, D. G., Eds.; Dekker Publishing: New York, 1971, p. 99.
73. Thurston, R.; Smith, W. T.; Cooper, B. P. Entomol. Exp. Appl. 1966, 9, 428.
74. Buhr, H.; Toball, R.; Schreiber, K. Entomol. Exp. Appl. 1958, 1, 209.
75. Harley, K. L. S.; Thorsteinson, A. J. Canad. J. Zoo. 1967, 45, 305.
76. Chakraborty, M. K.; Weybrew, J. A. Tob. Sci. 1963, 7, 122.
77. Thurston, R. J. Econ. Entomol. 1970, 63, 272.
78. Patterson, C. G.; Thurston, R.; Rodriguez, J. G. J. Econ. Entomol. 1974, 67, 341.
79. Yang, R. S. H.; Guthrie, F. E. Ann. Entomol. Soc. Am. 1969, 62, 141.
80. Stürckow, B.; Löw, I. Entomol. Exp. Appl. 1961, 4, 133.
81. Sinden, S. L.; Schalk, J. M.; Stoner, A. K. J. Am. Soc. Hortic. Sci. 1978, 103, 596.
82. Harley, K. L. S.; Thorsteinson, A. J. Can. J. Zool. 1967, 45, 305.
83. Aina, O. J.; Rodriguez, J. G.; Knavel, D. E. J. Econ. Entomol. 1972, 65, 641.
84. Staedler, E.; Hansen, F. E. Symp. Biol. Hung. 1976, 16, 267.
85. Aina, O. J.; Rodriguez, J. G.; Knavel, D. E. J. Econ. Entomol. 1972, 65, 641.

86. Duffey, S. S.; Isman, M. B. Experientia 1981, 37, 574.
87. Williams, W. G.; Kennedy, G. G.; Yamamoto, J. D.; Thacker,
     J. D.; Bordner, J. Science 1980, 207, 888.
88. Tingey, W. M.; Mackenzie, J. D.; Gregory, P. Am. Potato J.
     1978, 55, 577.
89. Charrier-Ladreix, Y. Physiol. Veg. 1977, 15, 619.
90. Wollenweber, E. Biochem. Syst. Ecol. 1975, 3, 35.
91. Wollenweber, E. Phytochem. 1976, 15, 438.
92. Wollenweber, E. Z. Pflanzenphysiol. 1974, 73, 277.
93. Wollenweber, E. Biochem. Sys. Ecol. 1975, 3, 47.
94. Biftu, T.; Stevenson, R. J. Chem. Soc. Perkin 1978, 1,
     360.
95. Wollenweber, E. Z. Naturforsch., C 1974, 29, 362.
96. Lotter, H.; Wagner, H.; Saleh, A. A.; Cordell, G. A.;
     Farnsworth, N. R. Z. Naturforsch., C: Biosci. 1979, 34,
     677.
97. Saleh, A. A.; Cordell, G. A.; Farnsworth, N. R. J. Chem.
     Soc., Perkin Trans. 1 1980, 5, 1090.
98. Suga, T.; Iwata, N.; Asakawa, Y. Bull. Chem. Soc. Japan
     1972, 45, 2058.
99. Wollenweber, E. Am. Fern J. 1978, 68, 13.
100. Nilsson, M. Acta Chem. Scand. 1961, 15, 211.
101. Ramakrishnan, G.; Banerji, A.; Chadha, M. S. Phytochem.
     1974, 13, 2317.
102. Wollenweber, E. Phytochem. 1976, 15, 2013.
103. Wollenweber, E. Z. Naturforsch., C 1977, 32, 1013.
104. Star, A. E.; Mabry, T. J.; Smith, D. M. Phytochem. 1978,
     17, 586.
105. Star, A. E.; Mabry, T. J.; Smith, D. M. Phytochem. 1978,
     17, 586.
106. Wollenweber, E. Flora 1979, 168, 138.
107. Nilsson, M. Acta Chem. Scand. 1961, 15, 154.
108. Star, A. E., Mabry, T. J. Phytochem. 1971, 10, 2817.
109. Wollenweber, E.; Dietz, V. H.; Smith, D. M.; Seigler, D. S.
     Z. Naturforsch., C 1979, 34, 876.
110. Wollenweber, E. Biochem. Syst. Ecol. 1975, 3, 35.
111. Wollenweber, E. Biochem. Syst. Ecol. 1975, 3, 47.
112. Suga, T.; Iwanatu, N.; Asakawa, Y. Bull. Chem. Soc. Japan
     1972, 45, 2058.
113. Wollenweber, E. Z. Naturforsch., C 1977, 32, 1013.
114. Wollenweber, E. Phytochem. 1977, 16, 295.
115. Popravko, S. A.; Kononenko, G. P.; Tikhomirova, V. I.;
     Wulfson, N. S. Bioorganiches Kaya Khimiya 1979, 5, 1662.
116. Saleh, A. A.; Cordell, G. A.; Farnsworth, N. R. Lloydia
     1976, 39, 456.
117. Wollenweber, E.; Egger, K. Tetrahedron Lett. 1970, 19,
     1601.
118. Wollenweber, E.; Lebreton, P.; Chadenson, M. Z.
     Naturforsch., B 1972, 27, 567.
119. Rabesa, Z. A. Doctoral Thesis, U. of Lyon, Lyon, France,
     1980.

120. Erdtmann, H.; Novotny, L.; Romanik, M. Tetrahedron, Suppl. 8, Pt. I 1966, 22, 71.
121. Rangaswami, S.; Iyer, R. T. Indian J. Chem. 1969, 7, 526.
122. Wollenweber, E. Phytochem. 1972, 11, 425.
123. Sunder, R.; Ayengar, K. N. N.; Rangaswami, S. Phytochem. 1974, 13, 1610.
124. Star, A. E.; Rösler, H.; Mabry, T. J., Smith, D. M. Phytochem. 1975, 14, 2275.
125. Wollenweber, E. Ber. Deutsch. Bot. Ges. Bd. 1976, 89, 243.
126. Jay, M.; Favre-Bonvin, J.; Wollenweber, E. Can J. Chem. 1979, 57, 1901.
127. Jay, M.; Wollenweber, E.; Favre-Bonvin, J. Phytochem. 1979, 18, 153.
128. Dietz, V. H.; Wollenweber, E.; Favre-Bonvin, J.; Smith, D. M. Phytochem. 1981, 20, 1181.
129. Favre-Bonvin, J.; Jay, M.; Wollenweber, E.; Dietz, V. H. Phytochem. 1980, 19, 2043.
130. Wollenweber, E.; Dietz, V. H.; Schullo, D.; Schilling, G. Z. Naturforsch., C 1980, 35, 685.
131. Nilsson, M. Acta Chem. Scand. 1959, 13, 750.
132. Wollenweber, E.; Dietz, V. H. Biochem. Syst. Ecol. 1980, 8, 21.
133. Wollenweber, E.; Dietz, V. H.; MacNeill, C. D.; Schilling, G. Z. Pflanzenphysiol. 1979, 94, 241.
134. Wollenweber, E.; Favre-Bonvin, J.; Jay, M. Z. Naturforsch., C 1978, 33, 831.
135. Khosa, R. L.; Wahi, A. K.; Mukherjee, A. K. Curr. Sci. 1978, 47, 624.
136. Gonzalez, A.; Betancor, C.; Hernandez, R.; Salazar, J. A. Phytochem. 1976, 15, 1996.
137. Bottari, F.; Marshill, A.; Morelli, I.; Pacchiani, M. Phytochem. 1972, 11, 2519.
138. Seigler, D. S.; Smith, D. M.; Mabry, T. J. Biochem. Syst. Ecol. 1975, 3, 5.
139. Jameson, G. R.; Reid, E. N. Phytochem. 1975, 14, 2229.
140. Lyttle, T. F.; Lyttle, J. S.; Caruso, A. Phytochem. 1976, 15, 965.
141. Khan, H.; Zaman, A.; Chetty, H. L.; Gupta, A. S.; Dev, S. Tetrahedron Lett. 1971, 46, 4443.
142. Ayengar, K. N. N.; Iyer, R. T.; Rangaswamy, S. Indian J. Chem. 1972, 10, 482.
143. Iyer, R. T.; Ayengar, K. N. N.; Rangaswamy, S. Indian J. Chem. 1973, 11, 1336.
144. Bardouille, V.; Mootoo, B. S.; Hirotsu, K.; Clardy, J. Phytochem. '1978, 17, 275.
145. Imal, S.; Toyosata, T.; Sakai, M.; Sato, Y.; Fujioka, S.; Murata, E.; Goto, M. Chem. Pharm. Bull. 1969, 17, 335.
146. Faux, A.; Galbraith, M. N.; Horn, D. M. S.; Middleton, E. J. Chem. Comm. 1970, 243.

147.   Murakami, T.; Kimura, T.; Tanaka, N.; Saiki, Y.; Chen, C.
       Phytochem. 1980, 19, 471.
148.   Wollenweber, E.; Favre-Bonvin, J. Phytochem. 1979, 18,
       1243.
149.   Huhtikangas, A.; Huurre, A.; Partanen, A. Planta Med. 1980,
       38, 62.
150.   Bardouille, J. V.; Cox, M. Guyana J. Sci. 1977, 5, 61.
151.   Star, A. E. Bull. Torrey Bot. Club 1980, 107, 16.
152.   Wollenweber, E. Biochem. Physiol. Pflanz. 1974, 166,
       419.
153.   Mabry, T. J.; Gill, J. E.; Burnett, W. C., Jr.; Jones, S.
       B. in "Host Plant Resistance to Pests", ACS Symposium
       Series, No. 62; Hedin, P. A., Ed.; American Chemical
       Society: Washington D. C., 1977, p. 179.
154.   Burnett, W. C.; Jones, S. B., Jr.; Mabry, T. J. Am. Midl.
       Nat. 1978, 100, 242.
155.   Narain, N. K. Spectrosc. Lett. 1978, 11, 267.
156.   Drew, M. G. B.; Hitchman, S. P.; Mann, J.; Lopes, J. L. C.
       J. Chem. Soc., Chem. Commun. 1980, 802.
157.   Bohlmann, F.; Gupta, R. K.; Jakupovic, J.; King, R. M.;
       Robinson, H. Liebigs Ann. Chem. 1980, 11, 1904.
158.   Narain, N. K. Can. J. Pharm. Sci. 1977, 12, 18.
159.   Rustaiyan, A.; Niknejad, A; Danieli, B.; Palmisano, G.;
       Jones, S. Fitoterapia 1977, 48, 266.
160.   Sticher, O.; Afifi-Yazar, F. U. Helv. Chim. Acta 1979, 62,
       530.
161.   Rodriguez, E.; Dillon, M.; Mabry, T.; Mitchell, J. C.;
       Towers, G. H. N. Experientia 1976, 32, 236.
162.   Narasimhan, T. R.; Anant, M.; Swamy, M. N.; Babu, M. R.;
       Mangala, A.; Rao, P. V. S. Curr. Sci. 1977, 46, 15.
163.   Towers, G. H. N.; Mitchell, J. L.; Rodriguez, E.; Bennet,
       F. D.; Subba Rao, P. V. Biochem. Rev. 1977, 48, 65.
164.   Kanchan, S. D. Curr. Sci. 1975, 44, 358.
165.   Sharma, R. N.; Joshi, V. N. Biovigyanam 1977, 3, 225.
166.   Picman, A. K.; Towers, G. H. N.; Subba Rao, P. V.
       Phytochem. 1980, 19, 2206.
167.   Mears, J. A. J. Nat. Prod. 1980, 43, 708.
168.   Gershenzon, J.; Mabry, T. J.; Abstr. Ann. Mtg. Phytochem.
       Soc. N. Am. 1981, p. 9.
169.   Rogers, C. E. "Proceedings, EUCARPIA Symposium, Oil and
       Leguminous Crops, Prague, Czechoslovakia", in press.
170.   Morimoto, H.; Sanno, Y.; Oshio, H. Tetrahedron 1966, 22,
       3173.
171.   Herz, W.; DeGroote, R. Phytochem. 1977, 16, 1307.
172.   Bohlmann, F.; Dutta, L. N. Phytochem. 1979, 18, 676.
173.   Herz, W.; Kumar, N. Phytochem. 1981, 20, 93.
174.   Herz, W.; Kumar, N. Phytochem. 1981, 20, 99.
175.   Herz, W.; Kumar, N. Phytochem. 1981, 20, 1339.
176.   Waiss, A. C., Jr.; Chan, B. G.; Elliger, C. A.; Garrett, V.

R.; Carlson, F. C.; Beard, B. Naturwissenschaften 1977, 64, 341.

177. Stipanovic, R. D.; D'Brien, D. H.; Rogers, C. E.; Thompson, T. E. J. Agri. Food Chem. 1979, 27, 458.
178. Panizo, F. M.; Rodriguez, B. Anales De Quimica 1979, 75, 428.
179. Bohlmann, F.; Jakupovic, J.; King, R. M.; Robinson, H. Phytochem. 1980, 19, 863.
180. Kasprzyk, Z.; Janiszowska, W. Phytochem. 1971, 10, 1946.
181. St. Pyrek, J. Pol. J. Chem. 1979, 53, 1071.
182. St. Pyrek, J. Pol. J. Chem. 1979, 53, 2465.
183. Bohlmann, F.; Jakupovic, J.; King, R. M.; Robinson, H. Phytochem. 1980, 19, 863.
184. Popescu, H.; Fagarasan, E. Clujul Med. 1979, 52, 171.
185. Rodriguez, E. Chem. Eng. News 1980, 58, 42.
186. Hertz, W. in "Effects of Poisonous Plants on Livestock", Keeler, R. F.; van Kampen, K. R.; James, L. F., Eds.; Academic Press: New York, 1978, p. 487.
187. Kelsey, R. G.; Shafizadeh, F. Biochem. Syst. Ecol. 1980, 8, 371.
188. Curtis, J. D.; Lersten, N. R. Amer. J. Bot. 1974, 61, 835.
189. Van Hoof, L.; Vanden Berghe, D. A.; Vlietinck, A. J. Biol. Plant 1980, 22, 265.
190. Ahluwalia, V. K.; Prakash, C.; Gupta, M. C.; Mehta, S. Indian J. Chem., Sect. B 1978, 16, 591.
191. Dell, B.; McComb, A. J. Aust. J. Bot. 1975, 23, 373.
192. Dawson, R. M.; Jarvis, M. W.; Jefferies, P. R.; Payne, T. G.; Rosich, R. S. Aust. J. Chem. 1966, 19, 2133.
193. Jefferies, P. R.; Ratajczak, T. Aust. J. Chem. 1973, 26, 173.
194. Wollenweber, E. Phytochem. 1974, 13, 753.
195. Bose, P. C., Adityachaudhury, N. Phytochem. 1978, 17, 587.
196. Siddiqui, S. J. Indian Chem. Soc. 1937, 14, 703.
197. Bhaskar, A.; Seshadri, T. R. Indian J. Chem. 1973, 11, 404.
198. Seshadri, T. R. Proc. Ind. Acad. Sci. 1948, 27A, 129.
199. Seshadri, T. R. Proc. Ind. Acad. Sci. 1949, 28A, 106.
200. Das, A. K.; Mitra, S. R.; Aditayachaudhury, N.; Patra, A.; Mitra, A. K.; Chattejee, Mrs. A. Indian J. Chem., Sect. B 1979, 18B, 550.
201. Mitra, S. R.; Das, A. K.; Kirtaniya, C. L.; Adityachaudhury, N.; Patra, A.; Mitra, A. K. Indian J. Chem., Sect. B 1980, 19B, 79.
202. Sakakibara, M.; DiFeo, D., Jr.; Nakatini, N.; Timmermann, B.; Mabry, T. J. Phytochem. 1976, 15, 727.
203. Bernhard, H. O.; Thiele, K. Planta Med. 1981, 41, 100.
204. Bohnstedt, C. F.; Mabry, T. J. Res. Latinoam Quim. 1979, 10, 128.

205. Rhoades, D. F. in "Creosote Bush"; Mabry, T. J.; Hunziker, J. H.; DiFeo, D. R., Jr., Eds.; Dowden, Hutchinson and Ross, Inc.: Stroudsburg, 1977.

206. Stubblebine, W. H.; Langenheim, J. H. J. Chem. Ecol. 1977, 3, 633.

207. Langenheim, J. H.; Stubblebine, W. H.; Lincoln, D. E.; Foster, C. E. Biochem. Syst. Ecol. 1978, 6, 299.

208. Chow, P. W.; Jefferies, P. R. Aust. J. Chem. 1968, 21, 2529.

209. Errington, S. G.; Ghisalberti, E. L.; Jefferies, P. R. Aust. J. Chem. 1976, 29, 1809.

210. Ghisalberti, E. L.; Jefferies, P. R.; Sefton, M. A. Phytochem. 1978, 17, 1961.

211. Lukefahr, M. J.; Fryxell, P. A. Econ. Bot. 1967, 21, 128.

212. Adams, R.; Geissman, T. A. Chem. Rev. 1960, 60, 555.

213. Seshadri, T. R. Proc. Indian Nat. Sci. Acad. Part A. 1971, 37, 411.

214. Berardi, L. C.; Goldblatt, L. A. in "Toxic Constituents of Plant Foodstuff", 2nd ed.; Liener, I. E., Ed.; Academic Press: New York, 1969, p. 211.

215. Heinstein, P.; Widmaier, R.; Wegner, P.; Howe, J. in "Biochemistry of Plant Phenolics"; Swain, T.; Harborne, J. B.; Van Sumere, C. F., Eds.; Plenum Press: New York, 1979, p. 313.

216. Abou-Donia, M. B. Residue Review 1976, 61, 125.

217. Berardi, L. C.; Golblatt, L. A. in "Toxic Contituents of Plant Foodstuff"; Liener, I. E., Ed.; Academic Press: New York, 1980, p. 183.

218. National Coordinating Group on Male Antifertility Agents (China) Gynecol. obstet. Invest. 1979, 10, 163.

219. King, T. J.; DeSilva, L. B. Tetrahedron Lett. 1968, 261.

220. Datta, S. C.; Murti, V. V. S.; Seshadri, T. R. Indian J. Chem. 1972, 10, 263.

221. Seshadri, T. R.; Sharma, N. N. Curr. Sci. 1973, 42, 821.

222. Gray, J. R.; Mabry, T. J.; Bell, A. A.; Stipanovic, R. D.; . Lukefahr, M. J. J. Chem. Soc., Chem. Comm. 1976, 109.

223. Stipanovic, R. D.; Bell, A. A.; O'Brien, D. H.; Lukefahr, M. J. Agric. and Food Chem. 1978, 26, 115.

224. Stipanovic, R. D.; Bell, A. A.; O'Brien, D. H.; Lukefahr, M. J. Tetrahedron Lett., 1977, 567.

225. Stipanovic, R. D.; Bell, A. A.; O'Brien, D. H.; Lukefahr, M. J. Phytochem. 1978, 17, 151.

226. Bell, A. A.; Stipanovic, R. D.; O'Brien, D. H.; Fryxell, P. A. Phytochem. 1978, 17, 1297.

227. Stipanovic, R. D.; Bell, A. A.; Lukefahr, M. J. in "Host Plant Resistance to Pests", ACS Symposium Series, No. 62; Hedin, P. A., Ed.; American Chemical Society: Washington, D. C., 1977, p. 197.

228. Stipanovic, R. D.; Bell, A. A.; O'Brien, D. H. Phytochem. 1980, 19, 1735.

229. Stanford, E. E.; Viehoever, A. J. Agric. Res. 1918, 13, 419.
230. Chan, B. G.; Waiss, A. C., Jr. Beltwide Cotton Prod. Res. Conf. Proc. 1981, 49.
231. Chan, B. G.; Waiss, A. C.; Jr.; Lurdin, R. E.; Asen, S. Abstr. Ann. Mtg. Phytochem. Soc. N. Am. 1981, p. 9.
232. Bell, A. A.; Stipanovic, R. D.; Howell, C. R.; Fryxell, P. A. Phytochem. 1975, 14, 225.
233. Stipanovic, R. D.; Bell, A. A.; Mace, M. E.; Howell, C. R. Phytochem. 1975, 14, 1077.
234. Stipanovic, R. D.; Bell, A. A.; Howell, C. R. Phytochem. 1975, 14, 1809.
235. Zaki, A. I.; Keen, N. T.; Erwin, D. C. Phytopath. 1972, 62, 1402.
236. Abdullaev, Z. S.; Karimdzhanov, A. K.; Ismailov, A. I.; Islambekov, S. H.; Kamaev, F. G.; Sagdieva, M. G. Khim. Prir. Soedin. 1977, 3, 341.
237. Halloin, J. M.; Veech, J. A.; Carter, W. W. Plant Soil 1978, 50, 237.
238. Bell, A. A.; Stipanovic, R. D. Beltwide Cotton Prod. Res. Conf. Proc. 1977, 244.
239. Margalith, P. Appl. Microbiol. 1967, 15, 952.
240. Vichkanova, S. A.; Goryunova, L. V. Antibiotiki 1968, 13, 828.
241. Vichkanova, S. A.; Oifa, A. I.; Goryunova, L. V. Antibiotiki 1970, 15, 1071.
242. Shaver, T. N.; Garcia, J. A. J. Econ. Entomol. 1973, 66, 327.
243. Vermel, E. M. Acta Unio Int. Cancrum 1964, 20, 211.
244. Dorsett, P. H.; Kerstine, E. E. J. Pharm. Sci. 1975, 64(6), 1073.
245. Minyard, J.P.; Tumlinson, J. H.; Hedin, P. A.; Thompson, A. C. J. Agric. Food Chem. 1965, 13, 599.
246. Minyard, J. P.; Tumlinson, J. H.; Thompson, A. C.; Hedin, P. A. J. Agric. Food Chem. 1966, 14, 332.
247. Minyard, J. P.; Tumlinson, J. H.; Thompson, A. C.; Hedin, P. A. J. Agric. Food Chem. 1967, 15, 517.
248. Minyard, J. P.; Thompson, A. C.; Hedin, P. A. J. Org. Chem. 1968, 33, 909.
249. Minyard, J. P.; Hardee, D. D.; Gueldner, R. C.; Thompson, A. C.; Wiygul, G.; Hedin, P. A. J. Agric. Food Chem. 1969, 17, 1093.
250. Hedin, P. A.; Thompson, A. C.; Gueldner, R. C.; Minyard, J. P. Phytochem. 1971, 10, 1692.
251. Hedin, P. A.; Thompson, A. C.; Gueldner, R. C.; Minyard, J. P. Phytochem. 1971, 10, 3316.
252. Hedin, P. A.; Thompson, A. C.; Gueldner, R. C.; Ruth, J. M. Phytochem. 1972, 11, 2119.
253. Hedin, P. A.; Thompson, A. C.; Gueldner, R. C. Phytochem. 1975, 14, 2087.

254. Thompson, A. C.; Hanny, B. W.; Hedin, P. A.; Gueldner, R. C. Amer. J. Bot. 1971, 58, 803.

255. Hedin, P. A.; Thompson, A. C.; Gueldner, R. C. Phytochem. 1972, 11, 2356.

256. Kumanoto, J.; Waines, J. G.; Hollenberg, J. L.; Scora, R. W. J. Agric. Food Chem. 1979, 27, 203.

257. Hanny, B. W.; Gueldner, R. C. J. Agric. Food Chem. 1976, 24(2), 401.

258. Hedin, P. A.; Minyard, J. P.; Thompson, A. C.; Struck, R. F.; Frye, J. Phytochem. 1967, 6, 1165.

259. Hedin, P. A.; Collum, D. H.; White, W. H.; Parrott, W. L.; Lane, H. C.; Jenkins, J. N. in "Regulation of Insect Development and Behaviour"; Sehnal, R.; Zabza, A.; Menn, J. J.; Cymborowski, B., Eds; Wroclaw Technical University Press: Wroclaw, 1981, 1071.

260. Hedin, P. A.; Miles, L. R.; Thompson, A. C.; Minyard, J. P. J. Agric. Food Chem. 1968, 16, 505.

261. Shaver, T. N.; Lukefahr, M. J. J. Econ. Entomol. 1969, 62, 643.

262. Hanny, B. W.; Meredith, W. R., Jr.; Bailey, J. C.; Harvey, A. J. Crop Sci. 1978, 18, 1071.

263. Dilday, R. H.; Shaver, T. N. Crop Sci. 1980, 20, 91.

264. Yang, H. C.; Davis, D. D. Crop Sci. 1976, 16, 485.

265. Hanny, B. W. J. Agric. Food Chem. 1980, 28, 504.

266. Cook, O. F. U.S. Dept. Agri. Bur. Plant Indus. Bull. 88 1906, 87.

267. Lukefahr, M. J.; Houghtaling, J. E. J. Econ. Entomol. 1969, 62, 588.

268. Shaver, T. N.; Lukefahr, M. J.; Garcia, J. A. J. Econ. Entomol. 1970, 63, 1544.

269. Seaman, F.; Lukefahr, M. J.; Mabry, T. J. Proc. Nat. Cotton Council, Atlanta 1977, 102.

270. Elliger, C. A.; Chan, B. G.; Waiss, A. C., Jr. J. Econ. Entomol. 1978, 71, 161.

271. Lukefahr, M. J.; Martin, D. F. J. Econ. Entomol. 1966, 59, 176.

272. Abou-Donia, M. B.; Taman, F.; Bakery, N. M.; El-Sebab, A. H. Experentia 1974, 30, 1151.

273. Meisner, J.; Kehat, M.; Zur, M.; Ascher, K. R. S. Entomol. Exp. Appl. 1977, 22, 301.

274. Lukefahr, M. J.; Stipanovic, R. D.; Bell, A. A.; Gray, J. R. Beltwide Cotton Prod. Res. Conf. Proc. 1977, 97.

275. Chan, B. G.; Waiss, A. C., Jr.; Binder, R. G.; Elliger, C. A. Entomol. Exp. Appl. 1978, 24, 94.

276. Waiss, A. C., Jr.; Chan, B. G.; Elliger, C. A.; Binder, R. G. Beltwide Cotton Prod. Res. Conf. Proc. 1981, 61.

277. Meisner, J.; Ascher, K. R. S.; Zur, M. J. Econ. Entomol. 1977, 70, 149.

278. Meisner, J.; Zur, M.; Kabonci, E.; Ascher, K. R. S. J. Econ. Entomol. 1977, 70, 714.

279. Meisner, J.; Navon, A.; Zur, M.; Ascher, K. R. S. Environ.
      Entomol. 1977, 6, 243.
280. El-Sebae, A. H.; Sherby, S. I.; Mansour, N. A. J. Environ.
      Sci. Health 1981, B16, 167.
281. Meisner, J.; Ishaaya, I.; Ascher, K. R. S.; Zur, M. Ann.
      Entomol. Soc. Am. 1978, 71, 5.
282. Meisner, J.; Ascher, K. R. S.; Zur, M.; Kabonci, E. J.
      Econ. Entomol. 1977, 70, 717.
283. Abou-Donia, M. B.; Dieckert, J. W. Toxicol. Appl.
      Pharmacol. 1971, 18, 507.
284. Chan, B. G.; Waiss, A. C., Jr.; Lukefahr, M. J. J. Insect
      Physiol. 1978, 24, 113.
285. Georghiou, G. P. in "Annual Review of Ecology and
      Systematics", Vol. 3; Johnston, R. F., Ed.; Annual Reviews
      Inc.: Palo Alto, 1972, p. 133.
286. Winteringham, F. P. W.; Hewlett, P. S. Chem. Ind. 1964,
      35, 1512.
287. Harborne, J. B. in "Herbivores - Their Interaction with
      Secondary Metabolites"; Rosenthal, G. A.; Janzen, D. H.,
      Eds.; Academic Press: New York, 1979, p. 619.

RECEIVED August 23, 1982

BIOCHEMICAL AND PHYSIOLOGICAL MECHANISMS

Regulation of Synthesis and Accumulation of Proteinase Inhibitors in Leaves of Wounded Tomato Plants

C. E. NELSON, M. WALKER-SIMMONS, D. MAKUS, G. ZUROSKE, J. GRAHAM, and C. A. RYAN

Washington State University, Institute of Biological Chemistry and Biochemistry and Biochemistry/Biophysics Program, Pullman, WA 99164

Two proteinase inhibitors, Inhibitors I and II, accumulate in leaves of tomato plants when attacked by chewing insects or mechanically wounded. The accumulation of these two antinutrient proteins is apparently a defense response and is initiated by the release of a putative wound hormone called the proteinase inhibitor inducing factor (PIIF). The direction of flow of PIIF out of wounded leaves is primarily towards the apex and transport occurs maximally about 120 min following wounding. After a single severe wound, the in vitro translatable tomato leaf mRNA specific for Inhibitors I and II increases to a maximum within four hours and remains constant for about five hours when it decreases rapidly to about 50% of the maximum. The rate of in vivo accumulation of both inhibitor proteins steadily increases, reaching a steady state after nine hours. However, a second wound at nine hours results in a tripling of the steady state rate of inhibitor accumulation over the next several hours. The data indicates that the second wound causes no change in the apparent translational efficiencies of the mRNA for Inhibitors I and II but causes increased rates of inhibitor accumulation by providing more translatable inhibitor messages when the plant's translation system is operating at high efficiency.

A severe mechanical wound on a single leaf of tomato plants initiates a complex series of extracellular and intracellular reactions which result in the synthesis and accumulation of two proteinase inhibitors, Inhibitors I and II, in leaf cells (1, 2). A second wounding, within a few hours, results in a 2-3 fold increase in the rates of accumulation initiated by the

0097-6156/83/0208-0103$06.00/0

first wound. A putative wound hormone, called the proteinase
inhibitor inducing factor, PIIF, is released at the wound site
and travels throughout the plant to initiate synthesis and
accumulation of the two proteinase inhibitors, even in unwounded
leaves several cm from the wound site. We view this process as
a primitive immune-like response in which the plant is responding to pest damage by producing powerful antinutrient proteins,
the proteinase inhibitors, to help the plant discourage persis-
tent or future attacks by pests.

This wound response has provided a novel system for studying
the regulation of the expression of the two proteinase inhibitors
by the factor PIIF, triggered by severe environmental stress.
In this chapter we report our recent data concerning the direction and time course of PIIF transport through the plant following wounding, and our progress in initiating a program to study
the molecular biology of inhibitor accumulation.

Direction and Rate of Flow of the Wound Signal, PIIF

PIIF activity has been isolated by various techniques from
tomato leaves to yield a single broad peak from Sephadex G-50
that exhibits a Mw range of about 5000 to 10,000 daltons and is
primarily carbohydrate in composition (3). Properties of highly
purified PIIF preparations, such as loss of activity upon either
prolonged acid hydrolysis or periodate oxidation, and its
monosaccharide composition suggested that it was similar to the
pectic polysaccharides found associated with the plant cell
wall. In collaboration with Dr. Peter Albersheim, of the
University of Colorado, we found that a sycamore cell wall-
derived polysaccharide, called rhamnogalacturonan I, was as
active as tomato PIIF in inducing proteinase inhibitor accumulation in young tomato plants (4). This work substantiated that
tomato PIIF was a fragment of the plant cell wall. In subsequent
experiments we were able to enzymically degrade PIIF into
oligosaccharides with molecular weights of about 400 to 2000
that retained the capacity to induce proteinase inhibitors in
detached tomato leaves (3).

A hypothesis was presented (3) in which PIIF is released as
a mixture of poly- and oligosaccharides fragmented from the cell
wall by hydrolytic enzymes that either are activated during
wounding or are introduced by invading pests. PIIF, or a
product induced by its presence, could then be transported
rapidly through the plant vascular system to target cells where
it induces the synthesis and accumulation of proteinase inhibitor
proteins.

Direction of flow of PIIF. The time-course of Inhibitor I
accumulation in leaves of young tomato plants at the four leaf
stage, wounded at 0 time and at 72 hr, is shown in Fig. 1.
Wounding of an upper leaflet (left) did not cause accumulation

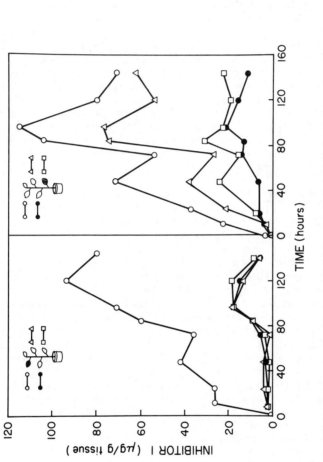

Figure 1. Time course of accumulation of Inhibitor I in terminal leaflets of young tomato plants. Leaves were wounded at the uppermost leaf (left) and the lowest leaf (right) by crushing across the midrib of the terminal leaflet with a hemostat at zero time and again at 72 h. The concentrations of Inhibitor I in leaves were determined immunologically.

of Inhibitor I in lower leaves, but a second wound (at 72 hr) on
the same leaf was weakly registered by the lower leaves and
resulted in some accumulation of the inhibitor. Wounding of the
lowest leaf (right) did cause upper leaves to accumulate Inhibitor I and a second wound at 72 hr significantly reinforced the
response, particularly in the two uppermost leaves. In Table I
the average levels of Inhibitor I in all four terminal leaflets
of tomato plants, singly wounded at the various leaflet positions, is shown. It is clear that inhibitors preferentially
accumulated in leaves near the apex of the plant and that the
transport of PIIF was primarily acropetally. The lower leaves
were not disposed to accumulate much Inhibitor I, even when the
lowest leaflets were wounded. These experiments suggest that
not only are lower older leaves much less responsive to PIIF,
but are apparently recipients of only a small amount of the
hormone that is released above them. The experiments also
demonstrated that all of the leaves are capable of releasing
PIIF when wounded, although the lowest wounded leaves do not
appear to be releasing as much PIIF as upper leaves.

Rate of flow of PIIF. A single slice of a sharp razor
blade through the leaf petiole apparently does not release
appreciable PIIF into the plant as evidenced by the lack of
accumulation of inhibitor in tissues of the excised leaf (5).
Thus, a single wound can be inflicted in a terminal leaflet and
then the entire leaf can be severed at a measured distance from
the wound site near the base of the petiole at various times to
determine how long it takes for PIIF to travel past the position
of the cut and out of the leaf, as judged by accumulation of
Inhibitor I in an adjacent upper leaflet 24 hr later. Our
previous experiments (6) suggested that about 2 hr was required
to maximally transport PIIF out of leaves (a distance of about 6
cm). We repeated these experiments herein with the purpose of
comparing the rate of PIIF transport with that of $[^{14}C]$glucose
that was applied to the wound immediately following the injury.

The time for transport of $[^{14}C]$glucose from its application
at the wound site to the base of the petiole (Fig. 2, right) was
nearly identical to that of PIIF that was released upon wounding.
The amount of glucose that was transported up to the next leaf,
however, was small compared to what passed through the petiole
of the wounded leaf. Very little radioactivity was detected in
the adjacent lower leaf. The maximum radioactivity was reached
at about 120 min following wounding, the same time that PIIF
transport through this region was maximal (Fig. 2, left). Thus,
both PIIF and $[^{14}C]$glucose were transported out of the wound
sites at approximately the same times and were both preferentially transported acropetally.

Mode of transport. A jet of hot air (80°C) focused on a
segment of the petiole near the main stem completely destroyed

TABLE I

INHIBITOR I ACCUMULATION IN LEAVES OF YOUNG TOMATO PLANTS
120 HOURS FOLLOWING WOUNDING OF INDIVIDUAL LEAVES AT
DIFFERENT LOCATIONS ON THE PLANTS

Young tomato plants having four leaves were wounded at the leaf
position shown below with a hemostat at 0 and 72 hr. After 120
hr following the initial wounding the terminal leaflet of each
leaf was assayed for Inhibitor I concentration.

Position of Wounded Leaf	Leaf #			
	1	2	3	4
	Inhibitor I Accumulation			
	(μg/g tissue)			
	94	12	12	17
	74	140	6	4
	105	96	56	20
	80	54	16	18
	9	0	0	0

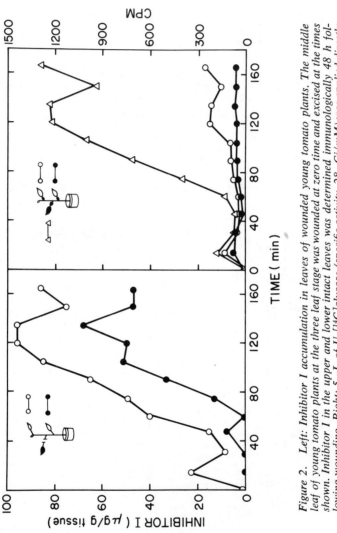
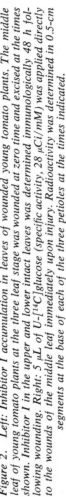

Figure 2. Left: Inhibitor I accumulation in leaves of wounded young tomato plants. The middle leaf of young tomato plants at the three leaf stage was wounded at zero time and excised at the times shown. Inhibitor I in the upper and lower intact leaves was determined immunologically 48 h following wounding. Right: 5 μL of U-[14C]glucose (specific activity, 28 μCi/mM) was applied directly to the wounds of the middle leaf immediately upon injury. Radioactivity was determined in 0.5-cm segments at the base of each of the three petioles at the times indicated.

the phloem but not the xylem, as evidenced by the drying of the treated area within an hour to form a fine strand of xylem with the leaf still intact with good turgor. This treatment caused the leaf itself to accumulate Inhibitor I over the next 24 hr, but it minimally affected the leaves of the rest of the plant, indicating that the destruction of the phloem did not initiate PIIF transport out of the leaf but only into the leaf itself. We subjected a segment of the petiole of the lowest leaf of several 3 leaf stage tomato plants, to a hot air jet to destroy the phloem. Within an hour the injured petiole segments had dried to form the thin strands of xylem. As before, this treatment did not release PIIF into the plant (Table II, treatment 3) while the leaf itself (leaf #1) accumulated considerable Inhibitor I. Subsequent wounding of leaves whose phloem had been destroyed (Table II, treatment 4) did not result in Inhibitor I accumulation in adjacent leaves, indicating that the phloem destruction had blocked its transport out of the wounded leaf.

While the evidence presented herein cumulatively supports the involvement of the phloem in translocating PIIF out of wounded tomato plants, the velocity of transport appeared to be much slower than expected for phloem transport. When carefully measured from point to point in petiole tissue the velocity of assimilate out of leaves, in general, is in the order of 1-5 cm/min (7, 8). Our techniques do not allow direct measurements with PIIF itself, but from our indirect measurements we calculate a gross movement of PIIF from the time of wounding to time of maximum transport of PIIF (as with ^{14}C glucose) to the base of the petiole, 6 cm from the wound site, of about 0.05 cm/min. This is over twenty times too slow for a process involving just phloem transport. Thus, if phloem transport is occurring, then a time of nearly 2 hr must be required to deliver a maximum quantity of PIIF into the phloem at or near the wound site. This rate is much slower than normal phloem loading, for example in healthy soybeans (9).

The data seems to indicate that the origins of PIIF and its introduction into the transport system of the plant is a complex system. We can however speculate about the nature of the process from available information. The 2 hr period required to reach maximum PIIF levels in the phloem could be a consequence of enzymic degradation of the cell wall following wounding, and/or the production and association of some type of chemical to the cell wall fragments, eventually entering the transport system. Thus, the phloem, which is involved in translocating carbohydrates, would be a logical candidate to transport small pectic fragments, preferentially to the upper leaves. This type of mechanism of PIIF production and release would require some type of loading mechanism that would allow the fragments to enter the phloem as, or after, they were produced. As a monitoring system for insect or pathogen damage, this process would be

TABLE II

EFFECT OF THE HOT AIR–DESTRUCTION OF PETIOLE PHLOEM ON THE
MOVEMENT OF PIIF OUT OF MECHANICALLY WOUNDED LEAVES

All treatments of young tomato plants (having three leaves) were
at the lowest leaf (#1) shown below. Hot air (80°C) was applied
to the base of the petiole through a window in a teflon shield
to destroy a segment of phloem tissue. A single wound, perpen-
dicular to the midrib was inflicted at the center of the terminal leaflet of leaf #1 three days after hot air treatment and
Inhibitor I levels were assayed in the leaves 24 hr later.

Treatment	Leaf #		
	1	2	3
	Inhibitor I Accumulation (μg/g tissue)		
1. No treatment	0	0	10
2. No phloem destruction leaf #1 mechanically wounded	133	126	86
3. Leaf #1 phloem destroyed, leaf #1 tissue unwounded	173	35	32
4. Leaf #1 phloem destroyed, leaf #1 tissue wounded	165	14	27

well suited, directing messages from wounded or damaged tissues
to the younger healthy tissues to initiate inhibitor accumulation.

In Vitro Synthesis of Pre-Proteinase Inhibitors with mRNA from Wounded Tomato Plants

The two proteinase inhibitors that accumulate in leaves of
wounded tomato leaves have been isolated and characterized.
Inhibitor I has a molecular weight of 41,000 and is composed of
subunits with molecular weights of about 8100 (10). It is,
therefore, a pentamer in its native state. Each subunit
possesses an active site specific for chymotrypsin, and the
apparent K_i for the inhibition of chymotrypsin is about $10^{-9}M$
(10). Inhibitor II has a molecular weight of about 23,000, is
composed of two subunits, and strongly inhibits both trypsin and
chymotrypsin with K_i values of about 10^{-8} and $10^{-7}M$ respectively
(10).
     Messenger RNA has been prepared from leaves of wounded and
unwounded tomato plants and only leaves of wounded plants
contain translatable mRNAs specific for Inhibitors I and II
(11). Both proteins have been shown to be translated in vitro
in a reticulocyte lysate system as preinhibitors, 2000-3000
daltons larger than those synthesized and accumulated in vivo
(11). The preinhibitors may be important in the compartmentali-
zation of the inhibitors as they are stored in the central
vacuole, or plant lysosome, of the plant cells (12). We have
now studied the time course of the increase in translatable mRNA
in leaves of wounded plants utilizing poly(A)$^+$ mRNA isolated at
various times following wounding.
     When young tomato plants are wounded, by chewing insects or
by a severe crushing of any type, the levels of total poly(A)$^+$
translatable mRNA for both proteinase Inhibitor I and Inhibitor
II rise rapidly during the first 4 hr after wounding (13). This
rise was measured by quantifying the immunoprecipitates that can
be recovered specifically from electrophoretic gels after
translation in a cell free rabbit reticulocyte lysate system
(11, 13). An example of such gels is shown in Fig. 3. In this
example the newly translated Inhibitors I and II were isolated
as immunoprecipitates with Inhibitor I IgG plus Inhibitor II
IgG and electrophoresed in 15% acrylamide gels in the presence
of SDS and mercaptoethanol. The analysis of radioactivity
($[^{35}S]$methionine) incorporated into each inhibitor in the gels
was taken as a measure of the concentration of inhibitor mRNAs.
Translatable mRNAs for Inhibitors I and II are present at near
maximum levels within 4 hr following wounding (Fig. 4A). The
levels remain high until 9 hr. After 9 hr the levels decrease
to less than half their original level (Fig. 4A). In the same
leaves, during the same time, the rates of in vivo accumulation
of Inhibitors I and II (Fig. 4B) steadily increase during the

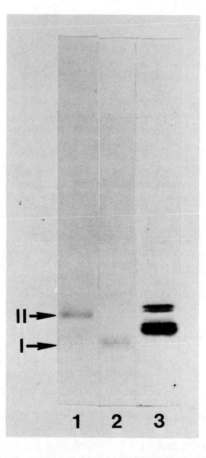

Figure 3.   SDS–urea electrophoresis of proteinase Inhibitors I and II. Key: Lane 1, proteinase Inhibitor II; Lane 2, proteinase Inhibitor I (both lanes stained with Coomassie blue); and Lane 3, [^{35}S]preproteinase Inhibitors I and II synthesized in an in vitro rabbit reticulocyte system immunoprecipitated for Inhibitors I and II.

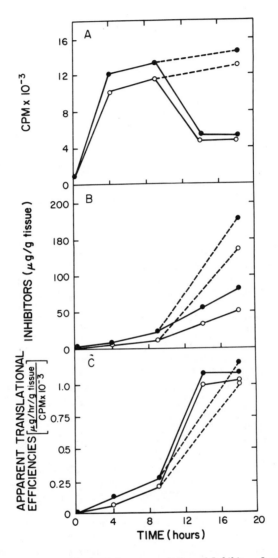

Figure 4.   Time course analysis of the accumulation of Inhibitors I and II protein, translatable mRNAs and apparent translational efficiencies in leaves of singly and doubly wounded tomato plants. Key: —●—, Inhibitor I, single wound; —○—, Inhibitor II, single wound; — — ● — —, Inhibitor I, double wound; and — — ○ — —, Inhibitor II, double wound.

A: In vitro translation of 4-μg quantities of tomato leaf mRNA and subsequent isolation of specific preinhibitors through the preformed antibody technique (11).

B: In vivo accumulation of Inhibitors I and II proteins in wounded tomato leaves.

C: Apparent translational efficiencies.

first 9 hr after wounding. By 14 hr the rates of accumulation have reached a steady state rate that remains constant for several hours. This suggests that some shift in the cell is occurring at about 9 hr that is reflected in both the in vitro translation of mRNA and in the in vivo accumulation of inhibitors. The cause of the shift in levels of translatable mRNA is unknown and could reside in either some structural feature of the mRNAs themselves that change their translational rates, or the differences could result in changes in their rates of synthesis or degradation. Nevertheless, the shift results in species of mRNA that are responsible for the more efficient steady state rate of synthesis of the two inhibitors after 9 hr.

A second wounding of the leaves 9 hr after the initial wounding, tripled the rate of accumulation of both Inhibitors I and II (Fig. 4B) over those of once wounded plants. This second wound also resulted in the maintenance of the mRNA levels present at 9 hr so that the decrease in mRNA noted in singly wounded plants did not occur. The translatable mRNA levels remained high through the 18th hr (Fig. 4A). Thus, high mRNA levels at 18 hr in doubly wounded plants is reflected in over a doubling of Inhibitors I and II synthesis and accumulation. As shown in Fig. 4C, a second wounding after 9 hr did not further increase the translational efficiency of Inhibitors I and II mRNA although it significantly increases the rates of inhibitor synthesis and accumulation. The second wound apparently provides more mRNA when the plant's translational system is already operating at high efficiency. The apparent increase in translational efficiency of mRNA after 9 hr is not the result of an increase in total poly(A)$^+$ mRNA. Two separate extractions of total mRNA from leaves of wounded tomato plants showed the opposite; that substantially less poly(A)$^+$ RNA was present at 14 hr (40 ± 10.6 μg/g of leaf) and 18 hr (55 ± 3.5 μg/g of leaf) after a single wound than at 9 hr (96 ± 4.2 μg/g of leaf tissue).

A similar time course for translatable mRNA has been reported (14) in barley aleurone layers, in response to gibberellic acid. Total poly(A)$^+$ RNA was found to increase dramatically in the first 12 hr after hormone application, followed by a rapid decline to 25% of the maximum at 18 hr. On the other hand, the accumulation of ovalbumin mRNA in response to progesterone in chick oviducts (15) is an example that does not appear to behave in this manner.

The possibility that the difference in apparent translation efficiencies described here might be due to the presence of a different, perhaps larger species of mRNA being present at 9 or 18 hr, was investigated by analysis of the sizes of mRNAs for Inhibitors I and II on linear sucrose gradients. Ultracentrifugation of 200 μg aliquots of mRNA from leaves of 9- and 18-hr single wounded plants and 18-hr double wounded plants demonstrated that all three mRNAS migrated identically (Fig. 5). To locate the position of poly(A)$^+$ RNAs specific for Inhibitors I

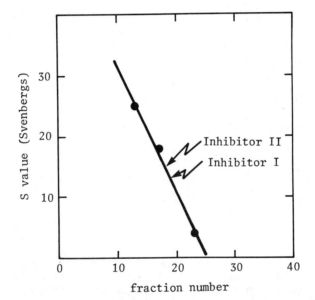

Figure 5. Size analysis of Inhibitors I and II specific mRNA from levels of 9- and 18-h singly wounded tomato plants and 18-h doubly wounded plants. Poly(A)⁺ RNA was applied to 15–30% linear sucrose gradients and was spun at 25,000 rpm. Twenty-five fractions were collected, the absorbency was measured, and the mRNA was precipitated by cold ethanol. In vitro translations were performed with each fraction in a rabbit reticulocyte system, and isolation of the preinhibitors with preformed antibody precipitates located the position of the two inhibitors. The gradients were calibrated by centrifugation of tomato leaf poly(A)⁻ RNA on an identical gradient. The locations of translatable mRNAs for Inhibitors I and II were identical with RNA obtained from 9- and 18-h singly wounded or 18-h doubly wounded plants.

and II the eluted gradient material was precipitated with ethanol
and in vitro translation performed. Fractions containing pre-
Inhibitors I and II were identified by specific immunoprecipita-
tion of the translation products (data not shown). Analysis of
the gradient (16) with known standards of tomato leaf RNA that
do not bind to oligo(dT)-cellulose columns (poly(A)⁻ material)
showed that Inhibitors I and II mRNAs always migrated with
sedimentation coefficients of 13S and 15S, respectively.

In order to further purify the specific messengers for the
inhibitors, additional 15-30% gradients were run on mRNA samples
that had been partially purified by oligo(dT)-cellulose. Fractions corresponding to the known position of Inhibitors I and II
mRNA were combined and recentrifuged in 10-25% linear sucrose
gradients (Fig. 6). By comparing the amount of radiolabel
specifically immunoprecipitated to the total number of counts
incorporated in the total translation reaction we calculated the
relative purity of the two messages (Table III). These physical
techniques have thus provided a 15-fold purification of the
specific mRNA for Inhibitor I and a 5-fold purification of
Inhibitor II mRNA.

The mRNA for both Inhibitors I and II appeared to be
typical of eukaryotic messengers that code for small proteins of
8-12,000 daltons having a poly(A)⁺ tail since both messengers
bind specifically to oligo(dT)-cellulose. There is no evidence
of translatable messengers for the two inhibitors in the RNA
fraction that did not bind to the oligo(dT) affinity resin (data
not shown).

The length of poly(A) segments present in poly(A)⁺ RNA from
pooled sucrose density gradient fractions rich in Inhibitors I
and II mRNA was determined by subjecting the RNA to pancreatic
and T_1 RNase digestion, and labeling the 5' terminal ends with
[γ-³²P]ATP and polynucleotide kinase. Results of the length
determination in polyacrylamide gels indicate that poly(A)
fragments are about 100 nucleotides long. The distribution of
lengths in the experiment represent a variety of mRNAs and the
tail lengths of the two Inhibitors are not known but are assumed
to be within this distribution.

In a further experiment we assayed for the presence of a
cap structure on the mRNAs for both Inhibitors I and II by
competitive inhibition by 7-methyl-guanosine 5'-monophosphate
($m^7G^{5'}p$) of the in vitro translation of these messengers.
Concentrations of 40 μM $m^7G^{5'}p$ inhibited by 50% the in vitro
translation of total tomato leaf poly(A)⁺ mRNA (Fig. 7A). This
level is 40-fold lower than that required to similarly inhibit
rabbit globin mRNA translated in a rabbit reticulocyte lysate
(17) and 4-fold lower than that required to inhibit the same
mRNA in a wheat germ system (18). It was of interest that the
translation of Inhibitor I is inhibited to 50% by 20 μM $m^7G^{5'}p$
while 50% inhibition of Inhibitor II requires less than 10 μM
(Fig. 7B). The basis of this difference is not understood but

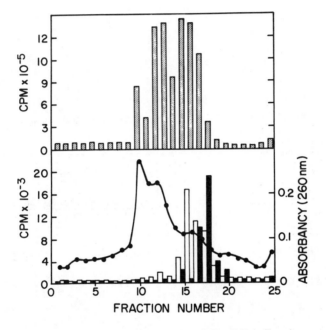

Figure 6. Partial purification of Inhibitors I and II mRNA. Fractions containing Inhibitors I and II mRNA determined by in vitro translation analyses were recovered from an initial 15–30% linear sucrose gradient, precipitated by cold ethanol, and applied to a 10–25% linear sucrose gradient. The sample was centrifuged for 36 h at 25,000 rpm. Fractions of the gradient were collected and subjected to in vitro translation analyses. The upper graph represents total [^{35}S]methionine incorporation assayed with 1 μL of the translation mixture as described (11). The bottom figure quantitates the radiolabel incorporated specifically into Inhibitor I (solid bars) and Inhibitor II (open bars).

TABLE III

PARTIAL PURIFICATION OF INHIBITORS I AND II mRNAs

Purification	% Counts Total Translatable Inhibitors Total Translatable Protein	
	Inhibitor I	Inhibitor II
Oligo(dT) Chromatography	0.27%	0.22%
First Sucrose Gradient	0.42%	0.35%
Second Sucrose Gradient	4.1%	1.2%

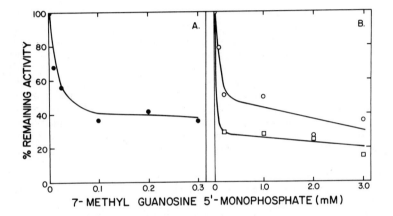

Figure 7. Inhibition of in vitro translation of tomato leaf mRNA by $m^7G^{5'}p$. Quantities of 4 μg mRNA from wounded tomato leaves were translated and analyzed.

A: Total incorporation of [^{35}S]methionine into trichloroacetic acid insoluble protein. Assays were done with 1-μL fractions of the translation reaction as described (11).

B: Incorporation of [^{35}S]methionine into pre-Inhibitors I (○) and II (□) as isolated by preformed antibody precipitates.

could be explained if there is a difference between the affin-
ities of the two proteinase Inhibitor mRNAs for the mRNA binding
site on the 40S initiation complex or between the in vivo cap
structures. Inhibition of translation by derivatives of $m^7G^{5'}p$
has been used as evidence that a specific mRNA contains a 5'
terminal cap structure (19), but this interpretation must be
treated with caution (20). However, the fact that the translation of Inhibitors I and II mRNA is several fold more strongly
inhibited by $m^7G^{5'}p$ than globin mRNA lends support to the
presence of capped structures at the 5' terminus of the inhibitor
mRNAs. The unequivocal identification of the cap structures
will require purification of the mRNAs for each inhibitor.
    Thus, studies of the properties of the mRNAs isolated at
different times after wounding do not reveal any differences
that might reflect the large changes in translational efficiencies observed in tomato leaves 9-14 hr after wounding. During
this time there may be changes in 5'-end capping, polyadenylation
and internal methylation. It is also possible that some cellular
component involved with synthesis and transport of inhibitors
into the central vacuole may be involved. The vacuole or lysosome is considered to originate with the Golgi apparatus.
Cellular events during the first hours following wounding may be
involved with the production of a component(s) to facilitate
ribosome-mRNA binding or some other process in the Golgi apparatus to specifically accommodate transport of Inhibitors I and II
into the vacuole.
    Alternatively, poor efficiencies of inhibitor mRNAs may be
due to their incorporation into ribonucleoprotein particles
(RNPs) such as found in sea urchin embryos (21). Newly made
mRNA in the embryos is found in RNPs and they apparently have
"weak" template activities while in these particles. The
presence of newly synthesized tomato mRNA in similar particles
might explain the apparently low translational efficiencies
noted herein. The use of chaotropic buffers in the preparation
of tomato leaf mRNA (11) would not differentiate between free or
polysome-bound mRNAs and those complexed in RNPs. If an RNP or
similar particle is involved, then its role must be a temporal
one since a second wound does not repeat the phenomenon (Fig. 4).
    Cloning of the mRNAs enriched in Inhibitors I and II
messages is underway to provide cDNA probes to more fully
explore the regulation of the tomato genes coding for Inhibitors
I and II in response to wounding. With such clones not only can
direct hybridization techniques be employed to probe the levels
of PIIF-induced mRNA, but more importantly, the genes for the
inhibitors can hopefully be isolated, the structural features of
the genes be studied and the molecular basis for the regulation
of gene expression in response to stress can be explored.

Summary

The wound-induced synthesis and accumulation of proteinase Inhibitors I and II in tomato leaves has provided a model system to study the regulation of proteinase inhibitor genes in plants. The simplicity of the phenomenon has made it possible to isolate the wound-factor, or hormone, and to study its release, direction and rate of transport in tomato plants. Messenger RNA has been isolated from leaves of wounded plants and contains translatable mRNAs for the two proteinase inhibitors. Studies with these mRNAs have provided a basis for the initiation of a program to clone inhibitor cDNAs for studies of the molecular basis of the wound-induced process of inhibitor synthesis.

Acknowledgements

We would like to thank Alan Rogers for excellent technical assistance and Richard Hamlin for growing the plants.

Literature Cited

1. Green, T.; Ryan, C.A. Science 1972, 175, 776-777.
2. Ryan, C.A. "Current Topics in Cellular Regulation"; E. Stadtman; B.L. Horecker, eds., Academic Press: New York, 1980; pp 1-23.
3. Bishop, P.D.; Makus, D.J.; Pearce, G.; Ryan, C.A. Proc. Natl. Acad. Sci. (USA) 1981, 78, 3536-3540.
4. Ryan, C.A.; Bishop, P.; Pearce, G.; Darvill, A.; McNeil, M.; Albersheim, P. Plant Physiol. 1982, 68, 616-618.
5. Ryan, C.A. Plant Physiol. 1974, 54, 328-332.
6. Green, T.R.; Ryan, C.A. Plant Physiol. 1972, 51, 19-21.
7. Canny, M.J. "Phloem Tranlocation"; Cambridge Univ. Press: Cambridge, 1973; pp 301-312.
8. Minchin, P.E.H.; Troughton, J.H. Ann. Rev. Plant Physiol. 1980, 31, 191-215.
9 Fisher, D.B. Plant Physiol. 1970, 45, 114-118.
10. Plunkett, G.; Senear, D.F.; Zuroske, G.; Ryan, C.A. Arch. Biochem. Biophys. 1982, 213, 463-472.
11. Nelson, C.E.; Ryan, C.A. Proc. Natl. Acad. Sci. (USA) 1980, 77, 1975-1979.
12. Walker-Simmons, M.; Ryan, C.A. Plant Physiol. 1977, 60, 61-63.
13. Nelson, C.E.; Ryan, C.A. Biochem. Biophys. Res. Commun. 1980, 94, 355-359.
14. Muthukrishnan, S.; Chanda, G.R.; Maxwell, E.S. Proc. Natl. Acad. Sci. (USA) 1979, 76, 6181-6185.
15. Mulvihill, E.R.; Palmiter, R.D. J. Biol. Chem. 1980, 255, 2085-2091.

16. McConkey, E.H. "Methods in Enzymology," Vol. 12, Pt. A; L. Grossman; K. Moldave, eds., Academic Press: New York, 1967; pp 620-634.
17. Fresno, M.; Vazquez, D. Eur. J. Biochem. 1980, 103, 125-132.
18. Hickey, E.D.; Weber, L.A.; Baglioni, C. Proc. Natl. Acad. Sci. (USA) 1976, 73, 19-23.
19. Filipowicz, W. FEBS Lett. 1978, 96, 1-11.
20. Takagi, S.; Mori, T. J. Biochem. 1979, 86, 231-238.
21. Rudensey, L.M.; Infante, A.A. Biochem. 1979, 18, 3056-3063.

RECEIVED August 23, 1982

Plant Polyphenols and Their Association with Proteins

JOHN McMANUS, TERENCE H. LILLEY, and EDWIN HASLAM

University of Sheffield, Department of Chemistry, Sheffield S3 7HF, United Kingdom

Plant polyphenols (syn. vegetable tannins) which complex strongly with proteins are based on two structural types. The first group, the proanthocyanidins, possess an oligomeric flavan-3-ol structure and the second class is composed of esters of gallic acid, (R) and (S)-hexahydroxydiphenic acid, and their derivatives usually with D-glucose. Polyphenols associate with proteins principally by intermolecular hydrogen bonding. A hypothesis is advanced to explain the propensity of polyphenols, as their molecular size increases, to precipitate proteins from solution. An important corollary of this theory is that simple phenols (M.Wt. < 200) should display similar behaviour if they can be maintained in solution at sufficiently high concentrations.

For a century and more the elucidation of the chemistry of a natural product(1, secondary metabolite) has been a dominant theme of organic chemistry. Today the emphasis has changed. The question - "what is the role of secondary metabolism in the life of plants and micro-organisms?" - is one to which investigators increasingly address themselves. The problem remains essentially unresolved although numerous ideas have been put forward. Several propositions centre on the suggestion that it is the process of secondary metabolism and not, in the general case, the secondary metabolites themselves which are of importance to the organism (2). These suggestions do not exclude the possibility, indeed probability, that the distinctive properties of individual secondary metabolites have, over the course of evolution, secured a particular niche for an organism in the living world. Contrasting with this viewpoint are those which suggest that an organism's capacity to defend itself against predation by some organisms and to attract others included the evolution of an ability to synthesise an array of secondary metabolites which in appropriate circumstances might repel or attract other organisms (3 , 4).

0097-6156/83/0208-0123$06.00/0

Plant Polyphenols

        Polyphenols are a distinctive group of higher plant secondary
metabolites (5). Two important classes - the proanthocyanidins
(1, Figure 1) and esters of gallic acid and (R) and (S)-hexadydro-
xydiphenic acid (Figure 5) are unique and their mode of biosynthesis in the plant enables molecules of molecular weight > 500
and in many cases of at least 2-3,000 to be formed. These molecules, in relatively low concentrations, moreover possess the important property of precipitating proteins, polysaccharides and
some alkaloids from solution. This association with proteins is
the basis of the use of polyphenols (syn. tannins) as agents for
the conversion of raw animal skins to durable, impermeable leather
(6). It has also been recognised as underlying many other phenomena associated with plant tissues. Thus it is held to be responsible for the astringency of unripe fruit, beverages such as
tea and wine, for the impaired nutritional characteristics of
some foods and for the inactivation of microbial enzymes and plant
viruses. The capacity of plants to metabolise these two classes
of polyphenol is a primitive characteristic that has tended to
become lost with increasing phylogenetic specialisation. According to Bate-Smith (7) their importance to the plant lies in their
effectiveness as repellants to predators brought about by their
ability to precipitate proteins whether this be an extracellular
microbial enzyme or the salivary protein of a browsing animal.

Proanthocyanidins and Procyanidins - In a classical study Bate-
Smith (8) used the patterns of distribution of the three principal classes of phenolic metabolites, which are found in the
leaves of plants, as a basis for classification. The biosynthesis of these phenols - (i) proanthocyanidins; (ii) glycosylated
flavonols and (iii) hydroxycinnamoyl esters - is believed to be
associated with the development in plants of the capacity to
synthesise the structural polymer lignin by the diversion from
protein synthesis of the amino-acids L-phenylalanine and L-tyrosine. Vascular plants thus employ one or more of the p-hydroxy-
cinnamyl alcohols (2,3, and 4), which are derived by enzymic
reduction (NADH) of the coenzyme A esters of the corresponding
hydroxycinnamic acids, as precursors to lignin. The same coenzyme
A esters also form the points of biosynthetic departure of the
three groups of phenolic metabolites (i, ii, iii), Figure 1.
        The two principal classes of proanthocyanidins found (10) in
plant tissues are the procyanidins (1, R = H) and the prodelphin-
idins (1, R = OH). Proanthocyanidins of mixed anthocyanidin
character (1, R = H or OH) have been noted. In any tissue where
proanthocyanidin synthesis occurs there is invariably found a
range of molecular species - from the monomeric flavan-3-ols
(catechins, gallocatechins) to the polymeric forms (1) and biosynthetic work (11) suggests a very close relationship between
the metabolism of the parent flavan-3-ol and the synthesis of
proanthocyanidins, Figure 4.

The majority of the structural and biosynthetic work in this field has been directed towards an understanding of the procyanidins (1, R = H) - the major group of natural proanthocyanidins, (11). ~ Fruit bearing plants have proved to be particularly rich sources of oligomeric condensed procyanidins (11, 12, 13) and studies have concentrated upon the four major naturally occurring dimeric procyanidins (B-1, B-2, B-3 and B-4, Figure 2) which have been isolated from fruit, fruit pods, seeds and seed shells, leaves and other tissues of a wide range of plants. The procyanidins occur free and unglycosylated and almost invariably they are found with one or both of the parent flavan-3-ols, (+)-catechin or (-)-epicatechin, Figure 2. The patterns of occurrence of the procyanidins are most effectively revealed by hplc or by two dimensional paper chromatography and they may be used as a taxonomic guide (11). The absolute stereochemistry in procyanidins B-1 and B-2 was determined as 4R and the C-C interflavan bond thus occupies a quasi-axial bond at C-4 on the "upper" flavan-3-ol heterocyclic ring (14). Conversely the absolute stereochemistry in procyanidins B-3 and B-4 was shown to be 4S and here the interflavan bond is thus in a quasi-equatorial position on the heterocyclic ring (14). Thus the principal procyanidin dimers (Figure 2) all possess a trans orientation of the C-3 hydroxyl group and the flavan substituent at C-4.

Despite the successful application of spectroscopic methods to the structure determination of the dimeric procyanidins anomalous features of the spectra led to speculation (12, 15, 16) that restricted rotation existed about the interflavan bond. A detailed study (14) confirmed this hypothesis and defined two different forms of hindered rotation associated with the two groups of procyanidin dimers (B-1 and B-2, 4R - absolute configuration) and (B-3 and B-4, 4S-absolute configuration). Examination of molecular models shows that, although several conformations are accessible by interflavan bond rotation, there exists one preferred conformation for each of the pairs of natural procyanidins. If these are viewed from different angles they bear an almost object to mirror image relationship; the structures are quasi-enantiomeric (Figure 3). Elaboration of the oligomeric procyanidins (1, R = H) by the addition of further identical flavan-3-ol units, bearing in mind the conformational restraints about the interflavan bond, leads to two general structures which may adopt helical conformations with opposite helicities (11). The central core of these polymers is composed of rings A and C of the flavan repeat unit and ring B - the ortho - dihydroxyphenyl ring - projects laterally from this core.

Earlier proposals (12, 13, 17) and the results of biosynthetic experiments (18) have been adumbrated into a scheme of biosynthesis for the procyanidins (Figure 4) in which it is suggested that they are formed as byproducts during the final stage of the synthesis of the parent flavan-3-ol structures, (+)-catechin and (-)-epicatechin (11, 18). A two step reduction of the flav-3-en-3-ol

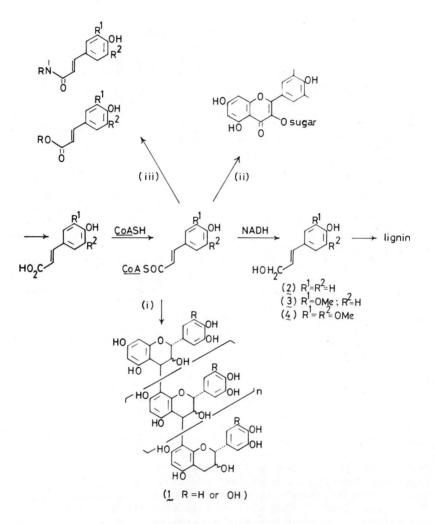

Figure 1.   Phenol biosynthesis in higher plants.

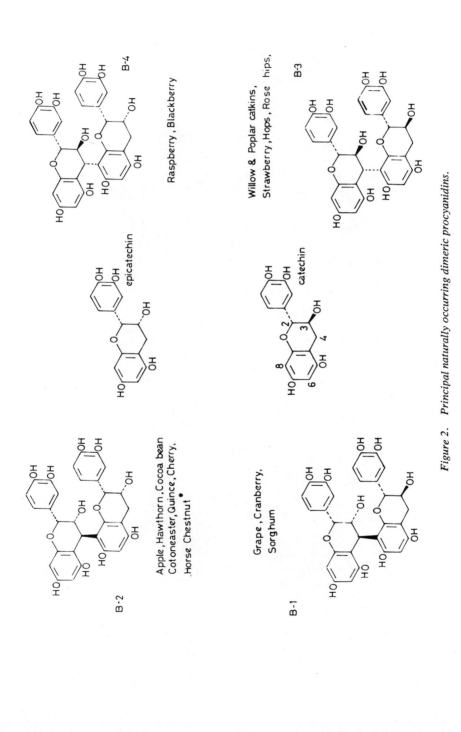

Figure 2. Principal naturally occurring dimeric procyanidins.

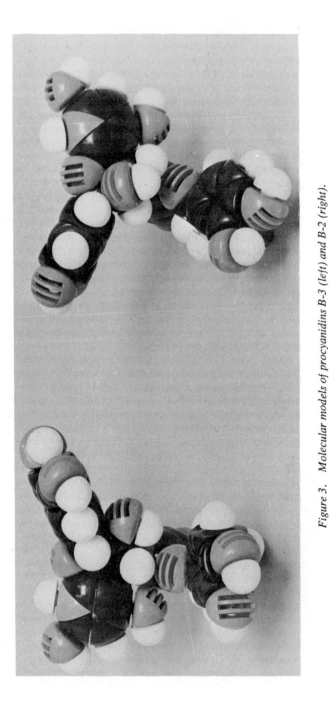

Figure 3. Molecular models of procyanidins B-3 (left) and B-2 (right).

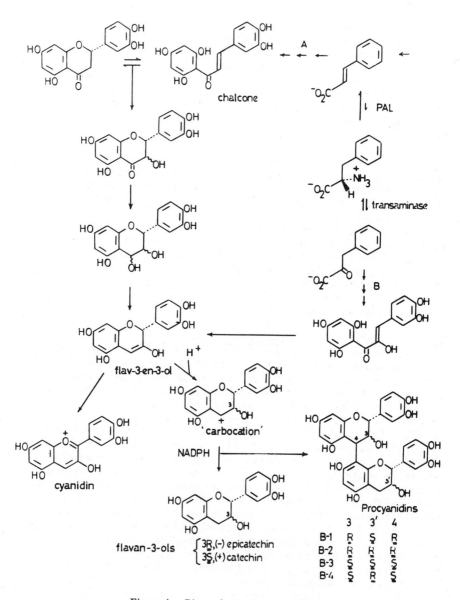

Figure 4.   Biosynthesis of procyanidins.

is postulated. If the supply of NADPH (say) is rate limiting the
intermediate C-4 carbocations (3S or 3R stereochemistry) might
escape from the enzyme active site and react with the reduction
product, the flavan-3-ol, to produce dimers, trimers ... and
eventually polymers. The hypothesis has been elaborated to
account for the various stereochemical features (11) and has been
supported by in vitro biomimetic experiments. The balance between
the metabolic flux to the flav-3-en-3-ol and the rate of supply
of the biological reductant (NADPH) probably determines the
balance between the lower and higher oligomeric forms of the procyanidins in each plant tissue. Thus tissues with a high flux
and poor supply of NADPH will contain predominantly the higher
oligomers - for example the seed coat of sorghum (19).

    This facet of proanthocyanidin metabolism was noted by early
workers (20) who recognised the presence in plants of leucoantho-
cyanins (proanthocyanidins) of quite different solubilities. With
increasing degrees of polymerisation the proanthocyanidins are
more difficult to solubilise in aqueous and alcoholic media and
those which may be coaxed into solution may have molecular weights
up to 7,000 - corresponding to structures such as (1) where
n = 18. Recent work has concentrated upon the structural examination of these polymers. Although Bate-Smith (8) generally
discounted a connection with lignin, an examination of the proanthocyanidin polymers, to ascertain if they do play a structural
role would be very valuable.

Esters of Gallic and Hexahydroxydiphenic Acids - An acute
observation made by Bate-Smith (8) in his taxonomic work was the
complete absence from plant tissues of 3,4,5-trihydroxycinnamic
acid. In situations where its occurrence might well have been
predicted, he found hexahydorxydiphenic acid and its derivatives
(Figure 5). Recent work has demonstrated the ubiquity in plants
of esters of R and S hexahydroxydiphenic acid with D-glucose and
of their presumed biogenetic precursors the galloyl-D-glucoses
(22, 23). Gallic acid (Figure 5) thus occupies a distinctive
position in phenolic metabolism in plants. Almost always it
occurs in ester form but in contrast to other natural phenolic
acids, which are invariably found as mono- and occasionally bis-
esters with polyols, gallic acid is encountered in a range of
esterified forms from the simple monoesters to the complex poly-
esters with D-glucose whose molecular weights extend to at least
2,000. The structural patterns discerned amongst gallic acid
metabolites are nevertheless highly reminiscent of those found in
other variants of secondary metabolism in other organisms - e.g.
polyketides in moulds and fungi.

    Gallic acid is most commonly associated with D-glucose. Many
plants metabolise simple esters and there appears to be a predis-
position for certain positions of esterification on the D-glucose
pyranose core (1 > 6 > 2 > 3,4). The metabolism of β-pentakis-
galloyl-D-glucose (Figure5) represents something of a metabolic

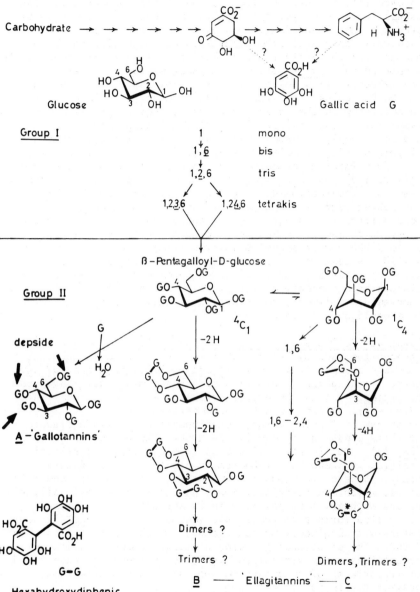

Figure 5. *Metabolism of gallic acid and hexahydroxydiphenic acid.*

watershed and from this point at least three different biosynthetic pathways diverge. From a taxonomic point of view these
pathways are quite distinct (24). One pathway proceeds to form
depside metabolites ('gallotannins', group IIA). Here further
gallic acid molecules are esterified as m-depsides to β-pentakis-
galloyl-D-glucose, preferentially at positions 3, 4 and 6. This
supports the view originally expressed by Emil Fischer (25) that
these esters are not only mixtures of isomers but also of
substances of differing empirical formulae.

A second and widely observed pattern of metabolism of β-
pentakis-galloyl-D-glucose is that (IIB, Figure 5) in which the
substrate is further transformed by oxidative coupling of pairs
of adjacent galloyl ester groups (2, 3 and 4, 6) on the D-gluco-
pyranose ring to give hexahydroxydiphenoyl esters.   Further
metabolism of these intermediates then also occurs by intermolecular C-O oxidative coupling to form 'dimers' - such as (5)
isolated from Rubus species - and possibly 'trimers'. Oxidative
coupling of galloyl ester groups takes place with the precursor
adopting the energetically preferred ($^4C_1$) conformation of the
sugar ring. Coupled with the observation that this is the
predominant mode of further oxidative metabolism of galloyl esters
in plants, it also seems reasonable to conclude that it is the
energetically preferred pathway.

A third and relatively minor pathway of metabolism of β-
pentakis-galloyl-D-glucose (IIC, Figure 5) occurs in some plants.
Here oxidative metabolism takes place via the energetically
unfavourable $^1C_4$ conformation of the substrate and coupling of the
galloyl ester groups occurs 1, 6 or 3, 6 to give hexahydroxydiphenoyl esters or 2, 4 to give the dehydrohexahydroxydiphenoyl
ester.   The compound geraniin is a key figure in this pathway of
metabolism (26).

Comparative Aspects of Polyphenol Metabolism - Proanthocyanidins
and the complex esters of gallic and hexahydroxydiphenic acid
show many structural similarities as plant metabolites. The shape
and size of the ester (5) is thus very similar to that of a
proanthocyanidin hexamer (1, n = 4). The most striking feature
of both structures however is the manner in which free phenolic
groups are distributed over the surface of the molecule providing
a structure with the inbuilt capacity for multidentate attachment
to other species by hydrogen bonding.

A curious but perhaps significant observation is that, although
several plant families retain the ability to biosynthesise
different types of complex polyphenol, rarely are both biosynthetic
capabilities displayed maximally. More often one particular
species specialises in one particular mode. Thus in the Ericaceae
many plants are rich sources of proanthocyanidins but Arcto-
staphylos uva-ursi has only a minimal proanthocyanidin biosynthetic capacity but combines with this a very high level of
gallic acid metabolism.

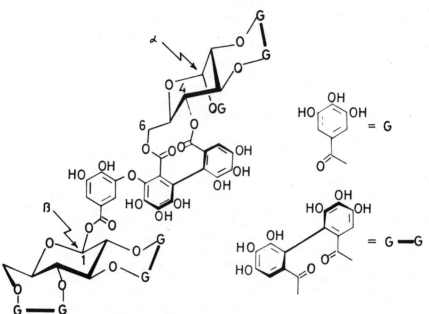

Compound 5

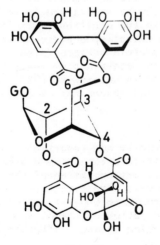

Geraniin

Polyphenol Interactions with Proteins

Studies of the association of polyphenols with proteins have a long history (27). Loomis (28) has succinctly summarised the conclusions of this earlier work. The principal means whereby proteins and polyphenols are thought to reversibly complex with one another are (i) hydrogen bonding, (ii) ionic interactions and (iii) hydrophobic interactions. Whilst the major thrust in earlier work was to emphasize the part played by intermolecular hydrogen bonding in the complexation, Hoff (29) has drawn attention to the possibility that hydrophobic effects may dominate the association between the two species.

Recent observations however now permit a hypothesis to be advanced to explain the propensity of complex polyphenols to precipitate proteins from aqueous solution (30). Two situations may be envisaged. At low protein concentrations the polyphenol associates - principally by hydrogen bonding via the ortho dihydroxyphenyl groups - at one or more sites on the protein molecule forming a relatively hydrophobic surface layer (Figure 6a). Aggregation and precipitation then ensue. Where the initial protein concentration is high the hydrophobic surface layer is formed by cross-linking of different protein molecules by the multidentate polyphenols (Figure 6b). Precipitation then follows as outlined, vide supra. This tendency to cross-link protein molecules explains the changing stoichiometry of the aggregates in relation to the initial protein concentration. An interesting corollorary of this hypothesis is that simple phenols such as pyrogallol and resorcinol should also be capable of precipitating proteins from solution if they can be maintained in solution at concentrations sufficient to push the equilibrium in favour of the protein-phenol complex and thus establish a hydrophobic layer of simple phenol molecules on the protein surface (Figure 6c). For many simple phenols the limit is determined by their own solubility but it can be achieved in water with BSA (3×10^{-5} molal) and pyrogallol (1 molal) and resorcinol (2 molal).

Postscript

Complex plant polyphenols readily and reversibly associate with proteins and they can precipitate them from dilute solution. This property is however a direct extrapolation of the characteristics of simple phenols themselves. The structural device represented by the plant polyphenols to a great extent obviates the need for a high molal concentration of phenol and it embodies the added feature that cross-linking between different molecular aggregates may be readily achieved.

The questions posed initially remain. What purpose do these molecules serve in plant metabolism? Does their presence point for example to the need by plants either now or during the course of evolution for such tailored molecules to reversibly coat the

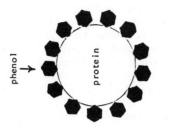

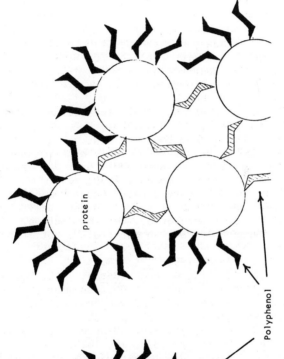

Figure 6. Protein–polyphenol association.

surface of proteins and hence modify their activity at particular phases of development? Answers to these problems will only emerge as and when a great deal more is known concerning the enzymology of processes governing the formation of these metabolites, how these processes are regulated and controlled, and their relationship to the network of primary metabolism from which they devolve. A truly formidable task but as Robert Louis Stevenson once wrote:

"To travel hopefully is a better thing than to arrive, and the true success is to labour".

El Dorado.

Literature Cited

1. Mothes, K. in "Secondary Plant Products" Encylopaedia of Plant Physiology New Series, Vol. 8; Ed. Bell, E.A. and Charlwood B.V.; Springer-Verlag, Berlin, 1980, p. 1

2. Bu'Lock, J.D. in "The Biosynthesis of Mycotoxins", Ed. Steyn, P.S., Academic Press, New York - London, 1980, p. 1

3. Janzen, K.H. in "Herbivores - Their Interactions with Secondary Metabolites"; Ed. Rosenthal, G.A. and Janzen, D.H.; Academic Press, New York - London, 1979, p. 331

4. Rhoades, D.F. in "Herbivores - Their Interactions with Secondary Metabolites"; Ed. Rosenthal, G.A. and Janzen, D.H.; Academic Press, Nre York - London, 1979, p. 3.

5. Harborne, J.B. in "Secondary Plant Products" Encylopaedia of Plant Physiology, New Series, Vol. 8; Ed. Bell, E.A. and Charlwood, B.V.; Springer-Verlag, Berlin, 1980, p. 329.

6. Howes, F.N., "Vegetable Tanning Materials"; Butterworths, London, 1939.

7. Bate-Smith, E.C., Food, 1954, 23, 124.

8. Bate-Smith, E.C., J. Linnaen Soc. (Bot.), 1962, 58, 95.

9. Zenk, M.H. in "Recent Advances in Phytochemistry" Vol. 12; Ed. Swain, T., Harborne, J.B. and van Sumere, C., Plenum Press, London and New York, 1978, p. 139.

10. Bate-Smith, E.C. and Lerner, N.H., Biochem. J., 1954, 58, 126.

11. Haslam, E., Phytochemistry, 1977, 16, 1625.

12. Thompson, R.S., Jacques, D., Tanner, R.J.N. and Haslam, E., J. Chem. Soc. (Perkin 1), 1972, 1387.

13. Weinges, K., Kaltenhaūser, W., Marx, H.-D., Nader, E., Nader, F., Perner, J. and Seider, D., Annalen, 1968, 711, 184.

14. Fletcher, A.C., Gupta, R.k., Porter, L.J. and Haslam, E. J. Chem. Soc. (Perkin 1), 1977, 1628.

15. Weinges, K., Marx, H.-D. and Perner, J., Chem. Ber., 1970 103, 2344.

16. Jurd, L. and Lundin, R., Tetrahedron, 1968, 24, 2652.

17. Geissmann, T.A. and Yoshimura, N.N., Tetrahedron Letters, 1966, 2669.

18. Opie, C.T., Porter, L.J., Jacques, D. and Haslam, E., J. Chem. Soc. (Perkin 1), 1977, 1637.
19. Gupta, R.K. and Haslam, E., J. Chem. Soc. (Perkin 1), 1978, 892.
20. Robinson, R. and Robinson, G.M., J. Chem. Soc., 1935, 744.
21. Porter, L.J., Czochanska, Z., Foo, L.Y., Newman, R.H., Thomas, W.A. and Jones, W.T., J. Chem. Soc. Chem. Commun., 1979, 375.
22. Okuda, T., Yoshida, T., Mori, K. and Hatano, T., Heterocycles, 1981, 15, 1323.
23. Haslam, E., Fortschr. chem. org. Naturst., 1982.
24. Magnolato, D., Gupta, R.K., Haddock, E.A., Layden, K. and Haslam, E., Phytochemistry, 1982.
25. Fischer, E., Ber., 1919, 52, 809.
26. Okuda, T., Nayeshiro, H. and Yoshida, T., Chem. Pharm. Bull. (Japan) 1977, 25, 1862.
27. Van Buren, J.P. and Robinson, W.B., J. Agric. Food Chem., 1969, 17, 772.
28. Loomis, W.D., Methods. Enzymol., 1974, 31, 528.
29. Oh, H.I., Hoff, J.E., Armstrong, G.S. and Haff, L.A., J. Agric. Food Chem., 1980, 28, 394.
30. McManus, J., Lilley, T.H. and Haslam, E., J. Chem. Soc., Chem. Commun., 1981, 309.

RECEIVED September 10, 1982

The Role of Natural Photosensitizers in Plant Resistance to Insects

T. ARNASON—University of Ottawa, Biology Department,
Ottawa, Ontario K1N 6N5, Canada

G. H. N. TOWERS and B. J. R. PHILOGÈNE—University of British Columbia,
Botany Department, Vancouver, British Columbia V6T 1W5, Canada

J. D. H. LAMBERT—Carleton University, Biology Department,
Ottawa, Ontario K15 5T6, Canada

Recent work has suggested that certain secondary metabolites from plants are capable of photosensitizing
insects. With the recognition of increasing numbers
of natural photosensitizers including polyacetylenes,
furanocoumarins, β-carbolines and extended quinones,
this unusual mechanism of plant defense now appears
to be present in a wide variety of plant families.
Polyacetylenes, a group, of over 500 diverse compounds
which are especially widespread in the Asteraceae are
powerful photosensitizers. At least one sulphur
derivative, α-terthienyl is more toxic to mosquito
larvae than DDT. The toxicity of some of these
compounds is mediated by the production of singlet
oxygen although free radicals may also be involved.
Furanocoumarins, furanoquinolines and β-carbolines on
the other hand interact with DNA and cause gross
chromosomal abnormalities in vivo. The furanocoumarin,
8-methoxypsoralen has been shown to be highly phototoxic to herbivorous Spodoptera larvae. It has also
been suggested that the leaf rolling habit of certain
microlepidopteran larvae on leaves containing photosensitizing furanocoumarins is related to light
avoidance. Although only limited information is available about the effects of naturally occurring compounds
on insects, work with synthetic dye photosensitizers
provides a broader basis for understanding the photosensitization of insects.

0097-6156/83/0208-0139$06.00/0

Light is often a forgotten or underestimated factor in the study of insects (1) and until recently little attention has been paid to its role in plant insect relations. In particular we wish to address the question of activation of plant secondary substances by light and their subsequent photosensitizing effects on insects. At least two examples of insect photosensitizing secondary metabolites, furanocoumarins and polyacetylenes have now been reported. These studies suggest that this mode of plant defense has particular adaptive advantages and may be more widespread than previously imagined. Fortunately the mode of action of photosensitizers from plants is understood from work with target organisms other than insects. In addition some detailed information on the effects of photosensitizers on insects is suggested from work with synthetic dye sensitizers. These latter aspects are crucial to the understanding of the effects of natural photosensitizers and are treated first in this review.

Photosensitization

For photosensitization to occur light must be absorbed by the photosensitizer and cause a reaction to take place in the biological system (2). Two main types of photosensitization are known to occur. Those involving O_2 are called photodynamic sensitizations and are mediated by the bulk of the 400 synthetic and natural photosensitizers known (3). The second type of photosensitization does not involve O_2 and is associated with the furanocoumarins.

Photodynamic sensitizations are photo-oxidations which proceed by one of two pathways (3,4). Type I photosensitizations (less common) yield superoxide radicals (O_2^-) through a series of electron transfer processes following light absorption. Anthraquinone dyes and flavins operate through this mechanism. In type II sensitizations, the sensitizer is excited from the ground state, S_0, to the first excited state, ^1S by absorption of a photon. Crossing over leads to the formation of the excited triplet state, ^3S which interacts with ground state oxygen, $^3O_2$. The transfer of excitation energy leads to the production of singlet oxygen ($^1O_2$) and the sensitizer returns to its ground state:

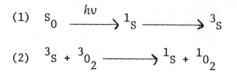

$$(1) \quad S_0 \xrightarrow{\ h\nu\ } {}^1S \xrightarrow{\hspace{2cm}} {}^3S$$

$$(2) \quad {}^3S + {}^3O_2 \xrightarrow{\hspace{2cm}} {}^1S + {}^1O_2$$

Because the ground state of the sensitizer is regenerated the reaction is catalytic. Once formed the activated species of oxygen, $^1O_2$ from type II reactions O_2^- from type I mechanisms cause biological damage by oxidation of biological molecules. They are also potentially interconvertible (5). Recent advances have allowed the identification of $^1O_2$ in photochemical damage of biological systems by the use of quenchers (6) or D_2O which enhances the lifetime of $^1O_2$ (7) Superoxide is detected in photo-oxidations by the use of the scavenging enzyme superoxide dismutase (8). Important target molecules are proteins, lipids (especially cholesterol) and nucleic acids (3, 4, 9) but the effects in vivo are largely dependant on the site to which the photosensitizer binds. Thus rose bengal (compound I) binds to and lyses membranes while acridine orange penetrates to the nucleus and causes damage to DNA (6, 10).

The second group of photosensitizers bind and create photo-chemical damage at the level of DNA without any O_2 requirement for activity. These include the well studied furanocoumarins (11) and more recently the furoquinoline alkaloids (12). Work with the furanocoumarins suggests that the first step is intercalation of the photosensitizer into DNA in such a way that one or two of the double bonds of the ring structure align with the pyrimidine double bonds. The excited state of the photosensitizer undergoes cycloaddition to the pyrimidine (usually thymine) forming a photoadduct (II). Compounds such as the furoquinoline, dictamnine (X) and hindered furanocoumarins form only monofunctional adducts (11, 12) while the linear furanocoumarins such as 8-methoxypsoralen (8-MOP) (IV) form difunctional adducts (III) leading to interstrand cross-linkage of DNA. Both types of damage but especially the latter have serious consequences for DNA transcription or duplication and can lead to cell death or mutagenesis (II).

Effect of Photosensitizers on Insects

Because of the availability of synthetic dye sensitizers, their effect on insects is somewhat better understood than photo-sensitizing plant secondary substances. The first report of the effect of photosensitizers on insects was made by Barbieri in 1928 (13), 28 years after the discovery of photosensitization by Raab. A 1972 review by Graham (14) reported only five investigations into the photosensitization of insects, but the number has more than doubled since that time, principally due to the work at Mississippi State. Some of these reports are summarized in TABLE I. Studies have concentrated mainly on the thiazine dye methylene blue (II) and a range of xanthene dyes of which rose bengal (I) was the most prominent. Four orders of insects are represented including eggs, larvae, pupae and adults. No unusually resistant species have been found and the results with the boll weevil and fire ant are particularly significant since

(I)

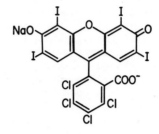

(II)

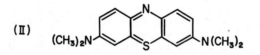

(III)

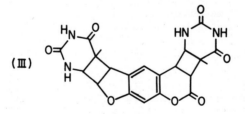

(IV)

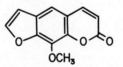

TABLE I

Studies of the Effect of Photodynamic Dye Sensitizers on Insects

Order	Organism	Photosensitizer	Mode of Administration	Reference
Diptera	Culex pipiens quinquefasciatus (mosquito larvae)	rose bengal	water treatment	(15)
	Aedes aegypti & Anopheles spp. (mosquito larvae)	uranin, erythrosin acridine red eosin, rhodamine, rose bengal	"	(16)
	Anopheles sp. & Aedes sp. (mosquito larvae)	xanthene dyes	"	(13)
	Aedes triseratus and Culex pipiens quinquefasciatus (mosquito larvae)	rose bengal	"	(17)
	Musca domestica (adult house fly)	xanthene dyes	diet	(18)
	Musca autumnalis (adult face fly)	xanthene dyes	diet	(19)
	Drosophila melanogaster	benzpyrene, methyl cholanthrene, dibenzanthrene	diet	(20)
Coleoptera	Anthonomus grandis (adult boll weevil)	xanthene dyes	diet	(21)
	Tenebrio molitor (meal worm larvae)	mythylene blue	injection	
Hymenoptera	Solenopsis richteri (fire ant)	xanthene dyes	diet	(22)
Lepidioptera	Pierris brassicae (cabbage butterfly larvae)	methylene blue	diet	(23)

they indicate that even hard bodied insects are susceptible to
photodynamic action. Mortality in most cases was found to be
directly proportional to the photon fluence and dye concentration. Except for the more recent work, toxicities are difficult
to compare from one study to the next because of the range of
conditions possible and the difficulty in estimating the fluence
of light absorbed with lamps of different spectral quality and
dyes with different absorption maxima. An attempt at quantita-
tion of the results has been made by derivation of second (22)
and third order (24) rate constants for photo-oxidation. Using
the second order rate constant insects can be ranked in order of
their susceptibility: Aedes triseratus (larvae) > Anthonomus
grandis (adult) > Musca autumnalis (adult) > Solenopsis richteri
(adult). With mosquito larvae susceptibility falls with the
number of instars. The rate constant can also be used to rank
dye toxicity (18). Both dye toxicity to house flies and quantum
yield for dye phosphorescence increase in the order: fluoroscein,
eosin yellow, phloxin B, erythrosin B and rose bengal. This
result is explained by the current view that phosphorescence and
the production of $^{1}O_2$ in the photosensitization reaction are
dependant on the proportion of dye molecules in their triplet
excited state (3).

At the physiological level it is well established that
vital dyes such as nile blue, neutral red and methylene blue
retard larval development under normal lighting conditions
(12L/12D with source unspecified) (25-27). Female but not male
pupal weights are also reduced. Unfortunately experiments were
conducted without dark controls so that it is difficult to
evaluate the role of photosensitization in these effects. As
house flies and fire ants succumb to photosensitization, they
lose motor control and become more excitable (28). This
suggested a neurotoxic effect and investigation of fire ant
acetylcholinesterase in vitro revealed that this enzyme was
sensitive to photo-oxidation. In vivo results, however,
revealed no effect on the enzyme which suggests another mode of
action. Epoxidation of cholesterol and membrane lysis may be
alternative primary sites. If this were the case ecdysone
metabolism of insects would probably also be effected.

Furanocoumarins

Although furanocoumarins are well studies for their effects
on human skin, recent work has suggested their *raison d'être* in
plants may be linked to their role as protective agents that
are effective against insects or pathogenic fungi (29, 30). This
group of compounds is reported in 8 families but find their
greatest diversity in the Apiaceae and Rutaceae. Berenbaum has
demonstrated their activity against insect herbivores in feeding
trials with a polyphagous herbivore , the fall armyworm
Spodoptera eriania which will feed on carrot which does not

contain furanocoumarins but not on parsnip that does contain
these compounds. Larvae were administered 0.1% 8-methoxypsoralen
(IV), a furanocoumarin compound which occurs widely in the
Apiaceae by treatment of artificial diet. These levels are comparable to those in plants and caused 100% mortality in larvae that
were also treated with near UV, the activating wavelength range
for these sensitizers. Dark controls and insects fed untreated
diets but irradiated with near UV showed a slight reduction in
survivorship. The effect of near-UV and 8-MOP together was highly
significant and is presumably due to the effect of the compound
on larval DNA. Thus to feeding generalists 8-MOP represents a
formidable barrier to grazing on the plants that contain it.
Since the precursor to psoralens, umbelliferone, is not toxic to
armyworms, Berenbaum hypothesizes that these plants have escaped
their enemies by the alteration of their chemical phenotype.
Insects have in some cases responded by overcoming these defenses.
For example aphids that feed exclusively on the Apiaceae take up
and bind 8-MOP but are unaffected by it, perhaps due to some
detoxification mechanism (29). Yu, S.J., Berry, R.E., and Terriere,
L.C.,(1979) showed that phytophagous insects possess enzymes that
are induced by plant secondary substances and that these enzymes
are involved in resistance to otherwise toxic compounds (31).
Microlepidoptera appear to have adapted to life on phototoxic
Apiaceae by a leaf rolling habit that screens out near UV. (30).
Cluster analysis of the fauna of the Apiaceae indicates that the
insect assemblages of plants containing phototoxic furanocoumarins
are similar and different from plants in the family that do not
contain phototoxic furanocoumarins (32). This lends support to the
hypothesis that a specialist group of insects exists that is
adapted to phototoxic Apiaceae.

Polyacetylenes

     Polyacetylenes are a very large group of secondary compounds
whose photosensitizing properties have recently been established
by our research team at U.B.C. (33). These compounds have conju-
gated double and triple bond systems (e.g. compound VII and VIII)
or may be biosynthetically cyclized into thiophene compounds such
as alpha terthienyl (α-T) (compound VI). Polyacetylenes and their
thiophene derivatives occur in several families but find their
greatest diversity in the Asteraceae, the largest plant family
(34). As in the case of the furanocoumarins, photosensitization
is mediated by the near-UV region of the spectrum but the mechanism of action does not involve cross-linking of DNA (35). Some
compounds such as α-T are clearly photodynamic in their mode of
action (36) but the situation is less clear for compounds containing triple bond systems such as phenylheptatriyne (PHT) (VIII).
Photo-oxidation of cholesterol and acetylcholinesterase has been
observed (37) with this and other polyacetylenes but lysis of red
blood cells and photosensitization E. coli were not O_2 dependant

(38). These results suggest a novel mechanism of action and are an active area of research.

Polyacetylenes are toxic to a broad range of organisms (39) but are especially toxic to insects. At 0.5 ppm, 9 of 14 compounds tested were toxic to first instar mosquito larvae (Aedes aegypti) in 30 min treatments with sources of near UV (15 $\overline{W/m^2}$) (40). The compounds were more active in sunlight. For example α-T killed second instar larvae instantaneously at 4 ppm. Compounds VII, α-T and PHT were especially active and were selected for further testing in dose response experiments (41). For similar near UV treatments, the LC_{50} for α-T was 19 ppb, 79 ppb for compound VII and 1.0 for PHT. Alpha T had some activity in the dark (LC_{50} = 0.74 ppm) indicating that it may bind to a sensitive cell site and direct photosensitization to this site. This compound was so active in the presence of near UV that its potential as a commercial larvicide was evaluated in simulated pond trials. For 500 larvae placed in 200 l of water in summer sunlight, 100% mortality was observed in 15 min at 200 ppb and 120 min at 20 ppb. A detailed action spectrum for photosensitization undertaken with narrowband interference filters indicated that there was close agreement between action and absorption. This suggested that the polyacetylene was the absorbing species and was not interfering with metabolism in such a way as to cause a photosensitizing product to be formed. Such mechanisms do occur with icterogens produced by the genus Tribolus which are ingested by range animals (42). The result with polyacetylenes also differs from that with psoralens where a DNA-psoralen complex is thought to result in the deviation of the action spectrum from the absorption spectrum (43).

We are currently investigating the effect of polyacetylenes and near-UV in sublethal doses during feeding trials with Euxoa messoria (Lepidoptera, noctuidae). Potential for further work also exists with the adapted insect, the soldier beetle (Coleoptera, Cantheridae), which apparently uses a polyacetylene as a defense compound (44).

Other compounds.

Despite the hypothesis that the evolutionary significance of phototoxic secondary substances may be linked to their ability to discourage insect herbivores, most research has been directed toward their effects on human skin and range animals (42). In an attempt to extend our knowledge of insect photosensitizers we have screened a number of plant secondary substances (TABLE II) for their photosensitizing activity to 4th instar mosquito larvae Aedes atropalpus under solar simulating lamps.

Hypericin (IX) is an extended quinone from St. John's wort (Hypericum spp), a common pasture weed throughout the world. This photodynamic compound causes a photosensitive disease called 'bighead' in sheep that consume it (42). Related compounds, the

(**V**)

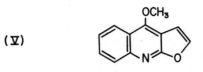

(**VI**)

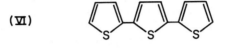

(**VII**)

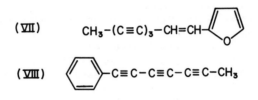

$CH_3-(C\equiv C)_3-CH=CH-$

(**VIII**)

$-C\equiv C-C\equiv C-C\equiv C-CH_3$

(**IX**)

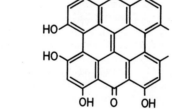

(**X**)

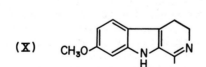

CH_3O-

TABLE II

Phototoxicity of Plant Secondary Metabolites
to Mosquito Larvae

Compound	Type	Phototoxic Activity
Hypericin	Extended quinone	+ +
Dictamnine	Furo quinoline alkaloid	+ +
Harmaline	β-Carboline alkaloid	+ +
Methoxyharmaline	β-Carboline alkaloid	+ +
Harman	β-Carboline alkaloid	−
Norharman	β-Carboline alkaloid	+
Harmalol	β-Carboline alkaloid	+
Berberine	Isoquinoline alkaloid	+

Note: Compounds were screened at several concentrations
in 24-hr acute toxicity test with fourth instar
Aedes atropalpus larvae. Tests were run with
parallel trials in the dark and under solar
simulating "vita lites[R]" with an intensity of
400 w/m^2. Activities were rated as follows
(−) no difference between light and dark trial,
(+)   enhancement of light toxicity over dark, (+ +)
larger enhancement of light toxicity over dark.

fagopyrins occur in buckwheat (fagopyrum sp).   In addition we
have included two groups of alkaloids, the β-carbolines (of which
harmaline (X) is an example) and the furoquinoline alkaloid
dictamnine (V) which were recently discovered by one of us
(Towers) to be phototoxic to yeast (45, 46).   The activity of
berberine was suggested by its fluorescence.   Obviously many
compounds are phototoxic or have enhanced toxicity in light as
compared to dark.   We believe this demonstrates the potential for
further work in this field and suggests that this is an area that
has been overlooked.

One of the reasons that many photosensitizing compounds have
been overlooked is because of their apparent lack of color.   For
example the polyacetylene PHT is a potent insect photosensitizer
in natural sunlight is completely colorless in solution.   Its
spectrum reveals (Figure 1) strong absorption bands in the near
UV which are well beyond the human visual limit (380 nm), but
within the range of wavelengths transmitted by the atmosphere
(generally > 300 nm) (47). In addition it should be noted that
the energy absorbed at these wavelengths is considerably higher
than in the visible range and may be a factor in the high
toxicity of compounds like alpha T.

In conclusion it is evident that many research opportunities
exist in the identification and characterization of new
substances and evaluation of their ecological evolutionary and
physiological significance.   In a practical sense it can be
hoped that some of these new compounds because of their novel
mode of action may be useful for the control of phytophagous
insects as part of integrated pest management programs.

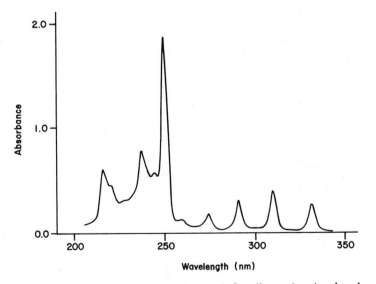

Figure 1.   Absorption spectrum of 1 ppm of phenylheptatriyne in ethanol.

ACKNOWLEDGEMENT

    This work was supported by NSERC and an Agriculture Canada
E.M.R. grant to one of us (Arnason).

Literature Cited

1. Philogène, B.R. Am. Nat. 1982, 118, in press.
2. Turro, N.J.; Lamola, A.A. in "The Science of Photobiology"
    K.S. Smith (ed) Plenum: New York. 1977, 63-86.
3. Spikes, J.D. in "The Science of Photobiology" K.C. Smith (ed)
    Plenum: New York. 1977, 87-112.
4. Grossweiner, L.I. Topics in Rad. Res. Quart. 1976, 11,
    141-199.
5. Singh, A. Photochem. Photobiol. 1978, 28, 429-433.
6. Ito, K. Photochem. Photobiol. 1978, 28, 493-508.
7. Nilsson, R.; Kearns, D.R. Photochem. Photobiol. 1973, 17,
    65-68.
8. Jahnke, L.S.; Frenkel, A.W. Photochem. Photobiol. 1978, 28
    517-523.
9. Kulig, M.J.; Smith, L.L. J. Org. Chem. P 973, 38, 3639-3642.

10. Lamola, A.A.; Yamane, T.; Trozzolo, A.M. Sci. 1973, 179,
    1131-1133.
11. Song, P.S.; Tapely, K.J. Photochem. Photobiol. 1979, 29,
    1177-1197.
12. Pfyffer, G.E.; Panfil, I.; Towers, G.H.N. Photochem.
    Photobiol. 1982, 35, 63-68.
13. Barbieri, A. Riv. Malariol, 1928, 7, 456-463.
14. Graham, K. Can. J. Zool. 1972, 50, 1631-1636.
15. Carpenter, T.L.; Heitz, J.R. Envir. Entomol. 1980, 9, 533-7.
16. Schildmacher, H. Biol. Zentralbl. 1950, 69, 468-477.
17. Pimprikar, G.D.; Norment, B.R.; Heitz, J.R. Environ.
    Entomol 1979, 8, 856-9.
18. Fondren, J.E.; Norment, B.R.; Heitz, J.R. Environ. Entomol.
    1978, 7, 205-208.
19. Fondren, J.E.; Heitz, J.R. Environ. Entomol. 1978, 7, 843-6.
20. Maltoltsy, G.; Fabian, G. Nat. 1946, 158, 877-8.
21. Callaham, J.J.; Broome, J.R.; Lindig, O.H.; Heitz, J.R.
    Environ. Entomol. 1975, 4, 837-841.
22. Broome, J.R.; Callaham, M.F.; Heitz, J.R. Environ. Entomol.
    1975, 4, 833-6.
23. Lavialle, M.; Dumortier, B. C.R. Acad. Sc. Paris. 1978, 287,
    875-8.
24. Fondren, J.E.; Heitz, J.R. Environ. Entomol. 1978, 7, 891-4.
25. Barbarosa, P.; Peters, M. Ent. Exp. & App. 1970, 13, 293-9.
26. Barbarosa, P.; Peters, M. J. Med. Ent. 1970, 7, 693-6.
27. Barbarosa, P.; Peters, M. Histochem. J. 1971, 3, 71-93.

28. Callaham, M.F.; Lewis, L.A.; Holloman, M.E.; Broome, J.R.; Heitz, J.R. Comp. Biochem. Physiol. 1975, 51C, 123-8.
29. Camm, E.L.; Wat, C.K.; Towers, G.H.N. Can. J. Bot. 1977, 54, 25-
30. Berenbaum, M. Sci. 1978, 201, 532-3.
31. Yu, S.J.; Berry, R.E.; Terriere, L.C. Pesticide Biochem. Physiol. 1979, 12, 280-4.
32. Berenbaum, M. Ecology. 1980, 62, 1254-66.
33. Camm. E.L.; Towers, G.H.N.; Mitchell, J.C. Phytochem. 1975, 94, 2007-11.
34. Bohlmann, F.; Burkhard, T.; Zdero, C. "Naturally Occurring Acetylenes" Academic Press, New York, 1973.
35. Wat, C.K.; Biswas, R.K.; Graham, E.A.; Bohm, L.; Towers, G.H.N.; Waygood, E.R. J. Nat. Prod. 1977, 42, 103-111.
36. Arnason, T.; Chan, G.F.Q.; Wat, C.K.; Downum, K; Towers, G.H.N. Photochem. Photobiol. 1981, 33, 821-4
37. Wat, C.K.; MacRae, W.D.; Yamamoto, E.; Towers, G.H.N.; Lam, J. Photochem. Photobiol. 1981. in press.
38. Arnason, T.; Wat, C.K.; Downum, K.; Yamamoto, E.; Graham, E.; Towers, G.H.N. Can. J. Microbiol. 1980, 26, 698-705.
39. Towers, G.H.N.; Wat, C.K. Rev. Latinoamer Quim. 1978, 9, 162,170.
40. Wat, C.K.; Prasad, S.K.; Graham, E.A.; Partington, S.; Arnason, T.; Towers, G.H.N.; Lam, J. Biochem. Syst. Ecol. 1981, 9, 59-62.
41. Arnason, T.; Swain, T.; Wat, C.K.; Graham, E.A.; Partington, S.; Towers, G.H.N.; Lam, J. Biochem. Syst. and Ecol. 1981, 9, 63-68.
42. Giese, A. Photophysiology, 1971, 6, 77.
43. Nakayama, Y.; Morihaw, F.; Fukuda, M.; Hamano, M.; Toda, K.; Patnak, M. in "Sunlight and Man", Fitzpatrick, T.B. (ed), University of Tokyo Press. 1974.
44. Meinwald, J.; Meinwald, Y.C.; Chalmers, K.M.; Eisner, J. Sci. 1978, 160, 890.
45. McKenna, D.J.; Towers, G.H.N. Phytochem. 1981, 20, 1001-4.
46. Towers, G.H.N.; Graham, E.A.; Spencer, I.D.; Abramowski, Z. Planta Medica. 1981, 41, 136-142.
47. Jagger, J.J. in "The Science of Photobiology", Smith, K.D. (ed), Plenum: New York. 1977.

RECEIVED September 30, 1982

Natural Inducers of Plant Resistance to Insects

MARCOS KOGAN

University of Illinois at Urbana-Champaign, Section of Economic Entomology, Illinois Natural History Survey and Office of Agricultural Entomology, College of Agriculture, Urbana, IL 61801

JACK PAXTON

University of Illinois at Urbana-Champaign, Department of Plant Pathology, Urbana, IL 61801

Many environmental factors may affect herbivore/plant interactions by increasing the level of resistance or susceptibility of the plant to the herbivore. Among the factors with a demonstrated capability to induce changes in levels of resistance are temperature, solar radiation, water stress, soil fertility, insecticides, herbicides, fungicides, growth regulators, pathogen infection, weed competition, and previous or concurrent herbivore attack. Most chemical factors, responsible for the resistance of plants to arthropods studied to date, have involved the genetically controlled, injury-independent accumulation of metabolites with allomonal activity. A completely different plant defense strategy has been demonstrated for many pathogens for which phytoalexin accumulation is the result of a post-challenge response. The aforementioned environmental factors, including pest-related injury, can induce de novo synthesis and accumulation of compounds with allomonal properties (phytoalexins) or change the relative concentration of both nutrient and non-nutrient compounds. The focus in this paper is on the post-infestation induction of resistance   by arthropods, in a manner up until recently known to result only from pathogen infection. The insect-resistance role of phytoalexins is discussed and reference is made to other possible natural sources of inducers of resistance.

Antiherbivory in plants has been ascribed mainly to the presence of physical defenses or to the injury-independent accumulation of secondary metabolites that have allomonal

0097-6156/83/0208-0153$06.00/0

properties. These metabolites are compounds that act as repellents, feeding deterrents, or toxins and are already present in the plant before herbivore attack. Another defense mechanism, probably equally important but more difficult to demonstrate, results from the absence of some required nutrients or compounds with kairomonal action, such as attractants and feeding excitants (1).

Antimicrobial defenses in most plants, on the other hand, have been shown to result from a post-challenge response of the plant elicited by the pathogen through products associated with the invading organism. Such products are extra-cellular polysaccharides of bacteria, or the β-glucans of Phytophthora mycellial walls (2).

Up until recently, it has been generally assumed that the two defense mechanisms have evolved under the specific selection pressure of animal herbivory or pathogenicity of microorganisms. Such mechanisms were studied independently by entomologists and plant pathologists with little or no effort invested to determine the commonalities that one might have expected to exist between the two systems. In theory, however, there would be great advantage for a plant to mobilize its defenses against herbivores only under the reasonable rise in risk of injury, such as at the initial stages of an insect infestation. The injury independent concentration of allomonal compounds may require a considerable amount of metabolic energy that needs to be diverted from other vital functions (3) (see also Mooney in this volume). If the herbivore attack does not occur at all, that energy is obviously wasted.

The post-challenge or injury-dependent response of plants is a manifestation of induced resistance. For the lack of previous definitions the following is proposed:

Induced resistance is the qualitative or quantitative enhancement of a plant's defense mechanisms against pests in response to extrinsic physical or chemical stimuli. These extrinsic stimuli are known as inducers or elicitors.

Inducers of Plant Resistance or Susceptibility Against Insects

Both physical and biological environmental factors have been shown to influence resistance or susceptibility of a plant to insects. Among these factors are: temperature, light, relative humidity, soil fertility and soil moisture, air pollutants (such as ozone, sulfur dioxide and others) (4). These are components of the plant's physical environment that occur independently of a man's intrusions. In addition, products made and intentionally applied by man may influence the quality of plants to insects; foremost among these are insecticides and fungicides, growth regulators and herbicides. Finally, plant pathogens and previous herbivory also influence the quality of the plant to insects.

Growth regulators, insofar as they are related to plant hormones, and plant pathogens and herbivores may be perceived as natural inducing factors. We considered herbicides together with growth regulators, because of the similarities in their chemistry and modes of action. We will concentrate our discussion on these types of inducers of plant defenses against insects.

Induction of Resistance by Growth Regulators and Herbicides

There have been many studies testing the effect of growth regulators on plants, and attempts have been made to measure this effect on insects. Table I provides a summary of some representative studies.

Table I.   Examples of Insect Responses to Applied Growth Regulators.

Inducer	Plant	Insect	Effect	Source
Maleic Hydrazide	Broad Bean	Pea Aphid	>Mortality <Fecundity	(5)
Chlormequat Chloride (=CCC, =Cycocel)	Broad Bean	Bean Aphid Cabbage Aphid	>Mortality	(6)
	Brussels Sprout	Pea Aphid Green Peach Aphid	<Fecundity	(8)
	Black Currant	Hyperomyzus	<Fecundity	(7)
	Oleander	Oleander Aphid	>Mortality	(9)
Daminozide	Pear	Pear Psylla	<Population	(10)
SADH	Petunia	Whiteflies	>Mortality	(11)
Gibberillin	Common Bean	Mites	<Fecundity	(12)
	Apple	Mites	<Fecundity	(13)
	Broad Bean	Pea Aphid	No-effect	(5)
		Bean Aphid	<Fecundity	(6)

Application of maleic hydrazide to broadbean plants increases mortality and reduces the fecundity of bean aphids (5). The growth retardant CCC (chlormequat) has been tested on several plant/insect systems, all with appreciable increases in resistance either due to reduced fecundity of females, or increased mortality of offspring; both resulting in an overall reduction in populations (6, 7, 8, 9). Occasionally, as is the case with many secondary metabolites, certain insect species are affected detrimentally by the application of a compound, but others are not. Gibberellin applied to broadbean reduces fecundity in the bean aphid (6), but has no effect on the pea aphid (5).

Herbicides also have been tested in their effect on insects (Table II). Results are even more ambiguous than those with growth regulators. For instance, 2-4D on barley reduced fecundity in two species of grain aphids, Rhopalosiphum padi and Macrosiphum avenae (14), but applied to corn increased fecundity of corn leaf aphids, Rhopalosiphum maydis (15), suggesting an improvement in the quality of food for the insect. Other herbicides have been tested with fewer insects, and results vary. Amitrol and Zitron applied to broadbean reduced fecundity and increased mortality in the pea aphid Acyrthosiphon pisi (16), but Banvel D, Barban or MCP applied to barley increased fecundity of the grain aphids M. avenae, R. padi, and Schizaphis graminum (17).

Table II. Insect Responses to Herbicides Applied to the Food Plant. (In the "Effect" column, R refers to resistance and S to susceptibility.)

Inducer	Plant	Insect	Effect	Source
				(18)
2,4-D	Wheat	Stem Sawfly	R>Mortality	(14)
	Barley	Grain Aphids	R<Fecundity	(15)
	Corn	Corn Leaf Aphid	S>Fecundity	
				(19)
	Broad Bean	Pea Aphid	S>Fecundity	
	Rice	Rice Stem Borer	S>Growth >Survival	(20)
Amitrole	Broad Bean	Pea Aphid	R<Fecundity >Mortality	(16)
Zytron				
Banvel D	Barley	Grain Aphids	S>Fecundity	(17)
Barban				
MCPA				

Induction of Resistance by Previous Pathogen Attack

MacIntyre, Dodds and Hare (21), and Hare (22) reported that some tobacco varieties hypersensitive to tobacco mosaic virus (TMV) can be protected against attack by other organisms. Local virus infections induced a systemic protection against the fungi Phytophthora parasitica var. *nicotiana* and Peronospora tabacina as well as the bacterium Pseudomonas tabaci (21). In addition, the reproductive rate of the green peach aphid, Myzus persicae, was reduced about 11% when females were allowed to feed and reproduce parthenogenetically on leaves at plant apices at least 6 nodes above the site of virus infection (21). Furthermore, it was found by Hare and coworkers (22) that growth rates of fourth instar tobacco hornworms, Manduca sexta, were reduced 27% when reared on tobacco leaves with local TMV lesions and 16% when reared on neighboring leaves without external symptoms. The symptomless leaves were on plants that had other leaves symptomatic of TMV. The whole plant was systemically induced to higher levels of resistance.

Induction of Resistance by Previous Herbivory

Feeding activities of herbivorous insects often result in physiological and morphological changes in the host plant. If these changes are in the direction of the accumulation of compounds with resistance properties, the herbivores themselves act as inducers. In Table III examples were drawn from several insect/ plant systems representing a whole spectrum of annual and perennial crops as well as vegetable crops, evergreen and deciduous leaf trees. In all these examples the induced resistance affects the inducing species itself. Thus, for instance, feeding by the pea aphid on alfalfa induces an increase in coumestrol that may affect additional feeding by other pea aphids (23). Changes may also occur in the levels of primary metabolites thus affecting the nutritional value of the leaves. For example, feeding by gypsy moth, Lymantria dispar, larvae on gray birch or black oak leaves produces physiological changes in the plants that may reduce the survival and growth rates of other larvae feeding on leaves several days after the initial attack (26). It is noteworthy that many changes that occur following herbivory on trees result from accumulations of phenolic compounds.

Table III.   Induction of Resistance by Previous Herbivory.

Inducing Herbivore	Plant	Induced Reaction	Source
Pea Aphid	Alfalfa	>Coumestrol	(23)
European Pine Sawfly	Pine	>Polyphenol Synt.	(24)
Sweet Potato Weevil	Sweet Potato	Ipomeamarone (Furanoterpenoid Phytoalexin)	(25)
Gypsy Moth	Gray Birch Black Oak	<Nutritional Value	(26)
Lep. Larvae	Birch	>Phenolics	(27)
Lygus disponsi	Sugar Beet Chinese Cabbage	>Quinones >Phenolics	(28) (29)
Cotton Bollworm	Cotton	>Phenolics	(30)
Striped Cucumber Beetle	Squash	>Cucurbitacins	(31)

A different type of induction occurs with the striped cucumber beetle, Acalymma vittata, feeding on squash (31). Feeding by this beetle results in an accumulation of cucurbitacins which are strong feeding excitants for the cucumber beetle itself. However, cucurbitacins are deterrent at high concentrations to the squash beetle, Epilachna tredecimnotata. In fact, the squash beetle seems to have evolved a behavioral adaptation to prevent accumulation of cucurbitacins in their feeding sites. Before starting the feeding process, they cut a circular trench encircling the leaf section upon which they will later feed. The trench prevents circulation of sap into the   enclosed area that remains free of the allomonal effect of the cucurbitacins, and suitable to feeding by the beetles (31).

From the above examples it is thus apparent that a plant challenged by herbivores, by plant pathogens, or otherwise stressed by specific chemicals does not remain biochemically indifferent. The plant responds with more or less drastic metabolic changes, some of which have a profound effect on additional herbivores or other plant pathogens. The chemical nature of these changes is now beginning to be elucidated.

Mechanisms of Induced Resistance

There seems to be a variety of mechanisms involved in resistance induced by the chemical and biological factors mentioned above. The mechanisms that we will discuss are the following: 1) changes in phenological synchronization between the plant and its complement of herbivores; 2) changes in the physiological state of the plant; 3) changes in nutrient concentration; 4) stimulation of compensatory mechanisms; 5) changes in concentrations of allelochemics--either increases in the concentration of allomones or reduction in the concentration of required kairomones; and 6) de novo synthesis of phytoalexins. This latter is perhaps the most exciting aspect of induced resistance and one upon which we will dwell in greater length.

Phenological Synchronization. Induced delays in budbreak in balsam fir reduce feeding by the spruce budworm, and defoliation and desquaring in cotton depress overwintering success of several major cotton pests. Table IV presents a summary of the work on these two systems.

TABLE IV. Growth Regulators Used in Induced Phenological Asynchronization.

Plant	Inducer	Insect	Effect	Source
Balsam Fir	Abscisic Acid	Spruce Budworm	Delayed budbreak	(32)
	Maleic Hydrazide Chlorflurenol			
Cotton	2,4-D (Amine)	Pink Bollworm	Defoliation	(33,34)
	Chlorflurenol	Heliothis spp.	New square shedding	(35,36)
	Chlorquemat (CCC)	Boll Weevil	Elimination of diapausing sites	(37)

When abscisic acid is applied to balsam fir, the break of buds in the spring is delayed and the emerging spruce budworm, Choristoneura fumiferana, larvae are forced to feed on old needles, which are a less desirable food. Attempts have been made, rather unsuccessfully, under natural forest situations, to manipulate budbreak with growth retardants for the control of the spruce budworm. However, this system seems to operate under controlled greenhouse conditions, and with some adjustments, it may have potential for practical applications in the field (32).

Practical results have been obtained in the control of several cotton pests by the timely application of defoliators or certain herbicides, such as 2-4D (amine), Chlorflurenol or Chlorquemat (CCC) (33, 34, 35, 37). Figure 1 shows the phenology of the cotton plant and the annual cycle of two major cotton pests: the boll weevil, Anthonomus grandis, and the boll worm, Heliothis zea. The majority of the bolls are formed and opened by the middle of September. Those that remain on the plant after that date do not contribute much to the final yield, but they do serve as food and reservoir for diapausing boll weevils, bollworms, and pink bollworms, Pectinophora gossypiella. It is therefore recommended that applications of defoliants be made starting in the middle of September to eliminate the overwintering sites or diapausing shelter for these pests. If this is done on a wide area, results are indeed outstanding. Even better results have been obtained for control of pink bollworm with the interruption of irrigation properly coupled with the application of defoliants.

Changes in Physiological State of the Plants. Resistance to fruit flies has been obtained with the proper application of growth retardants. Greany and coworkers in Florida reported that immature citrus fruits are resistant to fruit fly larvae (39). Resistance is apparently due to citrus peel oils which are mostly composed of monoterpenes and other terpenoids. Resistance is lost as the fruits senesce and change color. With timely applications of gibberllic acid, senescence of the peel is retarded, but the internal maturation processes are not affected. Thus the fruit concentrate sugars while the peel is still green and resistant to the fruit flies. Application of these growth regulators influences a whole range of post harvest pests, not only the fruit flies (39).

Changes in Nutrient Concentration. It has been reported since the 1960's that gibberellin, CCC and other growth regulators induce changes in total nitrogen and sugar levels in several plants. Rodriguez and Campbell (13), working with mites on apples, and Honeyborne (6), working with aphids on broadbeans and Brussels sprouts, recorded changes in pest populations following fluctuations in the concentrations of those nutrients.
Figure 2 (adapted after 13) shows that the increased concentration of gibberellin solutions applied at weekly intervals to apples produced a reduction in the concentration of total nitrogen and sugars with a consequent decline in mite populations.

Stimulation (or Inhibition) of Compensatory Mechanisms. An interesting effect of herbivore feeding on plants has been observed by several authors comparing hand defoliation with herbivory in grasses. One such report shows that the regrowth of

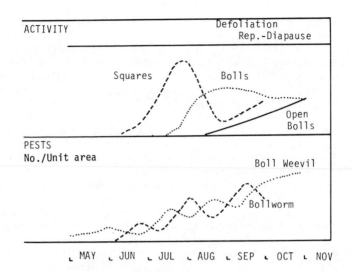

Figure 1. Phenological synchronization of cotton boll opening and diapause of boll weevils and bollworms (38).

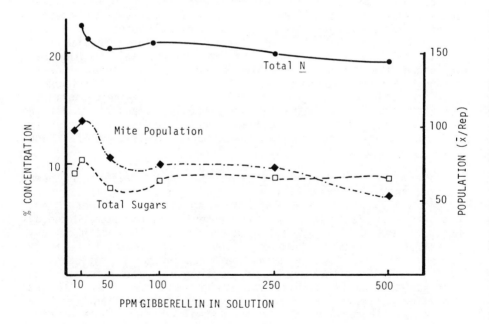

Figure 2. Effect of weekly sprays of gibberellin solutions on nitrogen (●) and total sugar (□) concentration in apple leaves and consequent effect on mite progeny production (■) (13).

grasses is stimulated by growth-regulator-type compounds in the saliva of ruminants (40). However, regrowth seems to be inhibited in grasses by grasshopper salivary gland and gut extracts at high defoliation levels, but it was apparently stimulated at low levels as is shown in Figure 3 (41). When 1/3 defoliation was implemented by actual feeding by grasshoppers, there was a substantial increase in the number of tillers. Tillering was much less in hand defoliated plants. However, when 100% defoliation was implemented, hand defoliation produced a greater number of tillers than grasshopper induced defoliation. The authors argued that wheat has evolved a mechanism whereby, at high defoliation levels induced by grasshoppers, tillering was inhibited. Such differential, density-dependent response by the grasses may be related to a feedback regulatory mechanism of both grass and grasshopper populations. Such mechanisms are, however, speculative, and there is still much to be elucidated in this area (41).

Changes in Allelochemic Concentration. Many allelochemic effects are related to concentrations of diphenolics and the activity of enzymes involved in their oxidation to quinones. Hori and Atalay (29) have shown that the small bug, Lygus disponsi, feeding on sugar beet leaves induced increases in the concentration of phenolics, and peroxidase and polyphenol oxidase activities. Similar increases were also observed immediately after feeding by Lygus disponsi on Chinese cabbage, but levels returned to near normal after three days. Leaf phenolics that remain at high concentrations after the initial feeding may influence the amount of herbivory and also pathogen attacks, although evidence to this effect was not available. Figure 4 (adapted after 29) shows the changes in polyphenol oxidase activity and in concentration in phenolic compounds following injury by Lygus disponsi on Chinese cabbage leaves. The activity and concentration are highest one day after injury, but fall and level off subsequently until 21 days.

De Novo Synthesis of Phytoalexins. Phytoalexins have been studied in great depth by plant pathologists. Excellent review papers are available in Hedin's ACS symposium volume, Host Plant Resistance to Plants (42), and more recently in the book edited by Horsfall and Cowling, Plant Disease (43). The antiherbivory effect of phytoalexins, however, is only now beginning to be fully appreciated. It is apparent that pathogen induced phytoalexins do have a definite effect on insect herbivores. There is mounting evidence that herbivore-inflicted injury may also result in the induction of phytoalexin production and accumulation.

The antiherbivory effects of phytoalexins have been studied in our laboratory using the Mexican bean beetle and the soybean looper (44). The Mexican bean beetle, Epilachna varivestis, is an oligophagous species that feeds preferentially on legume hosts.

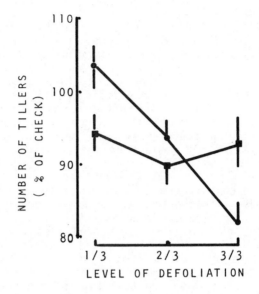

Figure 3.   Tiller production by wheat plants following defoliation by hand (■) or by grasshopper grazing (●) (41).

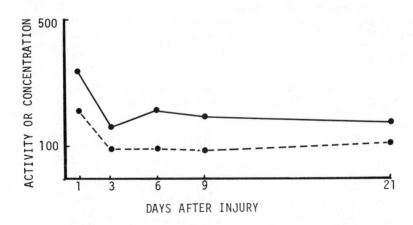

Figure 4.   Changes in polyphenol oxidase activity (– ● –) and phenolic compounds (●) in Chinese cabbage tissue after injury by Luygus disponsi *(29).*

We tested the effect of phytoalexins on the feeding preferences of
both adults and larvae. Soybean seedlings were germinated in the
greenhouse and, when cotyledons were fully opened, small discs
were cut and irradiated with UV-light to elicit phytoalexin
production. Another batch of discs obtained from the second
cotyledonary pair of each plant was not irradiated and was used as
a control. Discs were allowed to incubate for 48 hours, and then
they were exposed in pairs of treated and untreated discs to both
adults and larvae of the Mexican bean beetle. Results of the test
were analyzed using a preference index that gives a value of one
if there is no preference for either disc in the pair, and a
result smaller than one if the untreated or control disc is
preferred. Feeding preferences were measured by counting the
number of scars clearly visible on the cotyledon discs. It was
observed that the phytoalexin rich discs were probed, as shown by
fine mandibular markings, but were not fed upon. All the feeding
was done on the control discs. Feeding preferences on cotyledons
were tested also with the soybean looper, Pseudoplusia includens.
Results showed no feeding deterrency by the phytoalexins. The
effect of pure phytoalexins was tested also using the soybean
looper larvae feeding on artifical media. Due to the small amount
of pure phytoalexins available for the studies, we developed a
miniature system to test the effect of the diet on the growth
rates of newly born larvae up to about the 8th day of life. In
this case, results were not conclusive. When coumestrol was
tested at .1, .5, and 1% concentration in the diet, the percent
survival decreased but the weight gain at the 1% concentration did
not differ from the control and was twice as high as the weight
gain on the .1% diet (Table V). This is difficult to explain on
toxicological grounds. The survival of larvae on the 1%
glyceollin diet was drastically reduced (Table VI), but those
larvae that did survive gained on the average more weight than
those feeding on the control diet.

Table V. Soybean Looper Survival and Weight Gain to
8 Days on Artificial Media + Coumestrol (44).

Coumestrol Conc.	% Survival	% Wgt. Gain
0	96.0	100.0
0.1%	64.0	44.1
0.5%	68.0	54.1
1.0%	68.0	96.5

Table VI. Soybean Looper Survival and Weight Gain to 8 days on Artificial Medium + Glyceollin (44).

Glyceollin conc.	% Survival	% Wgt. Gain
0	83.2	100.0
0.1%	74.0	76.4
0.5%	75.5	139.2
1.0%	58.0	127.7

To test the validity of the bioassay itself we prepared a diet containing increasing amounts of rotenone, a compound derived from isoflavones and thus chemically not far removed from the soybean phytoalexins. Results in this case followed exactly the expected dose response curve (Table VII). Both survival and weight gain of larvae were drastically affected by increasing concentrations of rotenone. This experiment showed that the bioassay would be capable of detecting toxic effects of the phytoalexins on the soybean looper larvae, if such effects were acute. It showed also that the detoxification mechanisms in the soybean looper, a rather polyphagous insect, may permit it to adequately overcome the antibiotic effect of the isoflavonoid phytoalexins, but not that of the isoflavone rotenone.

Antiherbivory activity of phytoalexins has been demonstrated also in the field in studies by Sutherland and his coworkers (45, 46) with scarabaeid grubs feeding on the roots of several forage legumes. The phytoalexin vestitol was extracted from the roots of the forage legume Lotus pedunculatus, and it was shown to have a strong feeding deterrent effect on the grubs of Costelytra zealandica (46). Feeding deterrency was also demonstrated in our studies with the Mexican bean beetle and phytoalexin rich soybean cotyledons (44).

Table VII. Soybean Looper Survival and Weight Gain to 8 Days on Artificial Medium + Rotenone (Our Unpublished Data).

Rotenone conc.	% Survival	% Wgt. Gain
0	100.0	100.0
0.005%	87.5	46.1
0.05	79.2	16.0
0.5	70.8	9.0
5.0	29.2	3.9

In <u>vitro</u> studies of the effect of pure phytoalexins on the feeding of several phytophagous insects, on the other hand, have provided a diverse picture (Table VIII). For instance, vestitol and phaseollin have antifungal activity and also reduce feeding by <u>Costelytra zealandica</u> and <u>Heperonychus arator</u>. However, various other phytoalexins such as pisatin, genistein, and coumestrol may affect one species but not the other. Neither coumestrol nor genistein seem to inhibit soybean looper feeding or growth.

Table VIII. Effect of Legume Isoflavonoid Phytoalexins on Feeding by Insects.

Compound	Antifungal Activity	C. zealandica (Reduction in feeding)	H. arator	P. includens (Effect on growth rate)
(-)-Vestitol	+	+	+	
(-)-Phaseolin	+	+	+	
(+)-Pisatin	+	+	-	
Genistein	-	-	+	-
Coumestrol	-	-	+	-
Glyceollin	+			+

One may conclude from these studies that phytoalexins have a selective effect on herbivores. They inhibit feeding and growth on some, but seem to be innocuous to others. Given the coevolutionary origin of plant defenses, it is conceivable that some phytoalexins may even be kairomones for some insects. There is ample precedence to this dual role of several allelochemics (1). The example of cucurbitacins mentioned above is one of them.

Potential Uses of Induced Resistance in IPM

A better understanding of the role of phytoalexins in plant defenses and of the mechanisms of induced resistance may potentially open a powerful new approach to the control of insect pests of cultivated plants. If indeed, in light of the hypothesis of optimal defense strategies (3), a post-attack response is a more efficient line of defense than the attack-independent accumulation of allelochemics, the exploitation of phytoalexin-producing mechanisms may represent a fertile field for future investigations. Several uses of induced resistance may be conceived. Four of these approaches are briefly discussed.

Enhanced Response of Plants to Previous Herbivory by Means of Classical Breeding Methods. Since plants within the same species may differ in their ability to produce and concentrate phytoalexins in response to stimuli, the idea is to identify such plants and use their regulating genes in classical breeding programs.

Immunization of Plants by Attenuated Forms of Pathogens. As has been argued by Kuc and Caruso (47), plants can be immunized to achieve higher levels of resistance to pathogens. Similar mechanisms may conceivably provide a line of defense against phytophagous insects without the challenge-independent accumulation of defensive compounds.

Specific Control of Phytoalexin Accumulation by "Metabolite Shunting" of Biosynthetic Pathways. Graham and coworkers (personal communication), at the Monsanto Laboratories, St. Louis, have developed techniques to selectively shunt defensive metabolites, particularly of the shikimic acid cycle. Through various techniques, certain compounds are applied to plant aerial or root parts, and these compounds have the property of inducing specific accumulations of secondary metabolites. The directions of these accumulations are under known enzymic control (48), and the regulation of these enzymes is achieved by selecting appropriate inducers. Such inducers seem to provide a novel approach to the control of insects by magnifying the ability of plants to produce and concentrate antiherbivory compounds.

Improved Economic Injury Levels Through a Better Understanding of the Effects of Low Levels of Herbivory and Subeconomic Pathogen Infections. One of the critical factors in IPM is a realistic definition of economic injury levels for single pests and pest complexes. The understanding of the effect of low levels of herbivory and subeconomic infections by pathogens may help reassess the definition of economic injury levels, due to the evidence that these attacks may actually contribute to an increase in the ability of the plant to withstand future attacks.

Induced Resistance and IPM-Concluding Remarks

Seen from the perspective of a global pest control strategy, induced resistance represents a new dimension in control methodology. As one aspect of the overall field of plant resistance, induced resistance is parallel to genetic resistance and it may operate through the various modalities that are classically identified in genetic resistance. Thus, the escape mechanisms that were discussed as phenological asynchronies may be obtained or induced by cultural methods or by the timely application of growth regulators. Antibiosis, either through

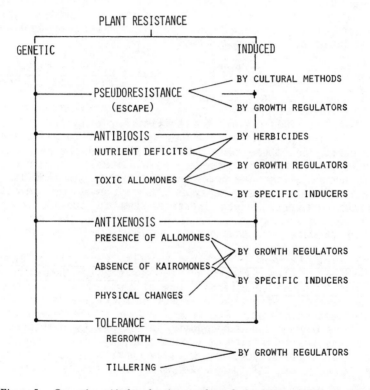

Figure 5.   Operation of induced resistance through classical modalities of resistance.

nutrient deficits or the accumulation of toxic metabolites, may be achieved by the application of herbicides, growth regulators or specific inducers, such as those mentioned under the concept of metabolite shunting. Antixenosis, or effects on the behavioral responses of insects, may be the result of accumulation of allomones, elimination of kairomones or promotion of physical changes in the plants, all of which can be effected by growth regulators or by specific inducers. Finally, tolerance may be obtained through the induction of regrowth or recovery of injured tissue or the tillering of grasses by growth regulators or by herbivory itself.

The increased evidence of the antiherbivory role of phytoalexins, and improvements in our ability to manipulate phytoalexin and other antiherbivory metabolite accumulations in plants by the use of growth regulators, specific inducers, or attenuated forms of pathogens may represent one of the most exciting developments in the fight against insects since the advent of the modern organo-synthetic insecticides. The difference is that we are enhancing the ability of the plants to produce their own defenses; therefore, we are not tampering with potentially hazardous chemicals, and we are not grossly impinging on the integrity of the environment. We are, however, bypassing the inherently contradictory objective of producing both high yielding plants and highly resistant plants. From the standpoint of optimal defense strategies, these two goals are often impossible to reconcile. Induced resistance represents an alternative. We may indeed be witnessing the birth of a fourth generation of insecticides.

Acknowledgements

This publication is a contribution of the Illinois Natural History Survey and Illinois Agricultural Experiment Station, College of Agriculture, University of Illinois at Urbana-Champaign. Supported in part by USDA Competitive Grant 59-2171-0-1-452-0, "Role of Phytoalexins in Soybean Resistance to Insect Pests"; the Regional Project S-157, and the US EPA (through Texas A&M University; CR-806277-02-0). The opinions expressed herein are those of the authors and not necessarily those of the funding institutions.

Literature Cited

1. Norris, D. M.; Kogan, M. in "Breeding Plants Resistant to Insects" (F. G. Maxwell and P. R. Jennings, ed.), Wiley: New York, 1980, pp. 23-61.
2. Cruickshank, I.A.M. in "Plant Disease: An Advanced Treatise" (J. G. Horsfall and E. B. Cowling, ed.), Academic Press: New York, 1980, pp. 247-67.
3. Rhoades, D. F. in "Herbivores: Their Interaction with

Secondary Plant Metabolites" (G. A. Rosenthal and D. H. Janzen, ed.), Academic Press: New York 1979, pp. 3-54.
4. Tingey, W. M.; Singh. S. R. in "Breeding Plants Resistant to Insects" (F. G. Maxwell and P. R. Jennings, ed.), Wiley: New York, 1980, pp. 87-113.
5. Robinson, A. G. Can. Entomologist 1960, 92, 494-9.
6. Honeyborne, C.H.B. J. Sci. Fd. Agric. 1969, 20, 388-90.
7. Singer, M. C.; Smith, B. D. Ann. Appl. Biol. 1976, 82, 407-14.
8. Van Emden, H. F. J. Sci. Fd. Agric. 1969, 20, 385-7.
9. Tahori, A. S.; Halevy, A. H.; Zeidler, G. J. Sci. Fd. Agric. 1965, 16, 568-9.
10. Westigaar, P. H.; Lombard, P. B.; Allen, R. B.; Strang, J. G. Environ. Entomol. 1980, 9, 275-7.
11. Tauber, M. J.; Shalucha, B.; Langhans, R. W. Hort Sci. 1971, 6, 458.
12. Eichmeier, J.; Guyer, G. J. Econ. Entomol. 1960, 53, 661-4.
13. Rodriguez, J. G.; Campbell, J. M. J. Econ. Entomol. 1961, 54, 984-7.
14. Adams, J. B.; Drew, M. E. Can. J. Zool. 1969, 47, 423-6.
15. Oka, I. N.; Pimental, D. Environ. Entomol. 1979, 3, 911-5.
16. Robinson, A. G. Can. J. Plant Sci. 1961, 41, 413-7.
17. Hintz, S. D.; Schulz, J. T. Proc. N. Cent. Branch Ent. Soc. Amer. 1969, 24, 114-7.
18. Gall, A.; Dogger, J. R. J. Econ. Entomol. 1967, 60, 75-7.
19. Maxwell, R. C.; Harwood, R. F. J. Econ. Entomol. 1960, 53, 199-205.
20. Ishii, S.; Hirano, C. Entomol. Exp. Appl. 1963, 6, 257-62.
21. MacIntyre, J. L.; Dodds, J. A.; Hare, J. D. Phytopathology 1981, 71, 297-301.
22. Hare, J. D. in "Impact of Variable Host Quality on Herbivorous Insects" (R. F. Denno and M. S. McClure, ed.), in press.
23. Loper, G. M. Crop Sci. 1968, 8, 104-6.
24. Tejelges, B. A. Can. J. Bot. 1968, 46, 724-5.
25. Akazawa, T.; Uritani, I.; Kubota, H. Arch. Biochem. Biophys. 1960, 88, 150-6.
26. Wallner, W. E.; Walton, G. S. Ann. Ent. Soc. Am. 1979, 72, 62-7.
27. Haukioja, E; Niemela, P. Ann. Univ. Turku (Finland) Ser. A, 1977, 59, 44-7.
28. Hori, K. Appl. Ent. Zool. 1973, 8, 103-12.
29. Hori, K.; Atalay, R. Appl. Ent. Zool. 1980, 15, 234-41.
30. Guerra, D. J. "Natural and Insect-Induced Levels of Specific Phenolic Compounds in Gossypium hirsutum L." M.S. Thesis, U. of Arkansas.
31. Carroll, C. R.; Hoffman, C. A. Science 1980, 209, 414-6.
32. Eidt, D. C.; Little, C.H.A. Can. Ent. 1968, 100, 1278-9.
33. Adkisson, P. L. J. Econ. Ent. 1962, 55, 949-51.

34. Kittock, D. L.; Mauney, J. R.; Arle, H. F.; Bariola, L. A. J. Environ. Qual. 1973, 2, 405-8.

35. Thomas, R. O.; Cleveland, T. C.; Cathey, G. W. Crop Sci. 1979, 19, 861-3.

36. Kittock, D. L.; Arle, H. F.; Henneberry, T. J.; Bariola, L. A.; Walhood, V. T. Crop Sci. 1980, 20, 330-3.

37. Bariola, L. A.; Henneberry, T. J.; Kittock, D. L. J. Econ. Entomol. 1981, 74, 106-9.

38. Hamer, J., ed. "Cotton Pest Management Scouting Handbook" Missisippi Coop. Extension Service (no date), 48 p.

39. Greany, P. D.; Hutton, T. T.; Rasmussen, G. K.; Shaw, P. E.; Smoot, J. J. J. Econ. Entomol. in press.

40. Detling, J. K.; Dyer, M. J. Ecology 1981, 62, 485-8.

41. Capinera, J. L.; Roltsch, W. J. J. Econ. Entomol. 1980, 73, 258-61.

42. Hedin, P. A., ed.; "Host Plant Resistance to Pests"; American Chemical Society: Washington, D.C., 1979.

43. Horsfall, J. G.; Cowling, E. B., ed; "Plant Disease: An Advanced Treatise"; Academic Press: New York, 1980.

44. Hart, S. V.; Kogan, M.; Paxton, J. D. J. Chem. Ecol. (submitted).

45. Russell, G. B.; Sutherland, O. R. W.; Hutchins, R. F. N.; Christmas, P. E. J. Chem Ecol. 1978, 4, 571-9.

46. Sutherland, O. R. W.; Russell, G. B.; Biggs, D. R.; Lane, G. A. Biochem. Systematics and Ecol. 1980, 8, 73-75.

47. Kuc, J.; Caruso, F. L. "Host Plant Resistance to Pests," American Chemical Society: Washington, D.C., 1979, pp. 78-89.

48. Hahlbrook, K.; Grisebach, H. Annu. Rev. Plant Physiol. 1979, 30, 105-30.

RECEIVED September 16, 1982

Cytochrome P-450 Involvement in the Interactions Between Plant Terpenes and Insect Herbivores

LENA B. BRATTSTEN

University of Tennessee, Department of Biochemistry and Graduate Program in Ecology, Knoxville, TN 37996

The cytochrome P-450-dependent microsomal mono-oxygenase system is important in several ways in insect herbivores that feed on terpene-containing plants. Cytochrome P-450 metabolises many terpenes to polar products that can be excreted, often after further conjugation reactions, or that may be more toxic to the insect. Many terpenes induce insect cytochrome P-450 to higher activity. Changes in cytochrome P-450 activity may influence hormone balance or pheromone production in the insect, implicating the plant allelochemicals as factors in the regulation of reproductive success of insect populations. Cytochrome P-450 is therefore an important factor in insect host-plant specialisations. Efforts at improving plant resistance to insect herbivory must take this enzyme system into consideration.

One or more of the large variety of terpenes, biosynthetic-ally related to each other as outlined in Figure 1, are present in almost all higher plants. The terpenes are all fairly to highly lipophilic compounds depending upon their state of oxidation and glycosylation. They therefore have a high potential for toxic interference with the basic biochemical and physiological functions of insect herbivores.

Although the role of the terpenes in the plants that produce them is still a matter of debate, evidence of an anti-herbivore function for many of them is accumulating. Modern DNA technology, gene splicing, and cloning techniques will undoubtedly make it possible to incorporate suitable defensive allelochemicals into selected crop plants in order to minimise crop devastation by insect herbivores. However, it is necessary to understand the fund-

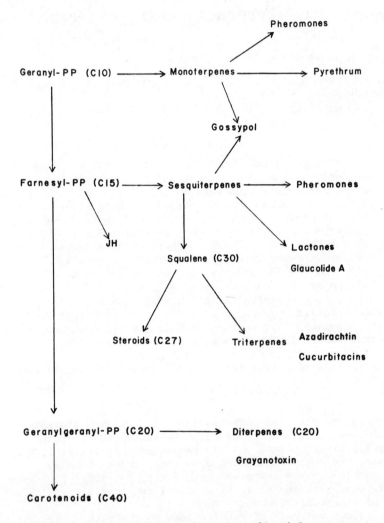

Figure 1.   Outline of the biosynthetic relationships of plant terpenes.

amental mechanisms by which plant populations and insect populations coexist in pristine ecosystems in order to take full advantage of this potential. It is clear that the insect herbivores
have developed many effective and highly specialised adaptations
for dealing with the unavoidable presence of these potentially
toxic chemicals in their food supply. Due to the great diversity
of specific solutions for survival that exist in nature, meaning-
ful efforts at mathematical modeling to establish predictive
models cannot be based simply on the gross behavior of experiment-
ally manipulated microcosms. Such models will at best be valid
only for the particular situation on which they are based. Modeling techniques are presently far ahead of the available basic
biological and ecological information upon which truly predictive
models should be based, in particular when toxicants of any kind
are involved.

In undisturbed ecosystems, plant and herbivorous insect populations coexist in a steady state condition dictated by external
biological and physical factors (parasites, predators, precipitation, temperature, soil quality, etc.). But this balance is also
regulated by myriads of very fundamental interactions, many or
even most of which are yet unknown, between the plant allelochemicals and the biochemical, physiological, and behavioral functions
of the insect herbivores (1).

The complex interactions of terpenes with insect growth and
reproduction and with the insects' ability to metabolise potentially toxic, lipophilic foreign compounds will be discussed briefly
in the following.

Toxicity of terpenes

As is the case with plant allelochemicals in general, the
terpenes are not usually acutely toxic. Table I provides a few
examples of terpene toxicities to mammals and shows that some are,
indeed, highly toxic, e.g. grayanotoxin and ergosterol. However,
the vast majority of the terpenes show acute mammalian toxicities
only in the order of several g/kg, i.e. they are not acutely toxic.
More data are available on phytotoxin toxicities to mammals than
to insects. It is not necessarily true that a compound which is
toxic to a mammal has a similar toxic effect in an insect. It is,
for instance, well known that the pyrethrins are highly toxic to
insects and are used with safety as selective insecticides.
Limonene, although not toxic to mammals, is acutely toxic to
Dendroctonus pine beetles (2, 3). On the other hand, chemicals
that are highly toxic to mammals may not be toxic to insects. An
example is ergosterol which is highly toxic to several mammals in
addition to the dog, but which lacks acute toxicity to larvae of
the southern armyworm, Spodoptera eridania (Brattsten, unpublished). This compound is probably used as a precursor in the
synthesis of cholesterol by at least some insects (4).

Table I
Mammalian acute toxicities of plant terpenes

Compound	LD_{50}	Animal, route
Grayanotoxin	1.2 mg/kg	mouse, ip
Ergosterol	4 mg/kg	dog, oral
Gossypol	550 mg/kg	pig, oral
Pyrethrum	200 mg/kg	rat, oral
Pulegone	120 mg/kg	rat, ip
Hymenovin	150 mg/kg	mouse, oral
α-pinene	several g/kg	rat, oral
Limonene	several g/kg	rat, oral
Carvone	several g/kg	rat, oral
Borneol	several g/kg	rabbit, oral
Cineole	several g/kg	rat, oral
Citral	several g/kg	rat, oral
Eugenol	several g/kg	mouse, oral
Geraniol	several g/kg	rat, oral
Menthol	several g/kg	rat, oral
Terpineol	several g/kg	rat, oral
Caryophyllene	several g/kg	rat, oral
Nepetalactone	several g/kg	rat, oral

All information from (46) except for hymenovin (47).

Terpenes as insect attractants and deterrents

Numerous terpenes are attractants for insects. Table II
shows some examples. The compounds are very often feeding
attractants. They can also be oviposition stimuli as, for
instance, α-pinene for the eastern spruce budworm or methyl iso-
eugenol for the carrot rust fly. The evolutionary details of
some of these relations, often reveal fascinating cases of insect
host-plant specialisations. A chemical that is a deterrent to
most insects can become an obligatory feeding cue for a specialist,
e.g. the spotted cucumber beetle's dependency on cucurbitacins for
food recognition. Even more intricate relationships exist as with
Ips and Dendroctonus bark beetles using the host tree (+)-α-pinene
as precursor for their own aggregation pheromone component, cis-
verbenol (5, 6).

A study with houseflies (7) shows clearly that very small
differences in the molecular structure can result in drastically
different biological effects as exemplified in Table III by the
optical isomers (-)-limonene, a fly attractant, and (+)-limonene,
a fly deterrent. A difference in oxidation state in the functional group as in citronellol, a fly attractant, and citronellal,
a fly deterrent, also causes different responses. A difference in
the length of the carbon chain as in farnesol (C15), a fly attractant, and geraniol (C10), a fly deterrent, also confers different

Table II
Plant terpenes as attractants for insects

Insect species	Compound	References
Eastern spruce budworm	α-pinene	48
Ips bark beetles	α-pinene	49
Scolytid bark beetles	α-pinene	50
	β-pinene	
	limonene	
	camphene	
	geraniol	
	α-terpineol	
Honeybee	geraniol	51
Japanese beetle	geraniol	52
	citronellal	
Pine beetle	α-terpineol	53
Pales weevil	eugenol	54
	α-pinene	
	anethole	
	citronellal	
Oriental fruitfly	methyleugenol	55
Carrot rustfly	mehyl isoeugenol	56
Boll weevil	α-pinene	57
	β-pinene	
	limonene	
	caryophyllene	
Lace wing	iridodiol	58
Spotted cucumber beetle	cucurbitacins	59
Willow beetle	salicin	60
Silk worm	terpinyl acetate	61
	linalool	
	hexenol	

biological properties. Table III also shows an example of the
concentration effect. In this case a low concentration of carvone
is an attractant, whereas a high concentration is a deterrent.
This is a very widely occuring phenomenon, and well known also in
human society, e.g. in the contexts of spices and perfumes.

Table IV gives a few examples of terpenes shown to be insect
deterrents. As with the previous examples of insect attractants
in Tables II and III, the structural diversity of the deterrent
compounds is remarkable. There are no clear and logical struc-
ture-activity relationships among the compounds with these behavioral effects. Specialised and unique effects and behavioral
adaptations are thus the rule in interactions between species
rather than the exception.

The Vernonia sesquiterpene lactone, glaucolide A, is a feeding deterrent for the southern armyworm, a broadly generalist
feeder. This compound is one of the few that the southern armyworm larvae reject. This insect and its close relative the fall
armyworm, S. frugiperda, are capable of feeding on a large variety
of plants (8). Both can also metabolize lipophilic foreign compounds, including plant allelochemicals and synthetic pesticides,
to excretable, polar metabolites by cytochrome P-450-dependent
oxidations.

Southern and fall armyworm growth on pulegone-laced diets

The mint monoterpene pulegone is not only a feeding deterrent
for the fall armyworm but is also toxic to this species (9). It
is, however, neither a feeding deterrent nor acutely toxic to the
southern armyworm at similar concentrations. Figure 2a shows that
a concentration of 0.1% pulegone in the diet (10) is toxic and
kills the fall armyworm larvae (11). In contrast, the southern
armyworm larvae feed freely on diets (12) containing either 0.01%
or 0.1% pulegone. Their weight increases and there is no delay
in their development (Figure 2b). However, further experiments
with the southern armyworm showed that pulegone can be a very
important factor in the successful coexistence of plants and herbivores (13).

Pulegone effects on the southern armyworm

The data in Table V (13) show that the southern armyworm
larvae attain higher maximal fresh body weights when pulegone is
present in their diet up to 0.1%. But the percentage of non-
water body constituents is reduced as is most obvious in larvae
feeding on a 0.2% pulegone diet. The latter diet also prolongs
the time the larvae spend in the sixth instar and they undergo the
pupal molt 8-10 days later than larvae fed control diets or diets
with lower pulegone concentrations. The 0.2% pulegone diet also
reduces the pupation success to 57% as shown in Table VI
(13), but adult emergence from these pupae occurs at almost normal
rate (85%).

Table III
Terpenes as housefly attractants and repellents (7).

Attractants		Deterrents	
(-)-limonene		(+)-limonene	
Citronellol		Citronellal	
Eugenol		Citral	
Farnesol		Geraniol	
Carvone	(low)	Carvone	(high)
		Camphene	
		Cineol	
		β -phellandrene	
		Carvacrol	
		Linalool	

Table IV
Plant terpenes as deterrents for insects

Insect species	Compound	References
Western pine beetle	Myrcene	62
	Limonene	
Gypsymoth larvae	Farnesol	63
	Geranial	
	Nerolidol	
Many species	Nepetalactone	64
	Azadirachtin	65
	Gossypol	66
Silkworm	Terpineol	61
	Geranylacetate	
	Geraniol	
African armyworm	Warburganal	67
	Xylomollin	68
Beet armyworm	Caryophyllene	69
Fall armyworm	Pulegone	9
	Glaucolide A	70
Southern Armyworm	Glaucolide A	70

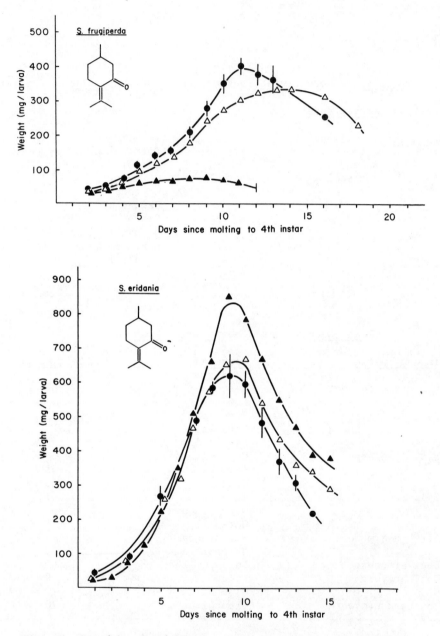

Figure 2. Growth from fourth instar to pupation on control or pulegone-containing diet of fall armyworm larvae (11) (top) and southern armyworm larvae (13) (bottom). Key: ●, control; △, 0.01% pulegone; and ▲, 0.1% pulegone.

Table V

Effect of dietary pulegone on body weight and accumulation of non-water body constituents in southern armyworm larvae and pupae (13)

Diet	Larvae		%dry mass	Pupae		%dry mass
	mg/fresh larval			mg/fresh pupa		
Control	622.4	111	12	282.8	22	21
0.01%	672.1	24*	10	278.1	20	20
0.1 %	784.8	96*	8*	323.0	53*	20
0.2 %	687.5	88	8*	296.2	36	15*

*Significantly different from control at P< 0.002 (T-test)
Data are mean ± S.D.
Larvae ate the 0.2% pulegone diet during the sixth instar only;
they ate the other diets from the fourth instar through to
pupation.

However, all levels of dietary pulegone affect the egg
production of the moths. The data in Table VI indicate a reduced
oviposition rate even at the lowest (0.01%) pulegone concentration
in the larval diet. The highest concentration (0.2%) reduces the
egg production to 10% of that in the control moths and hatching is
also reduced to 10% (13).
This case illustrates a possible, naturally occuring
mechanism whereby even a minute amount of a bioactive plant allelochemical may reduce the reproductive capacity of an insect herbivore population not drastically, but conceivably enough so that
continued co-existence is possible. The insect population may be
reduced only to the point where the plants can still grow and reproduce successfully, thereby insuring a continued food supply for
the insect herbivore.
An entirely different, in fact opposite, effect on insect reproduction by terpenes occurs with the desert locust. In
this case the monoterpenes α-pinene, β-pinene, limonene, and eugenol evaporating from desert shrubs about to bloom, precipitates
synchronised sexual maturation and mating activity in the locusts
(14). The spruce budworm is also stimulated to increased fertility levels by host tree monoterpenes (R.G. Cates, personal communication). It is possible that even opposite effects on reproduction in insects could occur depending on the specialisation of the
insect species to its environment, the diversity of the biological
activities of the compounds, and the high level of complexity of
the reproductive processes.

Table VI
Effects of larval dietary pulegone on development and oviposition
in the southern armyworm

Diet	%Pupation	Days to emergence	%Emergence	Days to oviposition	Eggs per female
Control	96	12 3	96	3.5	1900–2000
0.01%	92	12 3	93	3.5	1500–1800
0.1 %	82	12 3	93	3.0	1000–1200
0.2 %	57	15 3	85	2.5	100–200

Larvae ate the 0.2% pulegone diet during the sixth instar only;
they ate the other diets from fourth instar through to pupation.
The "Days to emergence" data indicate that moths emerge during a
6-day period with a peak on day 12 or 15 after pupation.

Terpene involvement in insect reproduction

Several different molecular mechanisms could be involved in
the reproductive inhibition observed in the southern armyworm.
For instance, many terpene derivatives mimic insect hormone action.
Juvabione (15) is the classical example of a juvenile hormone (JH)
mimic that prevents egg maturation in Pyrrhocoris bugs. Aromatic
terpene ethers (16), methylene dioxyphenyl terpene ethers (17),
and other farnesyl derivatives also have JH activity and the
latter ones (18) also cause sterility in Pyrrhocoris. For the most
part JH active terpenes are among the sesquiterpenes but several
monoterpenes also have insect sterilizing effects (19, 20). The
acyclic monoterpene citral reduces the fertility of rats by causing follicular degeneration (21).

The precocenes, another class of terpene derivatives with JH
antagonistic effects, inactivate JH synthesis by specific inhibition of corpora allata (CA) microsomal cytochrome P-450-dependent
mixed-function oxidases (22). These enzymes, of primary importance in lipophilic foreign compound metabolism, are essential in
the biosynthesis of JH as outlined in Figure 3 (23). They may
also contribute to the inactivation of JH as the metabolic scheme
in Figure 4 indicates (24, 25, 26) although epoxide hydrase and
esterase activities dominate here (26, 27). The presence of JH in
the female adult is necessary for vitellogenin synthesis (28, 29)
in many insects. Therefore, since the cytochrome P-450 oxidase
system is essential for the maintenance of balanced JH titers, even
slight changes in the P-450 system in response to external inducers and inhibitors could have profound effects on the dynamic
characteristics of insect populations.

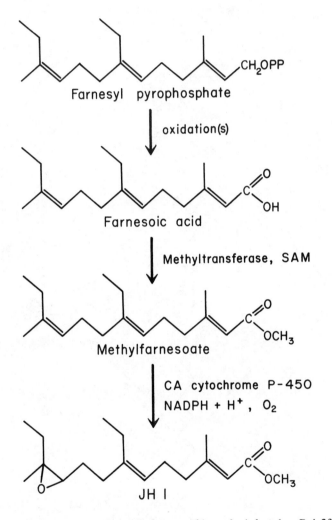

Figure 3. Outline of juvenile hormone biosynthesis based on Ref. 23.

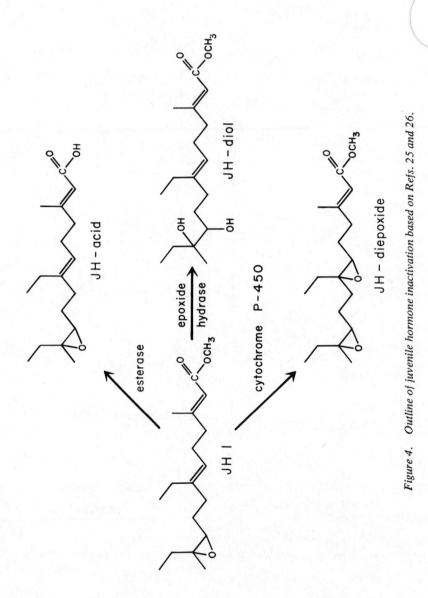

Figure 4.   Outline of juvenile hormone inactivation based on Refs. 25 and 26.

Interaction of terpenes and cytochrome P-450. Metabolism

The major metabolic fate of higher plant terpenes in mammals
is oxidation followed by conjugation usually to glucuronic acid
(30). However, specific data on even some of the most common compounds are not readily available. Limonene metabolism seems to
have been studied unusually intensively, maybe due to its thera-
peutic use to dissolve post-operatively retained gallstones.
Limonene is typically converted to transdiols via cytochrome
P-450-mediated epoxidation of either one of the two double bonds,
followed by epoxide hydration (31). It is also converted to several alcohols (32, 33) which may undergo further oxidation by de-
hydrogenases as indicated in Figure 5. All these products are
non-toxic and are excreted, in some cases after glucuronidation.
     The cases where terpene metabolism has been studied in in-
sects are very few indeed. Certain Ips and Dendroctonus bark
beetles convert monoterpenes such as α-pinene, β-pinene and myrcene to oxidation products, some of which have pheromonal activities (5, 6, 34, 35). A Dendroctonus bark beetle's cytochrome
P-450 converts α-pinene to several oxidized products after induction by α-pinene, and to at least one oxidized product without
prior induction (36). Rat liver cytochrome P-450 also converts
α-pinene to oxidation products (36) and this activity is induced
by phenobarbitol and β-naphthoflavone. There is also the interesting possibility that the bacterial flora in the bark beetles
may contribute to the oxidation of α-pinene to trans- and cis-
verbenol. A bacterium, Baccillus cereus, isolated from the hind-
gut of Ips paraconfusus catalyses these oxidations (37).
     The microsomal cytochrome P-450 system in the midguts of
southern armyworm larvae oxidises pulegone in vitro. The two
major products, 9-hydroxypulegone and 10-hydroxypulegone are formed by microsomes from larvae induced with either pentamethylbenzene or α-pinene. Microsomes from control (un-induced) larvae
only oxidise trace amounts of the compound. The 9-hydroxypulegone
rearranges spontaneously to menthofuran (38).
     In both these cases where an insect cytochrome P-450 system
has been shown to be responsible for the oxidation of α-pinene and
pulegone, the enzyme had to be induced to higher activity to effectively catalyse the reaction. This leads to the question of
whether insect P-450-dependent oxidations are sufficiently active
in natural situations to produce a significant amount of the
metabolites. Due to the importance of cytochrome P-450 oxidations
in pesticide metabolism, there are, fortunately, several studies
which show that the insect oxidase system is easily and rapidly
induced in response to a large variety of non-nutrient chemicals
in the food.

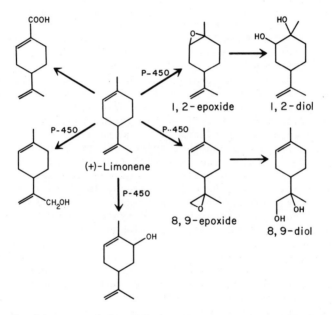

Figure 5.   Primary metabolites of limonene in mammals based on Refs. 31–33.

Interactions of terpenes and cytochrome P-450. Induction

     The classical examples of cytochrome P-450 inducers among the
terpenes are the hormonal steroids. Among the phytosterols,
sitosterol, stigmasterol (39), and ergosterol (Table VII) are inducers of the southern armyworm microsomal oxidases. The insect
molting hormones α-ecdysone and ecdysterone, widely occuring in
plants, are very potent inducers of housefly microsomal aldrin
epoxidase activity (40). Some of the most active inducers of insect cytochrome P-450 are among the monoterpenes, e.g. myrcene,
camphene (39) and other shown in Table VII. Menthol, menthone,
α-pinene, and β-pinene induce aldrin epoxidase activity in microsomes from the variegated cutworm larvae (41). However, in this
case, limonene was inhibitory (41) whereas both (+)- and (-)-
limonene are good inducers of the southern armyworm oxidase system
(39 and Table VII).

Table VII

Terpene induction of midgut microsomal cytochrome P-450 and
pyrethrum-dependent NADPH oxidation in southern armyworm
larvae (42)

Diet	Cytochrome P-450 (nmole/mg protein)	λ-max (nm)	NADPH oxidation (nmole/min,mg protein)
control	0.350	450.1	10.96
stigmasterol	0.396	450.3	12.13
ergosterol	0.371	449.8	16.57
α-pinene	0.881*	449.9	33.27*
β-pinene	0.741*	449.5	33.89*
(+)-limonene	0.683	449.8	32.55*
(-)-limonene	0.967*	449.8	31.20*
α-terpinene	0.876*	450.0	38.12*
γ-terpinene	1.028*	449.9	41.34*

*Significantly different from control at $P < 0.001$ (T-test).
Sixth instar larvae were fed *ad libitum* for 3 days on diets containing 0.2% of the terpene.

     Table VII shows the increase in cytochrome P-450 content in
microsomes from southern armyworm larval midguts resulting from
dietary exposure to several cyclic monoterpenes (42). It also
shows a closely corresponding increase in the rate of NADPH oxidation when pyrethrum is the substrate (R) being oxidised. The
microsomal cytochrome P-450 system is arranged as outlined in
Figure 6, consisting of a terminal heme-iron protein that in the
oxidised (Fe^{3+}) state binds the substrate (R). The complex undergoes two reductions during which bound molecular oxygen is converted to free radical species, one of which is inserted in the
substrate molecule, and the other one forms water. The reductions

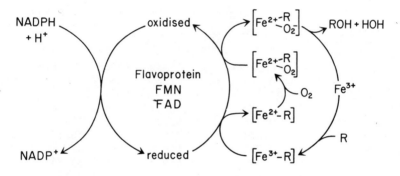

Figure 6. Outline of the microsomal cytochrome P-450 system.

are specifically dependent upon NADPH and a flavoprotein which
transports the reducing equivalents to the cytochrome. The entire
system is deeply embedded in the endoplasmic reticulum membranes
and depends on the membrane phospholipid fraction to assist in the
binding of the lipophilic substrate molecule. The NADPH-depen-
dence offers a convenient spectrophotometric method for measuring
the rate of oxygenation of substrates in cases where there is no
simple means of analysing the rate of product formation, or when
several different metabolites are formed as in the cases of
pyrethrum and pulegone.

The rates of pulegone-dependent NADPH oxidation in southern
armyworm microsomes are shown in Table VIII (38). In this case
microsomes from control diet-fed larvae show only trace activity
in agreement with results of the gas chromatographic metabolite
analyses. The southern armyworm cytochrome P-450 system apparently more easily oxygenates pyrethrum (Table VII) than pulegone. In
Table VIII the more potent inducer used is pentamethylbenzene and,
in accordance, the rate of pulegone metabolism is twice as high
with this inducer as when α-pinene is the inducer. All the plant
monoterpenes appear to induce a molecular form of the cytochrome
that is identical to the constitutive form as indicated by the
position of the absorption maximum of the cytochrome-CO difference
spectrum. Pentamethylbenzene, on the other hand, appears to induce a molecular form with some minute structural difference(s)
from the control form as indicated by a shift in the absorption
maximum to 449.0 nm. In the case of pulegone oxidation, the
Michaelis constants (K_m) indicate that both the pentamethylbenzene
and α-pinene-induced cytochrome have similar affinity for the compound. However, the pentamethylbenzene-induced form hydroxylates
pulegone in the C10 position three times more often than the
α-pinene-induced form (38).

The inducing effects on the southern armyworm cytochrome
P-450-mediated metabolism shown in Tables VII and VIII resulted
from 3 days' feeding on diets containing 0.2% of the terpene.
This concentration may occur in many plant species. However, a
much smaller dose of the inducer also effects a measureable
difference in the oxygenation rates as shown with α-pinene and
sinigrin induction of southern armyworm P-450 (39). The data in
Table IX show that a single dose of 100 μg/g of larval body weight
results in activities higher than those in control larvae. Considering the feeding behavior of the southern armyworm larvae,
consisting of feeding bouts of 10-20 minutes followed by equally
long resting periods (43), it seems reasonable to assume that the
metabolism of some bioactive leaf component may be increased
enough in a natural situation to contribute significantly to the
coevolutionary status of the insect with its host plant (39).
This would be a valid assumption whether a detoxified or an
activated polar product is formed.

Table VIII

Pulegone-dependent NADPH oxidation in relation to cytochrome P-450 content in midgut microsomes form southern armyworm larvae (38).

Diet	Cytochrome P-450 concentration (nmole/mg protein)	λ max (nm)	NADPH oxidation	
			V_{max} (nmole/min,mg protein)	K_m (µM)
control	0.346	450.0	Trace	
α-pinene	0.815	449.9	39.86 ± 4.88	10.28 ± 1.94
PMB	1.120	449.0	80.11 ± 11.52	11.64 ± 2.88

Larvae fed ad libitum for 3 days on control diet or diets containing 0.2% of the inducer.

Table IX

Rapid induction of microsomal oxidase activity
By terpenes in southern armyworm larvae

Inducer	Specific activity (nmole/min,mg protein)		Percent of control	Significance (2-tailed T-test)
control	1.66 ± 0.1	(7)	100	P < 0.001
d-carvone	2.22 ± 0.2	(4)	134	0.002>P>0.001
l-carvone	1.92 ± 0.2	(4)	115	0.002>P>0.001
caryophyllene	2.07 ± 0.2	(4)	125	0.01>P > 0.005
carotol	1.95 ± 0.2	(4)	117	

Larvae were individually fed a lima bean leaf disc loaded with a terpene dose of 100 ± 2 μg
per gram of their body weight. They were killed one hour after feeding. Activity of p-chloro
N-methylaniline N-demethylase was measured in post-mitochodrial supernatants as described earlier (71).

The data are mean ± S.D. of (N) experiments.

Conclusions

It would be charitable to say that the cytochrome P-450 system in insect herbivores does not entirely dictate their interactions with the host plants. There is a very active cytochrome P-450 system in the olfactory receptor region of the dog (44), indicating the tantalising possibility that such an enzyme system, if suitably located in the insect, is also contributing to the effects of plant allelochemicals on insect feeding behavior. The cytochrome P-450 may also be involved in pheromone biosynthesis in more species of insects (45) than the bark beetles. Cytochrome P-450 is directly involved in insect JH biosynthesis, of essential importance in both development and reproduction. The microsomal cytochrome P-450 system is of central and crucial importance in the metabolsim of lipophilic foreign compounds of all kinds, and may have derived its major biological significance in insect herbivores from exposure to the plant allelochemicals. It is beyond doubt that the cytochrome P-450 system strongly contributes to the great variety of unique solutions for survival that exist between insect herbivores and plants.

Acknowledgement

I thank C.K. Evans, C.A. Gunderson, and J.T. Fleming for excellent assistance. Previously unpublished work presented here was supported by U.S. Department of Agriculture, (SEA-GAMO) Competitive Research Grant Program grant No. 59-1471-1-1-695-0 and NSF grant PCM 81 00081.

Literature cited

1. Clayton, R.B., in "Aspects of Terpenoid Chemistry and Biochemistry" (T.W. Goodwin, ed.) Academic Press, New York, 1971, p. 1-27.
2. Smith, R.H. J. Econ. Entomol. 1965, 58, 509-510.
3. Coyne, J.F., Lott, L.H. Georgia Entomol. Soc. 1976, 11, 297-301.
4. Thompson, M.J., Svoboda, J.A., Kaplanis, J.N., Robbins, W.E. Proc. Roy. Soc. Lond. 1972, B180, 203-221.
5. Renwick, J.A.A., Hughes, P.R., Ty, T.D. J. Insect Physiol. 1973, 19, 1735-1740.
6. Vité, J.P., Bakke, A., Renwick, J.A.A. Can. Entomol. 1972, 104, 1967-1975.
7. Sharma, R.N., Saxena, K.N. J. Med. Entomol. 1974, 11, 617-621.
8. Tietz, H.M.: "An Index to the described Life Histories, early Stages and Hosts of the Macrolepidoptera of the Continental United States." A.C. Allyn Publ., Sarasota, 1972.
9. Zalkow, L.H., Gordon, M.M. Lanir, N. J. Econ. Entomol. 1979, 72, 812-815.

10. Shorey, H.H., Hale, R.L. J. Econ. Entomol. 1965, 58, 522-524.
11. Gunderson, C.A. M.S. Thesis, 1982, Univ. of Tennessee, Knoxville.
12. Brattsten, L.B., Price S.L., Gunderson, C.A. Comp. Biochem. Physiol. 1980, 66C, 231-237.
13. Gunderson, C.A., Samuelian, J.H., Evans, C.K., Brattsten, L.B., Submitted to J. Insect Physiol.
14. Carlisle, D.B., Ellis, P.E., Betts, E. J. Insect Physiol. 1965, 11, 1541-1558.
15. Bowers, W.S., Fales, H.M., Thompson, M.S., Uebel, E.C. Science, 1966, 154, 1020.
16. Bowers, W.S. Science, 1969, 164, 323-325.
17. Bowers, W.S. Science, 1968, 161, 895-897.
18. Masner, P., Slama, K., Landa, V. Nature, 1968, 219, 395-396.
19. Masner, P. Gen. Comp. Endocrinol. 1967, 9, 472.
20. Slama, K. A. Rev. Biochem. 1971, 40, 1079-1102.
21. Toaff, M.E., Abramovici, A., Sporn, J., Liban, E. J. Reprod. Fert. 1979, 55, 347-352.
22. Soderlund, D.M. Feldlaufer, M.F., Takahashi, S.Y., Bowers, W.S. in "Juvenile Hormone Biochemistry" (G.E. Pratts and G. T. Brooks, eds) Elsevier, 1981, p. 353-362.
23. Hammock, B.C. Life Sciences, 1975, 17, 323-238.
24. Yu, S.J., Terriere, L.C. J. Insect Physiol. 1974, 20, 1901-1912.
25. Yu, S.J., Terriere, L.C. Pestic. Biochem. Physiol. 1978 9, 237-246.
26. Slade, M., Wilkinson, C.F. Comp. Biochem. Physiol. 1974, 49B, 99-103.
27. Wing, K.D., Sparks, T.C., Lovell, V.M., Levinson, S.O., Hammock, B.D. Insect Biochem. 1981, 11, 473-485.
28. Pan, M.L., Wyatt, G.R. Science, 1971, 174, 503-505.
29. Reid, P.C., Chen, T.T. Insect Biochem. 1981, 11, 297-305.
30. Scheline, R.R. "Mammalian Metabolism of Plant Xenobiotics," Academic Press, New York, 1978.
31. Watabe, T., Hiratsuka, A., Isobe, M., Ogawa, N. Biochem. Pharmacol. 1980, 29, 1068-1071.
32. Regan, J.W., Bjeldanes, L.F. J. Agric. Food Chem. 1976. 24, 377-380.
33. Kodama, R. Yano, T., Furakawa, K., Noda, K., Ide H. Xenobiotica, 1976, 6, 377-389.
34. Hughes, P.R. Naturwiss. 1973, 60, 261-262.
35. Hughes, P.R. J. Insect Physiol. 1974, 20, 1271-1275.
36. White, R.A. Franklin, R.T., Agosin, M. Pestic Biochem. Physiol. 1979, 10, 233-242.
37. Brand, J.M., Bracke, J.W., Markovetz, A.J., Wood, D.L. Browne, L.E. Nature, 1975, 254, 136-137.
38. Brattsten, L.B., Fleming, J.T., Trammell, D.H., Zalkow, L.H. Submitted to Nature.
39. Brattsten, L.B., Wilkinson, C.F., Eisner, T. Science, 1977, 196, 1349-1352.
40. Terriere, L.C., Yu, S.J. Insect Biochem. 1976, 6, 109-114.

41. Yu, S.J., Berry, R.E., Terriere, L.C. Pestic. Biochem. Physiol. 1979, 12, 380-384.
42. Brattsten, L.B., Evans, C.K., Bonetti, S.J., Zalkow, L.H. Submitted to Comp. Biochem. Physiol.
43. Crowell, H.H. Ann. Entomol. Soc. Amer. 1943, 36, 243-249.
44. Dahl, A.R., Hadley, W.M. Hahn, F.F., Benson, J.M., McClellan, R.O. Science, 1981, 216, 57-59.
45. Brattsten, L.B. Drug Metab. Rev. 1979, 10, 35-58.
46. Duke, J.A. CRC Crit. Rev. Toxicol. 1977, 5, 189-237.
47. Ivie, G.W. Witzel, D.A. Herz, W., Kannan, R., Norman, J.O., Rushing, D.D., Johnson, J.H., Rowe, L.D., Veech, J.A. J. Agric. Food Chem. 1975, 23, 841-845.
48. Städler, E. Ent. exp. appl. 1974, 17, 176-188.
49. Renwick, J.A., Hughes, P.R., Krull, I.S. Science, 1976, 191, 199-200.
50. Rudinsky, J.A. Science, 1966, 152, 218-219.
51. Free, J.B. J. Agricult. Res. 1962, 1, 52-54.
52. Wilde, J. de, Ghent Landb. Hogesch. Mededel. 1957, 22, 335-347.
53. Kangas, E. Pertunnen, E. Oksanen, H. Rinne, M. Ann. Entomol. Fenn. 1965, 31, 61-73.
54. Thomas, H., Hertrell, G. J. Econ. Entomol. 1969, 62, 383-386.
55. Howlett, F.M. Bull. Entomol. Res. 1915, 6, 297-305.
56. Berüter, J., Städler, E. Zeitschr. Naturforsch. 1971, 26b, 339-340.
57. Folsom, J.W. J. Econ. Entomol. 1931, 24, 827-833.
58. Sakan, T. Issac, S., Hyeon, S. in "Control of Insect Behavior by Natural Products" (D.L. Wood, R.M. Silverstein, N. Nakajima, eds) Academic Press, New York, 1970, p. 237-247.
59. Benepal, P.S., Hall, C.V. Amer. Soc. Hort. Sci. 1967, 91, 353-359.
60. Kerns, H.G.H. Ann. Rep. Agric. Horticult. Stn., Long Ashton, 1931, p. 199.
61. Hammamura, Y. in "Control of Insect Behavior by Natural Products," (D.L. Wood, R.M. Silverstein, M. Nakajima, eds) Academic Press, New York, 1970, p. 55-80.
62. Smith, R.H. For. Sci. 1966, 12, 63-68.
63. Doskotch, R.W., Cheng, H.Y., Odell, T.M. Girard, L. J. Chem. Ecol. 1980, 6, 845-851
64. Eisner, T. Science, 1969, 146, 1318-1320.
65. Zanno, P.R., Miura, I., Nakanishi, K., Elder, D.L. J. Amer. Chem. Soc. 1975, 97, 1975.
66. Bottger, G.T., Patana, R. J. Econ. Entomol. 1966, 59, 1166-1168.
67. Kubo, L., Lee, Y.W., Pettei, M., Pilkiewicz, F., Nakanishi, K. J.C.S. Chem. Comm. 1976, 1013-1014.
68. Meinwald, J., Prestwick, B.D., Nakanishi, K., Kubo, L. Science, 1978, 199, 1167-1173.
69. Langenheim, J.H., Foster, C.E., McGinley, R.B. Biochem. Syst. Ecol. 1980, 8, 385-396.

70.   Burnett, W.C., Jr., Jones, S.B., Jr., Mabry, T. J. Amer.
      Midl. Nat. 1978, 100, 242–296.
71.   Brattsten, L.B., Wilkinson, C.F. Pestic, Biochem. Physiol.
      1973, 3, 393–407.

RECEIVED September 27, 1982

INSECT FEEDING MECHANISMS

Nonpreference Mechanisms:
Plant Characteristics Influencing Insect Behavior

J. A. A. RENWICK

Boyce Thompson Institute at Cornell University, Ithaca, NY 14853

The selection or avoidance of potential host
plants by phytophagous insects is guided by a
complex combination of physical and chemical
stimuli. Color, shape and olfactory cues may
play a role in the initial orientation, whereas
acceptance or rejection of a plant depends on
texture as well as chemical stimulants or deterrents. Initiation of feeding is stimulated or
deterred by the presence or absence of specific
chemicals or groups of chemicals, many of which
have been identified. The selection of a suitable
plant for oviposition is also crucial for survival of the progeny of most herbivorous insects,
but the chemical factors involved are known in
relatively few cases. Oviposition stimulants and
deterrents often appear to be quite different from
the chemicals that elicit or inhibit feeding
responses of larvae.

The importance of developing crop plants that are
resistant to major insect pests has created a need for
detailed examination of the mechanisms involved in resistance.
The widely recognized classification proposed by Painter (1)
appears to provide an acceptable break-down of the possible
bases of resistance for most purposes. However, some
modification of the terminology may be desirable before
beginning to analyze the individual mechanisms involved. The
term "nonpreference" refers to a behavioral response of the
insect to a plant, whereas "antibiosis" and "tolerance" refer
to plant characteristics. This anomaly has been addressed by
Kogan and Ortman (2), who suggested the term "antixenosis" to
describe the plant properties responsible for nonpreference.

0097-6156/83/0208-0199$06.00/0

ANTIXENOSIS   =   NONPREFERENCE
      .   .
      .   .
      .   .
Characteristic   Behavioral response
  of plant   of insect
(undesirable host)   (avoidance)

A problem may also arise in the separation of the
mechanisms since some overlap can occur between antixenosis
and antibiosis. These forms of resistance may be roughly
compared in the following manner:

ANTIXENOSIS   vs   ANTIBIOSIS
      .   .
      .   .
      .   .
Undesirability   Unsuitability
Avoidance by insect   Adverse effects
                              or Prevention of insect
                                 activity (e.g. feeding)

If a plant deters feeding by an insect, the mechanism of
resistance may be classified as antixenosis or antibiosis.
The critical question is whether the insect is completely
prevented from feeding, thus starving to death (antibiosis),
or would eventually feed on that plant when given no choice
(antixenosis). Since the answers to such questions are not
always known, this discussion will deal with the factors
affecting an insect's behavior in the selection of a host
plant, with emphasis on behavior.
       The choice of host plants is affected by a vast array
of positive and negative factors. These opposing forces
generally fall into one of the categories listed below:

Positive factors   Negative factors
Physical stimuli   Physical barriers
attractants   repellents
feeding stimulants   feeding deterrents
oviposition stimulants   oviposition deterrents

Whether a potential host plant is selected by an insect often
depends on a delicate balance which may be tipped in either
direction by the presence or absence of one of these factors.
       The subject of host selection has been reviewed in depth
by several authors (3, 4, 5), and the chemical factors
involved have been highlighted (6, 7). Reviews by Kogan (8)
and Hedin et al. (9) have provided comprehensive lists of
chemicals involved in the interactions between plants and
insects. Despite the vast number of plant-insect systems that

have been studied, it is surprising how few of the specific
chemicals responsible for eliciting particular behavioral
responses have actually been identified. No attempt will be
made here to further review the subject. Instead, a few
examples of the various types of interactions that could
contribute to plant resistance will be presented. Special
emphasis will be given to aspects that may have been
overlooked in the past, and one plant-insect system will be
examined in detail to demonstrate the interplay between the
various stimuli affecting insect behavior.

Physical Factors

Although we are primarily concerned here with plant
chemicals that influence insect behavior, it is important to
recognize the critical involvement of physical factors in the
host selection process. These may be simply classified as
visual and tactile stimuli. Among the visual cues affecting
orientation towards a host plant, color and shape appear to be
most important. A good example is provided by the apple
maggot fly, Rhagoletis pomonella, which is attracted to the
yellow hue of foliage for feeding and resting and to the form
of the fruit for mating and oviposition (10). The spruce
budworm lays its eggs on the needles of various conifers, and
the maximum number of eggs is found on twigs having a high
density of needles (11). The primary stimulus for oviposition
appears to be shape. Female moths readily lay eggs on paper
models of conifer twigs (12). In choice bioassays, we have
found no difference in the number of eggs laid on white
spruce, the preferred host, and English yew, a tree on which
the larvae cannot survive (13).
Morphological characteristics of potential host plants
may present barriers to insect feeding and oviposition.
Glandular hairs on plant leaves severely hamper the activities
of leaf hoppers and aphids (14, 15). The potato leafhopper,
for example, is restricted from feeding by glandular trichomes
of Solanum berthaultii and S. polyadenium, and mobility of the
insect is impaired by a sticky exudate (16). Some insects
also exhibit a preference for either smooth or rough surfaces.
The cowpea weevil, Callosobruchus maculatus, prefers smooth-
coated and well-filled seeds to rough and wrinkled varieties
for oviposition (17). However, the tobacco budworm moth,
Heliothis virescens prefers pubescent over smooth leaf cotton
plants for oviposition (18), and even in a no choice
situation, more eggs are laid on the smooth leaf cultivars
(19).

Chemical Factors

   Orientation. Chemical cues are involved in all three
phases of host selection behavior, i.e. orientation,
oviposition and feeding. Long range orientation of many
insects to their host plants is known to be guided by chemical
attractants, and reviews of the subject have provided lists of
insects and the sources of chemicals involved (9). But the
specific chemicals responsible for attraction are known in
very few cases. However, some of the best studied systems
appear to be found in the Diptera. Early work on fruit flies
revealed the involvement of essential oils in the host finding
process (20), and methyl eugenol was identified as an
attractant for the oriental fruit fly (21). Recently, the
combined effect of optical, chemical and tactile stimuli has
been demonstrated for the cherry fruit fly (22). Similar
results have been obtained with the onion fly, Hylemia
antiqua. The attractive properties of specific sulfur
compounds present in onions has long been recognized (23).
n-Propyl mercaptan and dipropyl disulfide are both attractants
and oviposition stimulants for the onion fly (24). However,
recent studies have pointed out the need for additional
compounds for maximum response (25). In addition, synergism
of visual and chemical cues occurs in the selection of
oviposition sites. A vertical image and yellow color
resembling an onion stem greatly enhance the effect of
volatiles in eliciting oviposition (26).
   The long range orientation of most other insects in
response to chemical attractants is not nearly as clear. The
olfactory orientation of the Colorado potato beetle has been
studied in considerable detail by Visser and coworkers.
Beetles are attracted by volatiles of several solanaceous
plants (27). The collection and characterization of compounds
emanating from potato plants resulted in the identification of
general green leaf volatiles such as hexanol, hexenols and
hexenal, which elicit a positive response (28). But the
specific compounds that enable the insects to recognize their
solanaceous hosts still remain a mystery.

   Feeding. The feeding behavior of phytophagous insects
has been studied much more widely than other aspects of the
insect/plant relationship. The reason for this probably lies
in the relative ease with which bioassays can be performed and
the results interpreted. Many insects can be reared on
artificial diets, and the effects of added plant constituents
can readily be determined. Some early studies by Dethier (29)
demonstrated a correlation between larval food choice and the
presence of specific chemicals in the umbelliferous host
plants of Papilio polyxenes. However, many of the compounds
typically found in the Umbelliferae are also present in other

plant families, and the Papilio larvae did in fact feed on
some of these species. Since that time, compounds or groups
of compounds that are typical of other plant families have
been identified as feeding stimulants for insects specializing
in these families (9). Secondary plant substances are
generally involved in such cases of specialization, even
though the primary function of these compounds in the
evolution of the plant is believed to be in defense against
herbivores (30). Many secondary plant substances are in fact
powerful feeding deterrents or antibiotic agents, thus
providing protection from generalist insects. The idea of
utilizing feeding deterrents as a means of protecting crops
from insect damage has received a lot of attention in the last
few years (31), and large screening programs have uncovered
several promising compounds (32). However, a balance often
exists between the repulsion of generalist insects and the
attraction of specialists. The characteristic bitter
substances of the cucurbits offer protection from mites and
other herbivores (33), but these cucurbitacins act as
kairomones for a group of diabroticite beetles (34). Also,
sinigrin, a glycoside found in most cruciferous plants, is
toxic to Papilio polyxenes larvae, which do not normally
attack crucifers. The southern armyworm, a generalist, is
inhibited by high concentrations of sinigrin, but feeding by
the imported cabbageworm is actually stimulated by this
compound (35).

The introduction of feeding deterrents into crop plants
through breeding programs would appear to be an ideal solution
to many pest problems. Some progress has been made in this
direction through the analysis of resistant and susceptible
varieties. Feeding deterrents have been isolated from sorghum
lines resistant to the greenbug, Schizaphis graminum, and
identified as p-hydroxybenzaldehyde, dhurrin and procyanidin
(36). Resistance to the European corn borer, Ostrinia
nubilalis in maize has been attributed to 2,4-dihydroxy-
7-methoxy-benzoxazin-3-one (DIMBOA) (37). This compound
inhibits normal development and increases mortality of the
larvae. However, recent studies have shown that other plant
factors are involved in resistance, and feeding deterrency may
play an important role (38). Breeding programs for incorporation of this resistance into commercially desirable varieties
appear to offer particular promise.

Oviposition. Although most of the research on host
selection has focused on feeding stimuli, most scientists
involved in this work acknowledge the fact that the choice of
food is largely predetermined by the gravid female at the time
of egg laying. The tiny hatchling larvae are usually
incapable of moving any distance to sample potential food
plants. Thus their survival depends on the judicious

selection of an oviposition site by the adult female. The
paucity of research in this area is probably due to the
problems involved in conducting behavioral experiments that
provide the insects with natural conditions for flight,
landing and sensory reception of the various stimuli.

It would seem logical to assume that similar chemical
stimuli are involved in larval feeding and adult oviposition.
This has in fact been demonstrated for Pieris brassicae, which
laid its eggs on green paper treated with sinigrin, a feeding
stimulant for the larvae (39). However, few other cases exist
where such a relationship can be definitely confirmed. A
distinct difference in the factors affecting oviposition by
specialist and generalist insects has been noted. The
polyphagous species may be stimulated to lay eggs by non-
specific cues such as moisture, sugars and amino acids. The
specialists (oligophagous and monophagous), on the other hand,
usually respond to specific secondary plant compounds.
However, the generalists may differ markedly in their
preference for plants containing particular allelochemicals.
In a comparison of cabbage looper and armyworm oviposition on
three species of Vernonia (Compositae) the cabbage looper
showed a distinct preference for V. gigantea and V. glauca,
which contain the sesquiterpene lactone, glaucolide A. The
armyworms laid more eggs on V. flaccidifolia, which lacks this
bitter compound (40).

The role of inhibitory stimuli in the choice of oviposition site by phytophagous insects has been emphasized by
Jermy and Szentesi (41). The acceptance or rejection of a
plant usually depends on contact with the plant surface or
through probing after landing. Specialists will oviposit if
the right stimulant is present, whereas acceptance by
generalists is governed to a large extent by the absence of
deterrents. Specialists may also be deterred by non-host
components which can interfere with the response to positive
signals. Three species of cabbage butterfly were deterred
from ovipositing on cabbage that was treated with extracts of
tomato and other non-host plants (42). On the other hand,
Pieris brassicae has been stimulated to oviposit on bean
plants (a non-host) by culturing these plants in a solution of
glucosinolates (43). So it appears that the balance between
inhibitors and stimulants is critical.

The oviposition behavior of Papilio butterflies has been
studied in some detail. Gravid females of the citrus
butterfly, Papilio demoleus, are attracted to host and
non-host plants almost equally by color. But the specific
attractant emitted from citrus plants increases the chances of
landing on these plants. Then contact chemical stimuli elicit
the ovipositional response of the butterflies (44).
Examination of another species, Papilio protenor demetrius, in
Japan has indicated that oviposition occurs in response to

contact stimuli released by the drumming action of the
forelegs on the leaf surface. Volatile components of the
plant do not appear to be involved in this case (45).
Variation in the behavior of individual females of Papilio
machaon has been noted. Different thresholds for acceptance
of alternative plants appear to occur, so that females
exhibiting a generalist strategy may actually lay eggs on
plants which are unsuitable as food for the larvae (46).

Butterflies in general seem to rely on a combination of
visual and chemotactile stimuli for oviposition. Field
observations coupled with laboratory experiments on Colias
butterflies have shown that chemical preferences for various
legume food plants are under genetic control. But in some
cases, chemical cues alone are not sufficient for females to
discriminate between species. Lupinus, a legume which is not
usually utilized by Colias, stimulates oviposition in the
laboratory, indicating its chemical similarity (47). But some
physical or environmental factors must play a role in nature.
Wiklund (48) has also concluded that adult and larval
preferences of Papilio machaon are determined by separate gene
complexes. Thus the possibility of oviposition on unsuitable
plants always exists.

The involvement of volatiles in the selection of oviposition sites appears to be particularly important in the
Diptera. Orientation and oviposition are closely tied to
attractant chemicals for the fruit flies and the onion fly,
which were discussed earlier. Recent work on the carrot fly,
Psila rosae, has resulted in the identification of both
volatile and non-volatile components of the recognition signal
(49, 50). The propenylbenzenes trans-methylisoeugenol and
trans-asarone, along with hexanal are sufficient to elicit
oviposition. In addition, a polyacetylene, falcarindiol, at
the leaf surface is highly stimulatory.

The importance of understanding the factors affecting
oviposition by phytophagous insects cannot be overemphasized.
This step in the life cycle is a key to survival of most
insect populations. Despite the problems associated with
behavioral studies on adult insects, considerable progress has
been made, and the potential for breeding plants that
discourage oviposition is gaining widespread recognition.

Environmental Factors

Host plant preferences of insect pests are often
influenced by environmental conditions. The effects of plant
stress on susceptibility to insect attack have been observed
in many crop plants. Stress factors affecting the
physiological state of the plant include drought, disease,
chemical pollutants, and high salt concentrations. Profound

differences in physiological conditions of the plant also occur with increasing age of the tissue.

The resistance of sorghum to grasshoppers has been related to the content of phenolics in the plant (51), and those cultivars with a high phenolic content suffer less damage from leaf-chewing insects in general (52). The levels of phenolics in healthy sorghum plants decrease as the plant matures, and environmental factors such as light intensity influence the concentration of phenolics (53). Thus wide variations in the feeding deterrent activity may be found within a particular cultivar. Similarly, host preferences of Heliothis zea have been found to change as the various host plants matured or entered a more attractive stage of development (54).

The stress produced by fungal pathogens on plants may have a distinct effect on their susceptibility to insects. Plant pathologists have noted the presence of large numbers of white flies on plants infected with Verticillium before the disease symptoms are apparent (H. Mussell, personal communication). The host selection behavior of grasshoppers feeding on wild sunflower is also affected by the presence of a pathogen. Leaf tissue infected with rust fungus, damaged by lepidopteran larvae or wilted by girdling beetles was preferred over healthy, green leaves (55).

The effect of plant age on the host selection process has been observed by comparison of two aphid species on brassica plants (56). A specialist, Brevicoryne brassicae prefers young leaves, which are higher in glucosinolates, whereas the generalist, Myzus persicae prefers older leaves, where amino acids are more important in the selection process.

Outbreaks of insect pests, particularly in forest ecosystems, have often been linked to air pollutants (57). Recent studies have shown that exposure of plants to sulfur dioxide can in fact affect their susceptibility to insect attack. Bean plants exposed to low levels of SO_2 were preferred for feeding by the Mexican bean beetle (58). Similar preferences were found with soybeans (59) both in the laboratory and in the field (60). Growth, rate of development and fecundity of the beetles were also increased on the treated plants.

The susceptibility of a plant to insect damage may also be affected by associated vegetation. The practice of mixed cropping is believed to minimize crop damage in many of the developing countries, where insecticide application is impractical. Effects on the population dynamics of insect pests have been demonstrated in mixed crops of Brassica/tomato and Phaseolus/weed grass (61). A recent study by Saxena and Basit (62) indicated that cotton can be protected from leafhoppers by planting non-hosts such as castor or sponge gourd. Oviposition on cotton was reduced in both cases, but

the mechanisms appear to be different. Volatiles from the
castor plants reduced the number of leafhoppers landing on
host plants, whereas sponge gourd was attractive to the
insects. Oviposition occurred on the gourd, but emerging
nymphs failed to develop on them and died.

Aggregation and Dispersion

Certain insects must aggregate on their host plants in
order to survive, and others depend on even distribution of
populations to prevent overcrowding of limited resources (63).
The mechanisms involved, in meeting these requirements often
depend on physical and chemical characteristics of the plant.
The aggregation of bark beetles on pine trees is mediated
by pheromones produced by the invading beetles. Attack en
masse is necessary to overcome the resin flow exuding from the
entrance holes. Many of the pheromones responsible for aggregation are produced by oxidation of terpene hydrocarbons
present in the resin (64, 65). Furthermore, the production of
a particular pheromone may depend on the optical rotation (or
absolute configuration) of a precursor. In the beetle Ips
paraconfusus, (-)-α-pinene is oxidized to cis-verbenol, a
component of the pheromone complex for this species (66),
whereas (+)-α-pinene is oxidized to trans-verbenol, which is a
pheromone for another species (67). Thus trees which lack the
precursors or right configuration of precursor might escape
invasion by these bark beetles.
Dispersion of insect populations may also depend on plant
constituents. Recent work in our laboratory has shown that
when cabbage looper larvae are feeding on a plant, the adult
females will avoid this plant for oviposition (68). The
deterrent is present in larval frass, but also in disrupted
plant tissue (69). Homogenized tissues from several host
plants of the cabbage looper were effective in deterring
oviposition. Both volatile and non-volatile components of the
plant appear to be involved (70). In another study, female
moths of the European corn borer were deterred from
ovipositing by the volatile emissions from injured plants
(71). Also, the olive fruit fly is deterred from further
oviposition on olives which are already attacked. In this
case, juice from the oviposition wound is active in signalling
occupancy to the gravid females (72).
These examples serve to illustrate the indirect role
that plant chemicals may play in the population dynamics of
insect pests. Work in this area has been very limited, but
the potential utilization of an insect's own spacing mechanism
may offer a new approach to pest management, and a good
understanding of the plant/insect relationship will be
essential.

Insect Pests of Crucifers

The insect community feeding on cruciferous crops has been widely studied over a long period of time. The reason for this interest stems not only from the commercial importance of these crops, but also from the fact that much is known about the chemistry of the insect-plant relationships. Furthermore, at least two of the major chemicals involved are commercially available.

Many insects have become specialists on crucifers and a few related plant families. These include flea beetles, leaf beetles, cabbage root fly, aphids, cabbage butterflies and the diamondback moth. At the same time, several polyphagous insects such as the cabbage looper, armyworms and aphids are major pests of crucifers. Comparative studies on these specialists and generalists have provided valuable information on host recognition and possible resistance mechanisms.

The Cruciferae are characterized chemically by the presence of glucosinolates, a group of glycosides which give rise to volatile hydrolysis products known as the mustard oils. The most common glucosinolate is sinigrin, or allylglucosinolate, which releases allylisothiocyanate upon hydrolysis. Both these compounds have been shown to be involved in the host selection behavior of several insects (73). Sinigrin stimulates feeding by the imported cabbageworm, Pieris rapae, the large white cabbage butterfly, P. brassicae, the diamondback moth, Plutella maculipennis, the mustard beetle, Phaedon cochleariae, and flea beetles, Phyllotreta spp.

When the effect of sinigrin on specialists and generalists is compared, varying responses are obtained (35). Feeding by Pieris rapae (a specialist) is stimulated. Larvae of Spodoptera eridania (a generalist) are unaffected by low concentrations, but feeding is inhibited by high sinigrin concentrations. Papilio polyxenes does not feed on crucifers, and all concentrations of sinigrin are toxic to larvae of this species.

An interesting study on the host plant selection of the horseradish flea beetle, Phyllotreta armoraciae, has shown that a flavonol glycoside stimulates feeding by this monophagous species (74). The combination of sinigrin and this compound, kaempferol 3-0-xylosylgalactoside, was more stimulatory than either chemical alone. This appears to be the first report of a crucifer feeding insect being stimulated to feed by an allelochemic which is not a glucosinolate. Feeding deterrents in other plants may also explain in part the specificity of the horseradish flea beetle. A comparison of monophagous and oligophagous flea beetles demonstrated selective feeding deterrence with cucurbitacins, cardiac glycosides and cardenolides (75). Cardenolides and cucurbitacins have been suggested as a second generation of protective compounds in Cruciferae.

Several insects appear to be attracted by mustard oil or allylisothiocyanate. These include the cabbage fly, <u>Phorbia floralis</u>, the vegetable weevil, <u>Listroderes obliquus</u>, the flea beetles, <u>Phyllotreta cruciferae</u> and <u>P</u>. <u>striolata</u>, the diamondback moth, <u>Plutella maculipennis</u>, and larvae of the imported cabbageworm, <u>Pieris rapae</u> (73). However, orientation towards host plants for oviposition by the <u>Pieris</u> butterflies seems to be guided primarily by visual cues (76).

Despite the implications and speculation that glucosinolates and/or mustard oil is involved in host selection by <u>Pieris rapae</u>, the mechanism of orientation and host recognition by this insect is not yet clear. The butterflies are attracted to green colors, and often land on non-hosts as well as hosts. Recognition of a suitable oviposition site seems to depend on contact stimuli. We have developed a bioassay to study the chemistry of the oviposition stimulant. Butterflies will readily land on green index cards, and if these are painted with a water extract of cabbage, oviposition will occur. Good discrimination between different concentrations of extract were obtained, but when sinigrin solutions were substituted for cabbage extracts, the ovipositional response was marginal. Sinigrin alone is therefore not sufficient to account for the stimulatory activity of cabbage. Preliminary fractionation has indicated that at least four compounds are involved, but the most active material is considerably less polar than sinigrin.

The effect of volatiles on oviposition by <u>P</u>. <u>rapae</u> was tested by placing whole cabbage leaves or a container of macerated cabbage tissue under the bioassay cards. There was no evidence for any increase in the number of landings on the cards with the volatiles, and it appears that oviposition is actually deterred to some extent by the presence of volatiles. The possibility still exists that very low concentrations might be attractive to the butterflies, but random landing does occur in the field, and so it seems reasonable to conclude that the non-volatile stimulant is the most important chemical cue for oviposition.

Conclusions

The behavior of insects in selecting a host plant for food and shelter is affected by a wide array of physical and chemical stimuli. Chemicals that play a role in resistance mechanisms may interfere with an insect's orientation, inhibit feeding, or deter oviposition. Most of the known mechanisms of resistance involve feeding deterrents, but the most vulnerable phase of the insect life cycle may prove to be oviposition. Environmental factors may influence the ability

of a plant to combat insect attack, and chemical constituents
of the plant may have indirect effects on the success or
failure of its attackers. The presence of specific chemicals
may offer protection from one insect pest, but may increase
the risk of invasion by other insects. Many assumptions have
been made in even the most thoroughly studied plant-insect
systems. But many more physical and chemical characteristics
need to be identified to explain the complex behavioral events
in host selection by individual insect species.

Literature Cited

1. Painter, R. H. "Insect Resistance in Crop Plants";
    University Press of Kansas: Lawrence/London, 1951; 520
    pp.
2. Kogan, M.; Ortman, E. F. Bull. Entomol. Soc. Am. 1978,
    24, 175-6.
3. Thorsteinson, A. J. Annu. Rev. Entomol. 1960, 5, 193-218.
4. Harris, P. Proc. Symp. IX Int. Congr. Plant Protection
    1979, 1, 105-9.
5. Kennedy, J. S. Ann. Appl. Biol. 1965, 56, 317-22.
6. Hsiao, T. H. Ent. exp. & appl. 1969, 12, 777-88.
7. Hedin, P. A.; Maxwell, F. G.; Jenkins, J. N. In: "Proc.
    Summer Inst. on Biol. Control of Plant Insects and
    Diseases" (F. G. Maxwell & F. A. Harris, eds.);
    University Press: Mississippi, 1974; p 494-527.
8. Kogan, M. Proc. Int. Congr. Entomol. 1976, 15, 211-27.
9. Hedin, P. A.; Jenkins, J. N.; Maxwell, F. G. In: "Host
    Plant Resistance to Pests" (P. A. Hedin, ed.); Am. Chem.
    Soc. Symp. Ser. 1977, 62, 231-75.
10. Prokopy, R. J., Owens, E. D. Ent. exp. & appl. 1978, 24,
    409-20.
11. Wilson, L. F. J. Econ. Entomol. 1963, 56, 285-8.
12. Städler, E. Ent. exp. & appl. 1974, 17, 176-88.
13. Renwick, J. A. A.; Radke, C. D. Environ. Entomol. 1982,
    11, 503-5.
14. Pearson, E. D. "The Insect Pests of Cotton in Tropical
    Africa"; Eastern Press: London, 1958; 355 pp.
15. Johnson, B. Plant Pathol. 1956, 5, 131-2.
16. Tingey, W. M.; Gibson, R. W. J. Econ. Entomol. 1978, 71,
    856-8.
17. Nwanze, K. F.; Horber, E.; Pitts, C. W. Environ. Entomol.
    1975, 4, 409-12.
18. Lukefahr, M. J.; Houghtading, J. E.; Graham, H. M.
    J. Econ. Entomol. 1971, 64, 486-8.
19. Robinson, S. H.; Wolfenbarger, D. A.; Dilday, R. H. Crop
    Sci. 1980, 20, 646-9.
20. Howlett, F. M. Trans. Entomol. Soc. London 1912, 412-8.
21. Howlett, F. M. Bull. Entomol. Res. 1915, 6, 297-305.

22. Levinson, H. Z. Z. angew. Entomol. 1977, 84, 1.
23. Peterson, A. J. Econ. Entomol. 1924, 17, 87-94.
24. Matsumoto, Y. In: "Control of Insect Behavior by Natural Products" (D. L. Wood, R. M. Silverstein, & M. Nakajima, eds.); Academic Press: New York, 1970; p 133-60.
25. Pierce, H. D., Jr.; Vernon, R. S.; Borden, J. H.; Oehlschlager, A. C. J. Chem. Ecol. 1978, 4, 65-72.
26. Miller, J. R.; Harris, M. O. Proc. 5th Int. Symp. Insect-Plant Relationships; Pudoc Press: Wageningen, 1982; in press.
27. Visser, J. H.; Nielsen, J. K. Ent. exp. & appl. 1977, 21, 14-22.
28. Visser, J. H.; van Straten, S.; Maarse, H. J. Chem. Ecol. 1979, 5, 11-23.
29. Dethier, V. A. Am. Nat. 1941, 75, 61-73.
30. Fraenkel, G. Science 1959, 129, 1466-70.
31. Munakata, K. Pure Appl. Chem. 1975, 42, 57-66.
32. Shalk, J. M.; Ratcliffe, R. H. FAO Plant Prot. Bull. 1977, 25, 9-14.
33. DaCosta, C. P.; Jones, C. M. Science 1971, 172, 1145-6.
34. Metcalf, R. L.; Metcalf, R. A.; Rhodes, A. M. Proc. Natl. Acad. Sci. USA 1980, 77, 3769-72.
35. Blau, P. A.; Feeny, P.; Contardo, L.; Robson, D. S. Science 1978, 200, 1296-8.
36. Dreyer, D. L.; Reese, J. C.; Jones, K. C. J. Chem. Ecol. 1981, 7, 273-84.
37. Klim, J. A.; Tipton, C. L.; Brindley, T. A. J. Econ. Entomol. 1967, 60, 1529-33.
38. Scriber, J. M.; Tingey, W. M.; Gracen, V. E.; Sullivan, S. L. J. Econ. Entomol. 1975, 68, 823-6.
39. David, W. A. L.; Gardiner, B. O. C. Bull. Entomol. Res. 1962, 53, 91-109.
40. Burnett, W. C.; Jones, S. B. Jr.; Mabry, T. J. Am. Midl. Nat. 1978, 100, 242-6.
41. Jermy, T.; Szentesi, A. Ent. exp. & appl. 1978, 24, 258-71.
42. Lundgren, L. Zool. Scr. 1975, 4, 253-8.
43. Ma, W. C.; Schoonhoven, L. M. Ent. exp. & appl. 1973, 16, 343-57.
44. Saxena, K. N.; Goyal, S. Ent. exp. & appl. 1978, 24, 1-10.
45. Ichinosé, T.; Handa, H. Appl. Ent. Zool. 1978, 13, 103-14.
46. Wiklund, C. Oikos 1981, 36, 163-70.
47. Stanton, M. L. Oecologia (Berl.) 1979, 39, 79-91.
48. Wiklund, C. Oecologia (Berl.) 1975, 18, 185-97.
49. Guerin, P. M.; Städler, E.; Buser, H. R. Proc. 5th Int. Symp. Insect-Plant Relationships; Pudoc Press: Wageningen, 1982; in press.
50. Städler, E.; Buser, H. R. Proc. 5th Int. Symp.

Insect-Plant Relationships; Pudoc Press: Wageningen,
1982; in press.
51. Woodhead, S.; Bernays, E. A. Ent. exp. & appl. 1978, 24,
123-44.
52. Woodhead, S.; Padgham, D. E.; Bernays, E. A. Ann. Appl.
Biol. 1980, 95, 151-7.
53. Woodhead, S. J. Chem. Ecol. 1981, 7, 1035-47.
54. Stadelbucher, E. A. Environ. Entomol. 1980, 9, 542-5.
55. Lewis, A. C. Proc. 5th Int. Symp. Insect-Plant Relationships; Pudoc Press: Wageningen, 1982; in press.
56. van Emden, H. F. In: "Phytochemical Ecology" (J. B.
Harborne, ed.); Academic Press: New York, 1972; pp 25-43.
57. Wenzel, K. F.; Ohnesorge, B. Forstarchiv 1961, 32, 177.
58. Hughes, P. R.; Potter, J. E.; Weinstein, L. H. Environ.
Entomol. 1981, 10, 741-4.
59. Hughes, P. R.; Potter, J. E.; Weinstein, L. H. Environ.
Entomol. 1982, 11, 173-6.
60. Hughes, P. R.; Weinstein, L. H.; Laurence, J. A.; Sacher,
R. F.; Dickie, A.; Johnson, L. Proc. 5th Int. Symp.
Insect-Plant Relationships; Pudoc Press: Wageningen,
1982; in press.
61. Perrin, R. M.; Phillips, M. L. Ent. exp. & appl. 1978,
24, 385-93.
62. Saxena, K. N.; Basit, A. Proc. 5th Int. Symp.
Insect-Plant Relationships; Pudoc Press: Wageningen,
1982; in press.
63. Prokopy, R. J. In: "Semiochemicals. Their Role in Pest
Control" (D. A. Nordlund, R. L. Jones, W. J. Lewis,
eds.); Wiley-Interscience: New York, 1981; pp 181-213.
64. Hughes, P. R. Z. angew. Entomol. 1973, 73, 294-312.
65. Hughes, P. R. J. Insect Physiol. 1974, 20, 1271-5.
66. Silverstein, R. M.; Rodin, J. O.; Wood, D. L. Science
1966, 154, 509-10.
67. Pitman, G. B.; Vité, J. P. Can. Entomol. 1969, 101,
143-9.
68. Renwick, J. A. A.; Radke, C. D. Environ. Entomol. 1980,
9, 318-20.
69. Renwick, J. A. A.; Radke, C. D. Ent. exp. & appl. 1981,
30, 201-204.
70. Renwick, J. A. A.; Radke, C. D. Proc. 5th Int. Symp.
Insect-Plant Relationships; Pudoc: Wageningen, 1982; in
press.
71. Schurr, K; Holdaway, F. G. Ent. exp. & appl. 1970, 13,
455-61.
72. Cirio, U. Redia 1971, 52, 577-600.
73. Schoonhoven, L. M. Recent Adv. Phytochem. 1972, 5,
197-224.

74. Nielsen, J. K.; Larsen, L. M.; Sørensen, H.; <u>Ent. exp. &</u>
     <u>appl</u>. 1979, <u>26</u>, 40–8.
75. Nielsen, J. K. <u>Ent. exp. & appl</u>. 1978, <u>24</u>, 41–54.
76. Hawkes, C.; Patton, S.; Coaker, T. H. <u>Ent. exp. & appl</u>.
     1978, <u>24</u>, 219–27.

RECEIVED August 23, 1982

Differential Sensory Perceptions of Plant Compounds by Insects

J. H. VISSER

Agricultural University, Department of Entomology, Binnenhaven 7, 6709 PD Wageningen, The Netherlands

The insect's perception of plant characters is directed towards the selection of a particular plant on which feeding ultimately results in growth and reproduction. By means of its gustatory receptors, an insect is informed about the nutritional quality of a plant. Plant odours are the chemical messengers for the insect's orientation. The olfactory orientation of the Colorado beetle in response to plant odours, and the specificity and sensitivity of its olfactory receptors are outlined. The differential perception of green odour , being composed of C6 alcohols, aldehydes and the derivative acetate, is a common feature in phytophagous insects. The sensitivity of olfactory receptors for the individual components of this complex vary in and between phytophagous insect species. Although leaf odours are characterized by the particular composition of their green odour, other volatile compounds are involved as well.

Most phytophagous insects exhibit specialized feeding habits; they feed on a restricted range of taxonomically related plant species, and are even specialized to feed on particular parts of these plants like leaves, stems, flowers, fruits or roots (1). The diversity in insect-plant interactions is overwhelming as each insect species shows a series of adaptations to its host plants. These adaptations involve morphological features like the insect's mouthparts, as well as behavioural and metabolic changes in order to cope with the physical and chemical characteristics of the plants to which phytophagous insects became adapted in evolutionary time. It is beyond the scope of the present paper to list all the adaptations of insects to plants, or even all the counter-adaptations of plants to insects. At the risk to generalize to an extent which over-simplifies the diversity in

0097-6156/83/0208-0215$06.00/0

insect-plant interactions, one might state that the chemical
diversity among plants is the principal factor underlying host
specificity of phytophagous insects (2,3).

The nutritional requirements of insect species exhibiting
different feeding habits like scavengers, parasites, predators
and phytophagous insects, are similar in a qualitative sense (4).
Each insect species needs, however, a particular quantitative
composition of nutrients in its diet to complete development (5).
The presence of toxic substances in plants, secondary plant
substances as they were formerly called by phytochemists, forms
a barrier which phytophagous insects have overcome by
specialization. Thus, an insect can tolerate or detoxify the
secondary plant substances present in its host plants, while the
majority of these substances being present in other plants still
acts as toxins (1). In this way phytophagous insects are adapted
to the metabolic qualities of their host plants, i.e. a
particular chemical composition of nutrients and secondary plant
substances.

At the same time insects are able to discriminate between
host and non-host plant species as they select plants on which
feeding ultimately results in growth and reproduction, and on the
other hand avoid poisoning or malnutrition on non-host plants. By
means of chemosensory sensilla, insects are able to perform the
difficult task, being well equipped analytical chemists, of
identifying the chemical composition of plants that insects meet
in their environment (6).

Gustation

Lepidopterous larvae bear on their mouthparts two pair of
styloconic sensilla (see Figure 1). The papilla of each sensillum
possesses one terminal pore which gives entrance to the dendritic
regions of four gustatory receptor cells. Besides, a fifth cell
in each sensillum acts as a mechanoreceptor in detecting
positional changes of the papilla (7,8).

The response spectra of the individual gustatory receptor
cells can be recorded by making use of electrophysiological
techniques (7,8). A capillary containing the test compound
dissolved in a saline solution is placed in contact with the
terminal opening of the sensillum and the responses of the
gustatory receptor cells to this compound are recorded by the
same electrode being connected to an amplifier, an oscilloscope
and X_t recorder. In case the test chemical evokes a response, a
train of action potentials is recorded. The response spectra of
the individual gustatory receptor cells in the sensilla
styloconica of Pieris brassicae larvae are shown in Table I.

Table I. Response spectra of gustatory receptor cells
in <u>Pieris brassicae</u> larvae (<u>7</u>,<u>8</u>).

medial sensillum styloconicum	{	cell	1 2 3 4	sugars feeding inhibitors glucosinolates salts
lateral sensillum styloconicum	{	cell	5 6 7 8	sugars glucosinolates amino acids anthocyanins
epipharyngeal sensillum	{	cell	9 10 11	sugars feeding inhibitors salts

This set of contact chemoreceptors enables the larvae to
perceive nutrients like sugars, salts and amino acids, as well as
secondary plant substances. The feeding inhibitor sensitive cell
in the medial sensillum styloconicum responds to a variety of
alkaloids and steroids, these are compounds which possess strong
feeding inhibitor action like strychnine, conessine and
azadirachtin. On the other hand the gustatory sensilla react to
glucosinolates, a class of secondary plant substances distributed
in the host plants of this insect i.e., <u>Brassica</u> species. Though
the medial, as well as the lateral sensillum styloconicum contain
a sugar sensitive and a glucosinolate sensitive cell, the sensilla
show different specificities. The sugar sensitive cell in the
medial sensillum responds to a number of carbohydrates, whereas the
receptor cell in the lateral sensillum responds restrictively to
sucrose and glucose. Aromatic glucosinolates are detected both
in the medial as well in the lateral sensillum; the responses to
aliphatic glucosinolates are restricted to the lateral sensillum.
It is remarkable that plant pigments like anthocyanins are
perceived as a taste by <u>P. brassicae</u> larvae.
        In addition to the sensilla styloconica, lepidopterous
larvae possess gustatory sensilla on the maxillary palps. Eight
basiconic sensilla are located on top of each palpus (see Figure
1). Five of them possess a terminal pore, and for that reason
these sensilla might be considered as contact chemoreceptors. The
remaining three show numerous small perforations all over the
cuticle, which indicates an olfactory function (<u>8</u>). The response
spectra of these sensilla are, however, still obscure.
        One pair of epipharyngeal sensilla located in the buccal
cavity completes the set of contact chemoreceptors in
lepidopterous larvae. In <u>P. brassicae</u> larvae, each of these

papilla-shaped sensilla contain three sensory cells: a) one cell
responds to sucrose and glucose, b) one feeding inhibitor
sensitive cell, and c) one salt sensitive cell (see Table I).

In biting their food and by means of a relatively small
number of gustatory receptor cells, the larvae are informed about
the composition of nutrients and secondary plant substances. Taste
perception, the integration of sensory information in the insect's
central nervous system, is not merely a process of summation.
Synergistic as well as antagonistic effects between individual
compounds can be observed in the food uptake of larvae on
artificial diets.

Sucrose incorporated in these diets enhances the food uptake
by P. brassicae larvae already at 0.002 Molar (see Figure 2). In
this respect D-glucose is less effective than sucrose.
Glucosinolates, amino acids and salts do not induce food uptake
by themselves. In the presence of suboptimal stimulating
concentrations of sucrose, the glucosinolate sinalbin increases
food uptake (Figure 2). The synergistic effect of sinalbin
already occurs below 10^{-5} Molar. In this way the interaction of
individual taste substances as a result of the central
integration of sensory information, defines the motor patterns
leading to feeding behaviour. The individual taste substances
can be described in terms of their action on food uptake and
their consecutive signal function. For P. brassicae larvae, taste
substances can be called either feeding stimulants (sucrose, D-
glucose), feeding incitants (glucosinolates), feeding co-factors
(amino acids, salts), or feeding inhibitors (strychnine,
conessine, azadirachtin).

Different insect species posses different gustatory receptor
cells, their response spectra being adapted to the perception
of chemical components distributed in their host plant species
(9). Taste perception in P. brassicae larvae forms a
representative example for phytophagous insects, which are able
to discriminate a number of compounds like sugars, amino acids,
salts, and secondary plant substances acting as feeding
inhibitors or feeding incitants (3,6,8,10).

Olfaction

Green odour. Food chemists are well aware of the complexity
of food odour blends being composed of numerous volatiles. The
chemical complexity of food odour blends is somewhat confusing
since analyses are frequently carried out on processed foods,
which includes canning and cooking of the original plant material
(11). During the heating process, a variety of volatile
substances is formed from non-volatile precursors. Therefore, one
might state that the present knowledge concerning the chemical
composition of odour blends which originate from living plants,
is but little. In order to trap the volatiles from the air over
plants, analyses can be carried out by making use of adsorbents
like carbon or Porapak Q (12).

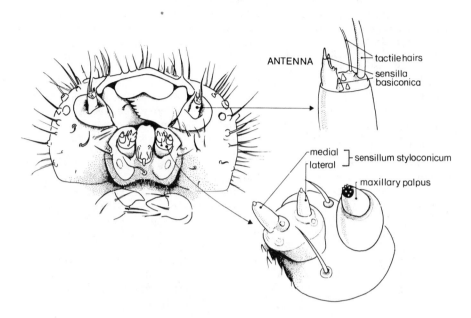

Figure 1.   *The head of a* Pieris brassicae *larva. The chemosensory sensilla are shown in detail* (7,8).

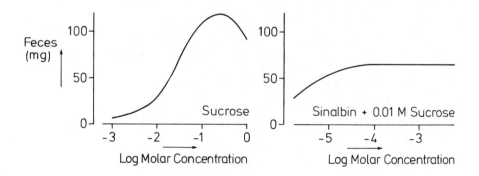

Figure 2.   *Feces production by* Pieris brassicae *larvae on artificial diets containing different amounts of sucrose and different amounts of sinalbin in the presence of sucrose* (7).

To study the odour over potato plant leaves, in 20 minutes 30 litres of air were drawn through a sample flask (volume: 5 litre) containing cut potato leaves (930 g), and the airborne components were trapped onto carbon (1-2 mg, see 13). The carbon trap was extracted with CS_2, and the extract was subjected to gas chromatography (Figure 3). Components were identified by using the gas chromatograph-mass spectrometer-computer system of the Central Institute for Nutrition and Food Research TNO (14). In the air space above leaves of fully grown potato plants, the following volatiles were detected: trans-2-hexenal, cis-3-hexenyl acetate, cis-3-hexen-1-ol and trans-2-hexen-1-ol. Besides, sesquiterpenes were obviously present at retention times of 23-29 minutes; their identities were not worked out in detail.

The class of straight chain C_6, saturated and unsaturated alcohols, aldehydes and the derivative acetate, forms a significant part of all leaf odour blends (15). These green odour components originate from oxidative degradation of the fatty acids in leaves i.e., linoleic and linolenic acid. Different plant species may show different compositions of their leaf odour blends in the proportions of the individual components of this green odour complex (15). In the air over cauliflower leaves, cis-3-hexenyl acetate is the predominant component (16), whereas in the air over leaves of fully grown potato plants cis-3-hexen-1-ol is the main component, followed by cis-3-hexenyl acetate, trans-2-hexenal and trans-2-hexen-1-ol (see Figure 3).

Differential sensory sensitivity. The insect's perception of plant odours differs essentially from their discrimination of non-volatile taste substances, as phytophagous insects may already perceive the odour at some distance from the plant. In adult phytophagous insects the antennae bear a large number of olfactory sensilla in order to detect the minute concentrations of the leaf odour components in the air downwind from a plant. The overall sensitivity of the antennal olfactory receptor system can be measured by making use of the electroantennogram technique (17). An electroantennogram (EAG) is the change in potential between the tip of an antenna and its base, in response to stimulation by an odour component. Such an EAG reflects the receptor potentials of the olfactory receptor cell population in the antenna.

The antennal olfactory receptor system in several phytophagous insects is very sensitive in the detection of the green odour components. In the Colorado beetle Leptinotarsa decemlineata, the threshold of response for trans-2-hexen-1-ol is circa 10^8 molecules per ml of air (17). In comparison, at 760 mm Hg and 20°C, 1 ml of air contains about 10^{19} molecules. The insects tested i.e., the migratory locust Locusta migratoria, the carrot fly Psila rosae (18), the cereal aphid Sitobion avenae (19), the Colorado beetle L. decemlineata (17), Leptinotarsa haldemani, the oak flea weevil Rhynchaenus quercus (20), the summer fruit tortrix moth Adoxophyes orana, and the large white butterfly P. brassicae, show differential sensory sensitivities for the individual green odour components (see Figure 4). For example, the antennal receptor system of the Colorado beetle is more sensitive for the alcohols than the corresponding aldehydes. Whereas in the carrot fly and in alate virginoparae of the cereal aphid the aldehydes cause higher responses than the corresponding alcohols. Each of the insect species shows certain traits in the character of their sensitivity spectra for the green odour components. This differential sensory sensitivity might represent a species specific adaptation of the set of olfactory receptors to the particular green odour composition of the host plants.

The sensitivity of the antennal olfactory receptor system differs even between Colorado beetle populations (see Figure 5). The beetles of the field population in Wageningen are relatively more sensitive for cis-3-hexenyl acetate when tested than those of the laboratory stock culture. Beetles of the Utah population are relatively less sensitive for trans-2-hexen-1-ol and trans-2-hexenal than the individuals of the field population in Wageningen, and those insects obtained from the laboratory stock culture. The functional significance of these differences for the geographic variation in host plant range of this insect species needs further elucidation (21,22).

Differential sensory perception. Phytophagous insects may use the particular green odour blend of their host plants to locate these suitable feeding and/or oviposition sites. The particular composition of the green odour affects the long-range olfactory orientation of Colorado beetles. When starved, these insects respond to the odour of their hostplant potato by walking upwind (odour-conditioned positive anemotaxis), and by increasing their speed of locomotion (direct chemo-orthokinesis), as shown in Figure 6 (23,14). The odours of several other solanaceous plant species induce these behavioural responses in Colorado beetles also (24). The group of solanaceous plant species tested, comprises host plants like Solanum carolinense and Solanum dulcamara, as well as non-host plants like Nicotiana tabacum, Petunia hybrida and Solanum nigrum. In general, the vapours of non-solanaceous plant species do not induce upwind locomotory responses in Colorado beetles, and can be regarded as "non-attractive" or neutral in this respect. At some distance from plants, leaf odour blends are the chemical cues which enable the Colorado beetle to discriminate between solanaceous and non-solanaceous plant species. In this way through olfactory orientation, the beetle's exploration is to some extent confined to a relevant part of the vegetation in which host plants occur.

The proportions of the individual green odour components in the leaf odour constitute a chemical message which, when perceived, directs the motor patterns of this insect. The

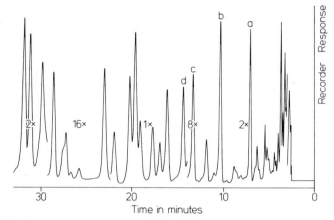

Figure 3. Components trapped from the air over cut potato leaves. Carbon traps were extracted with 60 μL of CS₂; 1 μL was used for GC (detector FID). GC conditions: WCOT Carbowax 20M column, 50 m long; temperature programmed 70–150 °C. Key: a, trans-2-hexenal; b, cis-3-hexenyl acetate; c, cis-3-hexen-1-ol; and d, trans-2-hexen-1-ol (14).

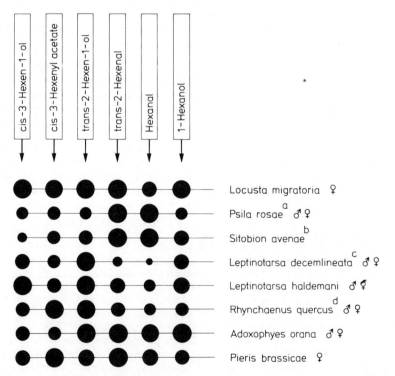

Figure 4. Sensitivity spectrum of the antennal olfactory receptor system in several phytophagous insect species to the green odor components. EAG amplitudes in response to the individual components are visualized in the areas of circles. Data were derived from Refs. 18 (a), 19 (b), 17 (c), and 20 (d).

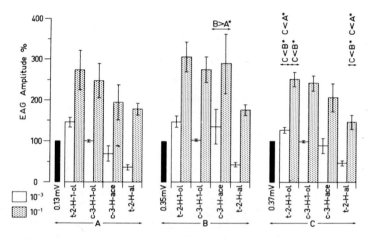

Figure 5.   Mean EAG responses of three male and three female Colorado beetles from different populations to trans-2-hexen-1-ol (t-2-H-1-ol), cis-3-hexen-1-ol (c-3-H-1-ol), cis-3-hexenyl acetate (c-3-H-ace), *and* trans-2-hexenal (t-2-H-al) *at two dilutions in paraffin oil, 10⁻³ and 10⁻¹ (v/v). Key: A, laboratory stock culture; B, field population Wageningen; C, field population Utah; vertical lines indicate 95% confidence intervals; and ∗, significant at* $P < 0.01$ *(Mann–Whitney U test).*

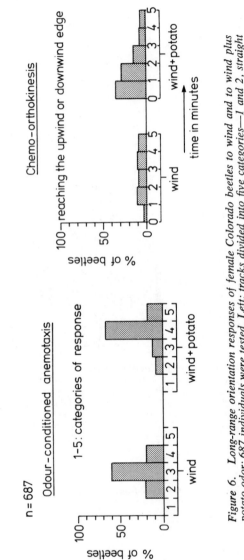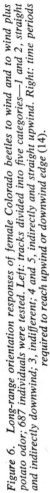

Figure 6. Long-range orientation responses of female Colorado beetles to wind and to wind plus potato odor; 687 individuals were tested. Left: tracks divided into five categories—1 and 2, straight and indirectly downwind; 3, indifferent; 4 and 5, indirectly and straight upwind. Right: time periods required to reach upwind or downwind edge (14).

chemical message is distorted as the concentration ratios of the green odour components in the leaf odour blend are changed artificially (25). On increasing the proportion of trans-2-hexen-1-ol or trans-2-hexenal in the potato leaf odour, the beetle's upwind locomotory response is "turned off". Thus, a change in the composition of the green odour impairs the attractiveness of a particular leaf odour blend (25).

The responses of single olfactory receptor cells in the antenna of the Colorado beetle to the green odour components and their isomers have been analysed (26). For the present purpose, these data are reduced to the elements acting in long-range olfactory orientation of this insect towards its host plant potato. The results from receptors responding with inhibition are discarded, as from their levels of spontaneous activity, these responses are suspected to represent artifacts. Moreover, the response thresholds of these receptors make their significance doubtful. The green odour components of potato leaves are considered to be the adequate stimuli for a number of olfactory receptor cells in the Colorado beetle antenna. The relative intensities of response in the receptor cells, which increase their neural activity upon stimulation by the individual green odour components of potato leaf, are shown in Figure 7 (26,14). Since not all single cells were tested for their responses to cis-3-hexenyl acetate, this compound is not discussed here.

The array of olfactory receptor cells reacts differentially to the green odour component stimuli of potato leaf, which is manifest by the continuum of their response spectra (see Figure 7). The observed interference with potato leaf odour caused by increasing the concentration of one green odour component occurs at concentrations near or even below the thresholds for an electroantennogram response (25). At these extremely low concentrations, the level of spontaneous activity in one single olfactory receptor cell is hardly changed. Therefore, in long-range olfactory orientation, the perception of green odour involves a concerted change of neural activity in numbers of olfactory receptor cells. An increase of the proportion of trans-2-hexen-1-ol or trans-2-hexenal heightens the neural activity in lines 1-11 more than in lines 12-23 (see Figure 7). This differential increment alters the contrast in the across-fibre pattern and, at the level of the central nervous system, modulates the beetle's orientation responses. Thus, green odour perception involves a differential response in the array of olfactory receptor cells (14).

Leaf odours. The total essence which emanates from growing leaves is not solely constituted of straight chain alcohols and aldehydes. In the insect's selection of a host plant, species-specific components might be involved. The leek moth Acrolepiopsis assectella is attracted by thiosulfinates, compounds isolated from leek. Cis-3-hexen-1-ol was also shown to be attractive (27).

The catches of carrot flies in yellow sticky traps are enhanced
by the release of hexanal and compounds isolated from the surface
wax of carrot leaves i.e., trans-methyliso-eugenol and trans-
asarone (28,29). The propenylbenzenes also stimulate oviposition
by this insect (29).

Electrophysiological results indicate that green odour
components are not the sole compounds involved in the perception
of leaf odour blends. EAGs of S. avenae to benzaldehyde (19),
and EAGs of A. orana to benzaldehyde, linalool, 1-octen-3-ol,
α-phellandrene and α-terpineol show circa the same size as EAGs
in response to green odour components. Whereas in P. brassicae
EAGs to allylisothiocyanate are small compared with the responses
to trans-2-hexenal (30). In P. brassicae larvae, the majority of
cells in the two large sensilla basiconica on the antenna (see
Figure 1) responds differentially to green odour components. Two
cell types react to allylisothiocyanate: the spontaneous activity
of neural discharges is increased in one cell and inhibited in
the second cell.

Most of the present-day information concerning the role of
leaf odour blends in host selection by phytophagous insects does
not proceed beyond the suggestion that some volatile compounds
might be called "attractants" and other "repellents" (31,32). In
case a volatile compound was shown to stimulate oviposition, it is
suggestive to describe the component as an "attractant": "although
a positive response is judged by the number of eggs laid, the
flies must be attracted to the odor before oviposition" (33). In
this line of thought, tests on long-range olfactory orientation
are not essential for assessing attractive features. J.S. Kennedy
has summarized the disadvantages of making use of the concept of
olfactory "attraction", and stated "this term (attractant) remains
no more than a blanket teleological term signifying and end-result,
conveying nothing about the component stimuli or reactions" (34).
In the same way, one might state that the term "repellent" is
misleading.

Phytophagous insects might be "attracted" over a long-range
in response to leaf odours, the insects showing an odour-
conditioned positive anemotaxis. The principal behavioural
response is directed to the wind, and is induced by the insect's
perception of the minute concentration of odour downwind from the
source. In the strict sense of the definitions an insect should
respond to an "attractant" by moving towards the source, which
coincides by moving upwind, and respond to a "repellent" by moving
away from the source. Thus, the response to a "repellent" should
result in the insect moving downwind, which makes no sense as
the insect does not escape from the "repellent" odour plume.
Moreover, it is hardly feasible phytophagous insects are equipped
with large numbers of olfactory receptors tuned to "repellents"
in order to detect the minute concentrations of these components
at long distances from the plant. For these reasons the effects
of "repellents" might expected to be restricted to close range,

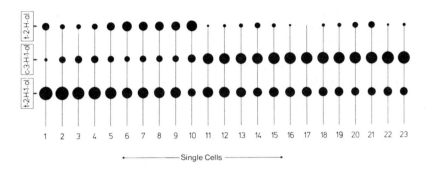

Figure 7. Intensity of neural activity in 23 olfactory receptor cells in the antenna of the Colorado beetle, in response to trans-2-hexen-1-ol (t-2-H-1-ol), cis-3-hexen-1-ol (c-3-H-1-ol), *and* trans-2-hexenal (t-2-H-al). *The increase of neural activity in response to individual green odor components is visualized in the areas of circles (14, 26).*

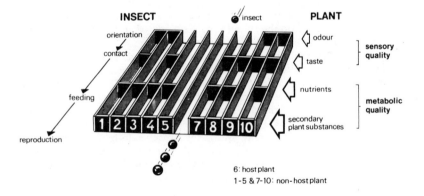

Figure 8. Host selection by phytophagous insects illustrated in a model of marbles. Row 6 is the host plant; rows 1–5 and 7–10 are non-host plants.

and different from deterrents as far as the volatile nature allows
these compounds to be detected just before an insect makes contact
with the plant.

Conclusions

The chemical factors underlying host selection by
phytophagous insects are outlined in Figure 8. Plants are
protected against herbivores mainly through secondary plant
substances. Phytophagous insects have to spend part of their
metabolic energy on detoxicating these noxious plant components,
and in evolutionary time developed a degree of tolerance to the
class of secondary compounds present in their host plant range.
Besides secondary plant substances, the proportions of individual
nutrients define the metabolic quality of a given plant which
is manifested in insect growth and reproduction.

In the co-evolution of insects and plants, the set of
chemosensory receptors in phytophagous insects adapted as to
perceive the metabolic qualities of plants. Thus, taste informs
an insect of the presence and concentrations of nutrients,
feeding incitants and feeding inhibitors. An insect depends
completely on the perceived taste quality in deciding for "take
it or leave it". The perception of plant odour enables an insect
to direct its motor patterns in order to explore efficiently the
surrounding vegetation. In this way host finding is not a matter
of "trial and error". Host plants meet the physiological
requirements of a given insect species, in the sense that an
appropriate sensory quality coincides with an appropriate
metabolic quality. Non-host and resistant plants are deficient
in at least one of the quality factors.

Host selection behaviour of phytophagous insects is a
catenary process. In the succession of the elements i.e.,
olfactory orientation, followed by contact and biting responses,
one observes an increment in specificity of both the chemical
plant signals as well as the chemoreceptors involved: (a) insect
olfactory receptor cells responding differentially to general
green odour components - the quality of an odour blend is coded
in an across-fibre pattern -, and (b) gustatory receptor cells
responding to plant species-specific feeding incitants and
inhibitors - the neural code is processed in labeled lines -.

Acknowledgements

The final drawings and glossy prints were prepared by W.C.T.
Middelplaats and J.W. Brangert respectively.

Literature Cited

1. Beck, S.D.; Schoonhoven, L.M. "Breeding Plants Resistant to
   Insects"; Maxwell, F.G., Jennings, P.R., Eds.;John Wiley:
   New York, 1980; p. 115-35.

2. Swain, T. Ann. Rev. Plant Physiol. 1977, 28, 479-501.
3. Schoonhoven, L.M. Ent. exp. & appl. 1982, 31, 57-69.
4. Southwood, T.R.E. "Insect/Plant Relationships"; van Emden, H.F., Ed.; Symp. Roy. ent. Soc. London, 1972, 6; Blackwell: Oxford; p. 3-30.
5. House, H.L. Ent. exp. & appl. 1969, 12, 651-69.
6. Staedler, E. Proc. XV Int. Congr. Ent. Washington D.C., 1976, p. 228-48.
7. Ma, W.C. Meded. Landbouwhogeschool Wageningen, 1972, 72-11.
8. Schoonhoven, L.M. "Insect/Plant Relationships"; van Emden, H.F., Ed.; Symp. Roy. ent. Soc. London, 1972, 6, Blackwell: Oxford; p. 87-99.
9. Drongelen, W. van J. comp. Physiol. 1979, 134, 265-79.
10. Schoonhoven, L.M. "Semiochemicals: their Role in Pest Control"; Nordlund, D.A., Ed.; John Wiley: New York, 1981; p. 31-50.
11. Straten, S. van "Volatile Compounds in Food", 3rd Suppl.; Central Institute for Nutrition and Food Research TNO: Zeist, 1979.
12. Pierce, H.D., Jr.; Vernon, R.S.; Borden, J.H.; Oehlschlager, A.C. J. chem. Ecol. 1978, 4, 65-72.
13. Grob, K.; Zürcher, F. J. Chromat. 1976, 117, 285-94.
14. Visser, J.H. "Olfaction in the Colorado Beetle at the Onset of Host Plant Selection"; Thesis, Agricultural University, Wageningen, 1979.
15. Visser, J.H.; Straten, S. van; Maarse, H. J. chem. Ecol. 1979, 5, 13-25.
16. Wallbank, B.E.; Wheatley, G.A. Phytochemistry, 1976, 15, 763-6.
17. Visser, J.H. Ent. exp. & appl. 1979, 25, 86-97.
18. Guerin, P.M.; Visser, J.H. Physiological Entomology 1980, 5, 111-9.
19. Yan, Fu-shun; Visser, J.H. Proc. 5th Int. Symp. Insect-Plant Relationships Wageningen; Visser, J.H., Minks, A.K., Eds.; Pudoc: Wageningen, 1982, in press.
20. Kozlowski, M.W.; Visser, J.H. Ent. exp. & appl. 1981, 30, 169-75.
21. Hsiao, T.H. Ent. exp. & appl. 1978, 24, 437-47.
22. Hsiao, T.H. Proc. 5th Int. Symp. Insect-Plant Relationships Wageningen; Visser, J.H., Minks, A.K., Eds.; Pudoc: Wageningen, 1982, in press.
23. Visser, J.H. Ent. exp. & appl. 1976, 20, 275-88.
24. Visser, J.H.; Nielsen, J.K. Ent. exp. & appl. 1977, 21, 14-22.
25. Visser, J.H.; Avé, D.A. Ent. exp. & appl. 1978, 24, 738-49.
26. Ma, W.C.; Visser, J.H. Ent. exp. & appl. 1978, 24, 520-33.
27. Thibout, E.; Auger, J.; Lecomte, C. Proc. 5th Int. Symp. Insect-Plant Relationships Wageningen; Visser, J.H., Minks, A.K., Eds.; Pudoc: Wageningen, 1982, in press.
28. Guerin, P.M.; Staedler, E. Proc. 5th Int. Symp. Insect-Plant

Relationships Wageningen; Visser, J.H., Minks, A.K., Eds.;
Pudoc: Wageningen, 1982, in press.

29. Staedler, E.; Buser, H.R. Proc. 5th Int. Symp. Insect-Plant
Relationships Wageningen; Visser, J.H., Minks, A.K., Eds.;
Pudoc: Wageningen, 1982, in press.

30. Behan, M.; Schoonhoven, L.M. Ent. exp. & appl. 1978, 24,
163-79.

31. Dethier, V.G. "Chemical Insect Attractants and Repellents";
Blakiston: Philadelphia, 1947.

32. Kogan, M. Proc. XV Int. Congr. Ent. Washington D.C., 1976,
p. 211-27.

33. Vernon, R.S.; Pierce, H.D., Jr.; Borden, J.H.; Oehlschlager,
A.C. Environm. Entomology 1978, 7, 728-31.

34. Kennedy, J.S. Physiological Entomology 1978, 3, 91-8.

RECEIVED August 23, 1982

Nutrient–Allelochemical Interactions in Host Plant Resistance

JOHN C. REESE[1]

University of Delaware, Department of Entomology and Applied Ecology, Newark, DE 19711

Once an insect has located a plant, the plant must be capable of supporting growth, development, and reproduction, if the plant is to serve as a host. The survival of plants, however, is due in part to their defensive strategies. Evidence is presented supporting the hypothesis that even susceptible plants are well-defended. Black cutworm larvae fed Pioneer 3368A corn seedlings grew to a weight of only 8.1% that of larvae fed an artificial diet. Some of the ability to inhibit growth can be extracted out of the plant and incorporated into the artificial diet demonstrating a chemical component, but physical and morphological factors appear to play an important role too. Data is presented indicating that neonate larvae are far more sensitive to the deleterious effects of the seedlings than larvae a few days old. Neonate larvae are also sensitive to handling, and using eggs instead of larvae to infest plants or inoculate diet seems advisable.

"As greater understanding of insect and plant biology, chemistry, and ecology is attained, we will be able to approach the goal of developing economic plants that are deliberately and foresightedly designed to be insect resistant" (1). One of the more interesting, and perhaps more poorly understood, areas of "... insect and plant biology, chemistry, and ecology ...," is the study of the interactions between allelochemics and nutrients. A suitable host plant must be capable of supporting growth, development, and reproduction, if the plant is to serve as a host. The feeding insect must ingest food "...that not only meets its nutritional requirements, but is also capable of being assimilated and converted into the energy and structural substances for normal

[1] Current address, Department of Entomology, Kansas State University, Manhattan, KS 66506

activity and development" (2,3). Thus, the concept of insect
dietetics includes a good deal more than nutrition in the narrow
sense; feeding behavior, nutrition, and post-ingestive effects
are also part of insect dietetics. The interactions between
nutrients and allelochemics at the behavioral, nutritional, and
post-ingestive levels may be very important parts of the mechanisms underlying the deleterious effects of plants on insect
growth and development. I propose that these deleterious effects
are probably not confined just to non-hosts or to highly resistant
crop varieties, but may be the rule; if a plant were truly unde-
fended against herbivore and pathogen attack, it would in all
likelihood not survive long enough to reproduce or produce a crop.

Defense of Susceptible Plants

The very survival of plants through evolutionary time is
"... due largely to their own defensive strategies ..." (4).
Most researchers probably accept this concept, and yet the preferred host plant of an insect species, or a crop that is decimat-
ed by a pest, is usually thought of as being quite susceptible
and rather poorly defended against insect attack.
I have started testing the hypothesis that even the most
susceptible plants are in fact remarkably well defended against
insect attack when compared to an artificial diet containing low
concentrations of defensive compounds, and having no morphological
means of defense. A few species of insects, in fact, have already
been observed to have greater fecundity and growth on artificial
diets, than on preferred host plants (5, 6, 7).
While working at the USDA Western Regional Research Center
in California with A. C. Waiss, Jr., I placed Heliothis zea larvae
on a glanded cotton variety, a glandless variety (presumably susceptible to bollworm larvae), and on artificial diet. Even the
weight of the glandless cotton group of larvae was only 25.1% of
the controls on artificial diet, indicating that even the suppos-
edly susceptible plant was chemically and/or morphologically quite
capable of reducing insect growth.
Black cutworms (Agrotis ipsilon) were fed Pioneer 3368A corn
seedlings (not known to be resistant) and artificial diet (8)
(Table I). By the sixth day the weight gain of larvae fed seedlings was a small fraction of that for larvae reared on diet.
Even though black cutworms can be severe pests of corn seedlings,
their growth is dramatically reduced by them. The artificial diet
may have been a richer source of nutrients than corn seedlings.
Therefore, I had samples of diet and seedlings analyzed for nitrogen content (as one overall measure of nutritional quality) at the
University of Delaware Soil Chemistry Laboratory. The artificial
diet was actually much lower in nitrogen than the seedlings.
Amino acid analysis showed no significant differences between
corn seedlings and the artificial diet.

Table I. Effects of Pioneer 3368A corn seedlings and artificial diet (8) on mean weights of black cutworm larvae.

Mean Larval Weights (mg) + SD

Larval Age (Days)	Reared on Plants	Reared on Diet	% of Control (Diet)
2	0.3 + 0.1	0.8 + 0.2	37.5% ***
3	0.4 + 0.2	2.4 + 1.2	16.7% ***
4	0.6 + 0.2	4.2 + 2.6	14.3% ***
5	1.0 + 0.5	11.7 + 5.3	8.5% ***
6	1.7 + 1.2	21.1 + 4.6	8.1% ***

*** indicates statistical significance between plant- and diet-reared larvae at $P<0.001$ level. N = 20.

The reduced growth of black cutworm larvae fed corn seedlings may be due to chemical and/or physical factors. In terms of chemical factors, the plant may be low in some essential nutrient and/or have growth inhibiting compounds present. These compounds may interact with nutrients directly or influence the overall nutritional physiology of the insect in some way. A bioassay searching for the absence of something is empirical at best. Assessing the importance of physical differences between a plant and an artificial diet is no simple matter either. Therefore, I have started to fractionate the plant in search of the presence of growth-inhibiting factors to see how important chemical factors are before trying to assess physical factors. Extracts were adsorbed on the alphacel portion of the diet in a way similar to the techniques of Chan et al (9) and using a modification of the black cutworm diet (8). Freeze-dried, powdered, corn seedling tissue was extracted in a soxhlet extractor with increasingly polar solvents. The residue was tested by substituting it for the alphacel portion of the diet. The greatest biological activity was found in the acetone fraction (Table II). When the effects of various concentrations (20 larvae per treatment) were bioassayed, there was a significant correlation between concentration and weight gain of the larvae in terms of percent of the controls (r = 0.85, 6 df). Methanol and benzene had similar dosage response curves (r = 0.69, 15 df and 0.75, 7 df, respectively).

Using single solvent extracts at 100% plant equivalency does not show enough activity to account entirely for the growth-inhibiting abilities of the intact seedlings. The recombining of different fractions may show additive or synergistic effects; it may also be that the physical characteristics of the intact and structurally complex plant tissue form a major defense.

Table II. Effects of extracts of Pioneer 3368A corn seedlings
on 10 day weights of black cutworm larvae. For
weights, each extract bioassayed at 100% plant equivalency (PE) (same concentration as in intact plant).
Residue bioassay at 90% of PE due to difficulty of incorporating into diet. Correlation coefficient is
given for experiments performed over a range of concentrations (25 - 100% PE).

Extract	Weight (% of Controls)$\pm$ SD	r (degrees of freedom)
Benzene	77.7 + 21.0	0.75 (7)*
Acetone	67.5 + 10.3	0.85 (6)**
Methanol	75.2 + 20.5	0.69 (15)**
Water	90.7 + 20.5	0.10 (12) N.S.
Residue	95.0 + 2.8	

* indicates statistical significance at P< 0.05 level.
** indicates statistical significance at P < 0.01 level.

Neonate Sensitivity

In developing sensitive bioassays to investigate effects
of plants on insect growth and development, black cutworm larvae
were placed on artificial diet and on corn seedlings. Neonate
larvae placed on corn seedlings weighed 1.7 mg 7 days later,
while neonate larvae put on artificial diet for 2 days and then
fed corn seedlings for 5 days weighed 10.2 mg. Thus, even though
the ones started on diet spent 71% of the experimental period on
plants, they weighed 7 times as much as the ones on plants from
the start. To see if this really suggests neonate sensitivity,
a 48 hour feeding period on artificial diet was inserted at
various points during the experiment. All of the treatments
yielded mean larval weights that were highly significantly
different from the controls (Table III), but what is particularly
important is that larvae given a boost on artificial diet during
the first 48 hrs. of the experiment were 3.3 times heavier at the
end of 8 days than those given diet during the second 48 hrs.
(significant at P<.001 level). This certainly suggests that neonate larvae are far more sensitive to plant allelochemics and/or
morphological defenses than older larvae. Judging by the magnitude of the differences in sensitivity, I suspect that the observed growth inhibition is due to a combination of both chemical
and morphological factors. Changes in sensitivity to allelochemics may revolve around the induction of detoxification systems (10), while morphological aspects may involve such aspects
as sclerotization of mandibles and thus the greater ability to
deal with such factors as silica. The magnitude of the observed
sensitivity may indicate that both chemical and morphological

factors contribute to growth reduction. Evidence suggesting neonate sensitivity has been found in a few other species (11, 12, 13), but has never been exploited very effectively in a bio-assay system. Such sensitivity to different experimental regimes may help explain difficulties sometimes encountered by different investigators trying to reproduce each others results.

Table III. Effects of being fed artificial diet at various times during experimental period. Black cutworm larvae fed on Pioneer 3368A corn seedlings when they were not on diet.

	Mean Weight (mg) at 8 Days $\pm$ SD	% of Controls
Plants Throughout	1.8 + 1.4	1.9% ***
Diet 1st 48 hrs	14.6 + 6.2	15.0% ***
Diet 2nd 48 hrs	4.4 + 3.0	4.5% ***
Diet 3rd 48 hrs	2.6 + 1.7	2.7% ***
Diet last 48 hrs	2.4 + 1.9	2.4% ***
Diet Throughout	97.6 + 29.4	100.0%

*** indicates statistically different from larvae fed diet throughout experiment, at P<.001 level. N = 17.

Neonate black cutworm larvae are not only particularly sensitive to the growth inhibiting effects of plants, but are also remarkably sensitive to various types of handling. This is a fact that is rarely taken into account in designing bioassays (most workers place larvae on diet or plants with a brush or with an inoculator using larvae mixed with corn cob grits). It is also an aspect of insect physiology about which very little is known. In an attempt to see if the hours between eclosion and being placed on diet or on plants are stressful to larvae, eggs were taken every 4 hrs from a single crystallizing dish of trypsinized eggs (eggs are routinely trypsinized off of the cheesecloth substrate and filtered onto a piece of filter paper in a Buchner funnel; the egg-covered filter paper is placed in a crystallizing dish that contains a layer of agar to maintain uniform humidity without excessive condensation) and placed onto diet. After enough larvae had eclosed in the crystallizing dish, larvae were placed on diet. Weight and survival data taken at the end of the experimental period indicate not only a striking reduction in both survival and weight (especially weight) after eclosion, but also a very intriguing relationship between age of the eggs at the time they were placed on diet and growth of re-sulting larvae (Table IV).

Table IV. Weights of black cutworm larvae which were placed on diet as eggs or larvae at various times after trypsinizing.

Hours After Trypsinizing	Stage When Placed on Diet	Mean Weight (mg) $\pm$ SD
8	Egg	29.6 ± 10.4
		**
28	Egg	47.6 ± 14.9
		***
52	Egg	70.6 ± 16.1
60	Egg	64.0 ± 34.0
		***
64	Larval	18.8 ± 4.5
		***
68	Larval	2.2 ± 2.7

Asterisks indicate statistical significance between two adjacent means. ** indicates statistical difference at $P < 0.01$ level. *** indicates statistical difference at $P < 0.001$ level. Larvae were weighed 7 days after eclosion. N = 15.

One factor in the large reduction in weight of larvae placed on diet as larvae instead of as eggs may be the physical movement of the larvae. An experiment was performed in which larvae were transferred from one plug of diet to a fresh one at various times during the experiment, beginning 6 hrs after eclosion (Table V). Clearly, there was much more sensitivity to being transferred during the first few hours after eclosion. Although this appears to be due to handling, the possibility that the diet is being conditioned in some way by the larvae can not be ignored. The experiment has been repeated several times with both fresh diet and with the larvae being transferred back to the same piece of diet with identical results in each case. Thus, the observed weight reduction is apparently due to being handled.

Insect Dietetics

As alluded to above, even a crop that sustains losses from an insect pest may actually be capable of inhibiting insect growth. A number of aspects of the insect dietetics may be interacting to produce such a situation. The feeding insect must ingest food "that not only meets its nutritional requirements, but is also capable of being assimilated and converted into the energy and structural substances required for normal activity and development" (2, 3).

Table V. Weight at 10 days of black cutworm larvae transferred from one plug of diet to another at various times after eclosion.

Time of Transfer (Hours after Eclosion)	Mean Weight (mg) ± SD	% of Control
6	110.7 + 54.4	58.7 ***
12	123.2 + 63.1	65.3 **
24	127.1 + 51.0	67.4 **
48	140.8 + 53.4	74.6 *
96	158.8 + 61.9	84.2
144	167.1 + 49.6	88.6
192	170.6 + 64.7	90.4
Control	188.6 + 65.2	100.0

*** indicates statistical difference from controls at P<0.001 level. ** indicates significance at P<0.005 level. * indicates significance at P<0.01 level. N = 20.

The role of allelochemic–nutrient interactions in insect dietetics has been investigated only rarely. Examples of such interactions abound in vertebrate literature (14, 15) and may supply useful leads for researchers working with insects. Many of the deleterious physiological effects of plant allelochemics may be due primarily to various interactions between these allelochemics and essential nutrients. In other words, it is important to not only consider the presence of nutrients, but also the "bio-availability" of these nutrients to the phytophagous insect.

Some of the more interesting examples of nutrient – non-nutrient interactions include some of the compounds that are analogs of nutrients. Mattson et al (16) found that cholesterol absorption decreased when various plant sterols were added to the diets of rats. A number of plant amino acids are not ordinarily required by herbivores and are usually not incorporated into proteins. For example, the structure of 3,4–dihydroxyphenyl-alanine (L-dopa) is similar to that of tyrosine. L-Dopa may play a role in favism (17), as well as having a number of other deleterious effects (18, 19, 20). Essential amino acids themselves can be deleterious if they are ingested in excessive quantities or if they are not in balance with other amino acids (21, 20).

Many more examples of nutrient allelochemic interactions could be cited from the vertebrate literature (see Liener (15) for a recent update of this topic); these interactions in insects

have been investigated at the molecular level in only a few
cases. However, the inhibition of assimilation or the efficiency
of conversion of assimilated food demonstrates an interaction
with the overall nutritional status of the organism. These indices have been employed to quantify host suitability and to
assess the interactions between plant compounds incorporated
into artificial diets and the nutritional physiology of the
insect. Such techniques have shown, for example that a phytophagous insect does not grow equally well on various plants, even
if no apparent behavioral barriers to feeding occur.

   Three of the most useful nutritional indices are assimilation (AD), efficiency of conversion of assimilated food (ECD),
and efficiency of conversion of ingested food (ECI). These and
a number of other indices were recently discussed by Scriber and
Slansky (22). If growth is inhibited, then it must either be
reflected in the amount eaten or in one or more of these indices
or both.

   As with any experimental technique, there are certain inherent sources of error in the nutritional index technique. Some
of these errors may be greatly magnified due to the calculations
involved and due to the exact ways in which the techniques are
employed. In my laboratory, Schmidt (unpublished data) has
succeeded in identifying and quantifying three major sources of
error. The first is the separation of uneaten diet from fecal
material. If the larva masticates or bores into the diet, this
becomes very difficult. Secondly, even if great care is used,
there is probably always some error in the taking of weight, including those used to calculate the percent dry matter of the
diet. These errors relate to the dry weight eaten by each
larva, which in turn affects the three indices. Most importantly, he found that very slight errors can be magnified tremendously through the mathematics involved in calculating the indices. If the amount of the diet eaten by a larva is small
relative to the amount given to it, then small errors in such
things as percent dry matter of the diet are magnified to an un-
acceptable level. If the larva eats 80% or more of the diet
given to it, this error magnification is reduced tremendously.
Thus, the single most important error-reducing technique is to
ensure that larvae are given an amount of diet such that they
will consume most of it during the experiment.

   The effects on nutritional indices can demonstrate an interaction between an allelochemic and the nutritional status of the
insect, but the number of actual compound-to-compound interactions that have been elucidated in insects has been small. One
of the most prominent examples is the structural analog of
L-arginine, L-canavanine (see paper by G. A. Rosenthal in this
symposium). Briefly, canavanine is similar enough to arginine to
be incorporated into the proteins of most insects, but these
canavanyl proteins do not function properly (23). Thus, it can
act as a competitive inhibitor of arginine metabolism (24, 25),

as an inhibitor of insect survival (26, 27, 20), reproduction (28), and metamorphosis (29).

Tannins are usually cited as examples of substances that can block the availability of proteins by forming complexes. Thus, when Feeny (30) found that oak leaf tannin reduced the growth of winter moth larvae (*Operophtera brumata* (L.)) and subsequently showed that oak leaf tannin forms a hydrolysis-resistant complex with casein in vitro (31), it was widely assumed that the growth-inhibiting effects of tannins in insects are due to the formation in the gut tract of tannin-dietary protein complexes that are not readily digested. It is also widely assumed that many digestive enzymes may be complexed, further reducing the rate of assimilation across the gut wall. Contrary to these assumptions, Fox and Macauley (32) found that tannins do not appreciably reduce the availability of nitrogen in *Paropsis atomaria* Oliv. larvae when fed *Eucalyptus* spp. having a wide range of condensed tannin concentrations. Bernays (33, 34, 35) performed a series of experiments using some rather high levels of tannin, but found little evidence for a reduction in digestion. Chan et al (36) have isolated a condensed tannin from cotton with a molecular weight of about 4850. Although this tannin inhibits the growth of *Heliothis virescens*, experiments with condensed tannin-casein or condensed tannin-polyamide complexes showed no reduction in biological activity (36). This suggests that the ability of tannin to inhibit growth involves something other than a reduction in assimilation due to complexing with gut tract proteins. In experiments with H. zea (37), cotton condensed tannin was found to be a relatively potent growth inhibitor, but no evidence was found for a reduction in assimilation. Instead, the primary growth inhibiting mechanism was an inhibition of ingestion. Thus, behavioral, as well as post-ingestive processes may affect the availability of nutrients.

In the case of the winter moth, tannins may very well reduce assimilation by complexing with dietary proteins. However, the assumption that this is how tannins inhibit growth in species other than the winter moth needs to be carefully re-examined. In the experiments cited above, mechanisms other than a reduction in assimilation appear to be operating.

There are mechanisms other than the complexing of proteins that may prevent nutrients from passing across the gut wall. Protease inhibitors decrease the availability of nutrients preventing the break-down of proteins into their component amino acids. The effects of protease inhibitors on insects have been reviewed (38, 39, 40; paper by C. A. Ryan in this symposium). Birk and Applebaum (41) have studied the adverse effects of soybean trypsin inhibitors on development and protease activity in *Tribolium castaneum*. In *Sitophilus oryzae*, high doses of soybean trypsin inhibitor caused adult mortality (42). The wound-induced accumulation of these inhibitors is discussed by C. A. Ryan elsewhere in this symposium.

Other enzymes may be inhibited, too. Liener (43) recently
summarized the literature on amylase inhibitors in wheat and
their effectiveness against stored product pests.

While not often listed as an essential nutrient, water is
critical to life and interacts with all other nutrients. For
insects, the moisture level of the host may have profound effects
on the nutritional physiology of the insect. Although stored
products insects may have remarkable abilities to conserve water,
many insects which live on growing plant tissue require relatively high moisture levels (44, 45, 46). On the other hand, water
dilutes the nutrients of the diet. For example, Celerio
euphorbiae larvae tend to eat more as the nutrients become more
dilute (47). Dilution of the diet of Prodenia eridania caused
an increase in efficiency of conversion (48, 49). Similarly,
Feeny (4) found that the efficiency of conversion of assimilated
food decreased with decreasing moisture levels of the food plants
of various lepidopterous larvae. In Hyalophora cecropia larvae,
moisture level and efficiency of conversion of both the nitrogenous and the caloric contents of the food were directly related
(50). The same decreasing efficiency of conversion with decreasing moisture level was found for black cutworm larvae (51).
The optimal moisture level for growth was quite different from
that for efficiency of conversion, due to the interaction between efficiency and the actual amount of dry material the larvae
ingested. This kind of information has important implications
for the interactions of nutrients and water which can help explain many of our observations of seasonal trends, herbivore
success on hosts, etc. (22).

Concluding Remarks

Many aspects of nutrient-allelochemical interactions are
probably key factors in the suitability of a given plant species
as a host for a particular insect. At best, this may be a less
than optimal situation for the insect, since even what appears
to be a susceptible plant is likely to be fairly well defended
against insect and pathogen attack. If the plant contains the
essential nutrients for the insect, but the utilization of these
nutrients is blocked in some way by allelochemics or by too much
or too little water, then growth may be slowed. If, due to behavioral modifiers, the insect will not feed on the plant, then
plant nutrients are not available to the insect.

Acknowledgement

Published with the approval of the Director of the Delaware
Agricultural Experiment Station as Miscellaneous Paper No. 988 ,
Contribution No. 519 of the Department of Entomology and
Applied Ecology, University of Delaware, Newark, Delaware.

Research supported by University of Delaware Research Foundation grant, Plant Defense to Insect Attack and Hatch Project 215, Host Plant Resistance to Black Cutworms in Corn.

I thank Meredith D. Field and Robert E. Johnson for conducting many of the experiments reported here.

Literature Cited

1. Beck, S. D. Ann. Rev. Entomol. 1965, 10, 207–232.
2. Beck, S. D.; Reese, J. C. Rec. Adv. Phytochem. 1976, 10, 41–92.
3. Beck, S. D. "Insect and Mite Nutrition," J. G. Rodriquez, Ed. North-Holland Publ., Amsterdam, 1972; pp. 1–6.
4. Feeny, P. P. "Coevolution of Animals and Plants," Gilbert, L. E.; Raven, P. H. Eds. University of Texas Press, Austin, 1975; pp. 3–19.
5. Beck, S. D. "Proceedings of the Summer Institute on Biological Control of Plant Insects and Diseases," Maxwell, F. G.; Harris, F. A. Eds. University Press of Mississippi, Jackson, 1974; pp. 290–311.
6. Whitworth, R. J.; Poston, F. L. Ann. Entomol. Soc. Am. 1979, 72:253–255.
7. Whitworth, R. J.; Poston, F. L. Proc. N. Centr. Br. – Entomol. Soc. Am. 1979, 33:22.
8. Reese, J. C.; English, L. M.; Yonke, T. R.; Fairchild, M. L. J. Econ. Entomol. 1972, 65:1047–1050.
9. Chan, B. G.; Waiss, A. C., Jr.; Stanley, W. L.; Goodban, A. E. J. Econ. Entomol. 1978, 71:366–368.
10. Brattsten, L. B. "Herbivores: Their Interaction with Secondary Plant Metabolites," Rosenthal, G. A.; Janzen, D. H. Eds. Academic Press, New York, 1979; pp. 199–270.
11. Freedman, B.; Nowak, L. J.; Kwolek, W. F.; Berry, E. C.; Guthrie, W. D. J. Econ. Entomol. 1981, 72:541–545.
12. Reese, J. C. "Current Topics in Insect Endocrinology and Nutrition," Rodriquez, J. G., Ed. Plunum Press, New York, 1981; pp. 317–336.
13. Shaver, T. N.; Parrott, W. L. J. Econ. Entomol. 1970, 63:1802–1804.
14. Reese, J. C. "Herbivores: Their Interaction with Secondary Plant Metabolities," Rosenthal, G. A.; Janzen, D. H. Eds. Academic Press, New York, 1979; pp. 309–330.
15. Liener, I. E., ed.; "Toxic Constituents of Plant Foodstuffs," Academic Press, New York, 1980; p 502.
16. Mattson, F. H.; Volpenhein, R. A.; Erickson, B. A. J. Nutr. 1977, 107:1139–1146.
17. Bell, E. A. "Toxicants Occurring Naturally in Foods," 2nd Ed., Nat. Acad. Sci., Washington, D. C., 1973; pp. 153–169.
18. Reese, J. C.; Beck, S. D. Ann. Entomol. Soc. Am. 1976, 69 68–72.

19. Rehr, S. S.; Janzen, D. H.; Feeny, P. P. Science 1973, 181: 81-82.
20. Janzen, D. H.; Juster, H. B.; Bell, E. A. Phytochem. 1977, 16:223-227.
21. Strong, F. M. "Toxicants Occurring Naturally in Foods," 2nd Ed., Nat. Acad. Sci., Washington, D. C., 1973; pp. 1-5.
22. Scriber, J. M.; Slansky, F., Jr. Ann. Rev. Entomol. 1981; 26: 183-211.
23. Rosenthal, G. A. Q. Rev. Biol. 1977, 52:155-178.
24. Vanderzant, E. S.; Chremos, J. H. Ann. Entomol. Soc. Am. 1971, 64:480-485.
25. Dahlman, D. L.; Rosenthal, G. A. J. Insect Physiol. 1976, 22: 265-271.
26. Isogai, A.; Chang, C.; Murakoshi, S.; Suzuki, A.; Tamura, S. J. Agr. Chem. Soc. Japan 1973, 47:443-447.
27. Harry, P.; Dror, Y.; Applebaum, S. W. Insect Biochem. 1976, 6:273-279.
28. Hegdeker, D. M. J. Econ. Entomol. 1970, 63:1950-1956.
29. Isoga, A.; Murakoshi, S.; Suzuki, A.; Tamura, S. J. Agr. Chem. Soc. Japan 1973, 47:449-453.
30. Feeny, P. P. J. Insect Physiol. 1968, 14:805-817.
31. Feeny, P. P. Phytochem. 1969, 8:2119-2126.
32. Fox, L. R.; Macauley, B. J. Oecologia 1977, 29:146-162.
33. Bernays, E. A. Entomol. Exp. App. 1978, 24:244-253.
34. Bernays, E. A. Ecol. Entomol. 1981, 6:353-360.
35. Bernays, E. A.; Chamberlain, D. J.; Leather, E. M. J. Chem. Ecol. 1981, 7:247-256.
36. Chan, B. G.; Waiss, A. C., Jr.; Lukefahr, M. J. Insect Physiol. 1978, 24:113-118.
37. Reese, J. C.; Chan, B. G.; Waiss, A. C., Jr. J. Chem. Ecol. 1982, In Press.
38. Ryan, C. A. Ann. Rev. Plant Physiol. 1973, 24:173-196.
39. Ryan, C. A. "Herbivores: Their Interaction with Secondary Plant Metabolites," Rosenthal, G. A.; Janzen, D. H. Eds.; Academic Press, 1979; pp. 599-618.
40. Ryan, C. A.; Green, T. R. Recent Adv. Phytochem. 1974, 8: 123-140.
41. Birk, Y.; Applebaum, S. W. Enzymologia 1960, 22:318-326.
42. Su, H. C. F.; Speirs, R. D.; Mahany, P. G. J. Georgia Entomol. Soc. 1974, 9:86-87.
43. Liener, I. E. "Toxic Constituents of Plant Foodstuffs," 2nd ed., Liener, I. E. Ed., Academic Press, New York, 1980; pp. 429-467.
44. Waldbauer, G. P. Entomol. Exp. Appl. 1962, 5:147-158.
45. Waldbauer, G. P. Entomol. Exp. Appl. 1964, 7:257-269.
46. Waldbauer, G. P. Adv. Insect Physiol. 1968, 5:229-288.
47. House, H. L. Canad. Entomol. 1965, 97:62-68.
48. Soo Hoo, C. F.; Fraenkel, G. J. Insect Physiol. 1966a, 12: 693-709.

49. Soo Hoo, C. F.; Fraenkel, G. J. Insect Physiol. 1966b, 12: 711–730.
50. Scriber, J. M. Oecologia (Berl.) 1977, 28:269–287.
51. Reese, J. C.; Beck, S. C. J. Insect Physiol. 1978, 24:473–479.

RECEIVED September 30, 1982

Chemical Basis for Host Plant Selection

JON BORDNER, DAVID A. DANEHOWER, and J. D. THACKER
North Carolina State University, Department of Chemistry, Raleigh, NC 27650

GEORGE G. KENNEDY, R. E. STINNER, and KAREN G. WILSON
North Carolina State University, Department of Entomology, Raleigh, NC 27650

The host range of the tobacco hornworm (Manduca sexta) is limited to selected members of the family Solanaceae. In an effort to better understand the chemical basis for the host plant selection process, we have undertaken an examination of both hornworm preferred and non-preferred members of the Solanaceae. Our investigations have shown this tc be a complex system involving the subtle interaction between such behavioral modulators as: (1) Ovipositional stimulants; (2) Feeding stimulants and imprinters; (3) Anti-feedants; (4) Repellants; (5) Insecticides. The results of these investigations will be discussed.

In addition to examining secondary plant substances and their role in modulating insect behavior towards host plant selection, we have been concerned with the role of the major plant metabolites in insect physiology after the selection process has been completed. It has been observed that the Mexican bean beetles (Epilachna varivestis) feeding upon soybean (Glycine max) in the southeastern regions of the U.S. are less well able to withstand the stress of desiccation than do beetle larvae feeding upon lima beans (Phaseolus lunatum). We are presently investigating the role that dietary lipids may play in this phenomena.

Through evolutionary time, phytophagous insects have tended towards specialization in their food selection process. The ability of insects to select a suitable hostplant amid numerous species of nonhosts is widely believed to be attributed to the insect's ability to discern the chemical difference between host and non-host. However, the host plant recognition process is not simply a matter of an insect recognizing a particular kairomone or avoiding a particular allomone. Rather the process is the result of the integration of numerous chemical and non-chemical factors, the complexity of which we are just beginning to appre-

0097-6156/83/0208-0245$06.25/0

ciate. Schoonhoven succinctly summarized contemporary thought on
this subject into 5 separate theorems (1).

Numerous attempts to isolate and identify host specific
kairomones and allomones for a variety of plant and insect species
have been reported (2). However, only in a few cases, particularly where the insect predator is monophagous, has a comprehensive
picture of the chemical basis for plant-insect interaction begun
to emerge. In the more general case of the oligophagous insect-
host plant system there has been a great deal of research effort
expended to unravel the chemical rationale for host plant specificity, but the results have been fragmentary and/or incomplete.
Although there remains little doubt that indeed insects respond to
chemical cues produced by hosts and non-hosts, the relative
paucity of hard chemical evidence contributing to a comprehensive
picture of insect-plant interaction makes any generalizations
tenuous and speculative at best. In fact, the only generalization
to be made may be that there is no simple dogma to the method in
which insects select their respective host plants. As M. S. Blum
(3) pointed out in his eloquent discussion of the mechanisms
insects use to detoxify, avoid or tolerate allomones, each insect
species and allomone must be regarded as a unique entity, i.e. it
appears that the available methods for dealing with toxins are as
numerous as the insect species themselves. Indeed, the mechanisms
of host plant recognition may be equally unique and diverse.

If the chemical rationale for insect-host plant interactions
is to be understood, then it is imperative that additional insect-
host plant systems be comprehensively examined with the assurance
that any chemical inferences represent biological fact. Implicit-
ly, any chemical methodologies that result from such investigations should serve as guidelines for investigations into other
systems as well. With these precepts in mind, we have undertaken
a thorough investigation into the chemical basis for host-plant
selection process of Manduca sexta which confines its host range
to selected members of the plant family Solanaceae.

The scope of our investigations into the M. sexta/Solanaceae
model system are best summarized in Table I. With three exceptions noted in Table I, we are currently investigating the
chemical factors responsible for the observed behaviors in both
hornworm preferred and non-preferred members of Solanaceae.
However, for the sake of brevity, we will discuss only the allomone isolated from the wild tomato, Lycopersicon hirsutum f. glabratum and the kairomonal factors found in horsenettle, Solanum
carolinense. (See Table I)

The isolation of the allomonal factor of L. hirsutum was
relatively straightforward and unremarkable as this factor was
typical of many organic molecules in its solubility properties
and was not apparently prone to thermal, photolytic or oxidative
degradation. The biologically active principle could be removed
from the leaf surface with a chloroform wash and could be selectively removed from unwanted pigments and waxes by redissolving

Table I. The M. sexta - Solanaceae Model

Genus/Species	Preference	Behavior
Solanum carolinense (horsenettle)	Preferred	Feeding Oviposition
L. hirsutum f.glabratum (wild tomato)	Non-preferred	Repellant
L. esculentum (domestic tomato)	Preferred	Feeding, Oviposition Orientation
Nicandra physaloides	Non-preferred	Repellant (4)
N. physaloides var. Albiflora	Preferred	Feeding
Datura stromonium (Jimson weed)	Preferred	Orientation (5)
S. tuberosum (potato)	Preferred	Feeding, Oviposition
Nicotiana spp. (tobacco)	Preferred	Feeding, Oviposition
Petunia spp. (petunia)	Non-preferred	Repellant (6)

the dried extract in ethanol. Classic column chromatography using
a stationary support of silicic acid and elution with a heptane to
chloroform gradient afforded a biologically active fraction that
was 85% pure by GLC analysis. Final purification was effected by
HPLC using a PAC bonded phase column (Whatman, Inc.) and elution
with a 5% gradient from heptane to methylene chloride.

The structure of this allomonal factor was established as
2-tridecanone using conventional spectral data. Having identified
the compound apparently responsible for the observed toxicity, it
then became necessary to quantitate the natural levels of 2-tride-
canone and ideally to use a single leaf of L. hirsutum in the
assay procedure. Fortunately, 2-tridecanone was amenable to GC
methodologies. Using this analysis on resistant (L. hirsutum)
and susceptible (L. esculentum) species of tomato, it was determined that foliage of L. hirsutum contained an average of 44.6 μg
2-tridecanone per cm^2 of leaflet surface area, whereas foliage of
the cultivated tomato, L. esculentum, averaged only 0.1 μg 2-tri-
decanone per cm^2. This information, together with the results of
in vitro studies showing the LC_{50} value 2-tridecanone for 1st
instar M. sexta larvae to be 17.1 μg/cm^2 of treated surface, suggests that 2-tridecanone is the allomonal factor responsible for
the mortality of hornworm larvae on L. hirsutum. A more comprehensive discussion of this research appears on references 7 and 8.

Turning our attention to kairomonal factors, it has been
observed that hornworm larvae show a feeding preference for the
host plant to which they were exposed during the 1st instar (9).
Yamamoto developed a bioassay utilizing a neutral substrate
(dandelion, Taraxacum officinale, leaf disks) which could be in-
fused with extracts of the preferred host plant in order to
detect a feeding preference in hornworm larvae. After screening
numerous members of the Solanaceae, we selected Solanum
carolinense (horsenettle) as it appeared to initiate the strongest
preference and have higher levels of the kairomonal factor in leaf
tissues.

Isolation of the feeding factor for M. sexta was a far more
difficult task. Whereas 2-tridecanone is a simple, stable molecule soluble in organic solvents, the feeding factor is water
soluble, occurs at trace levels in plant tissue, and is easily
hydrolyzed under mild alkaline or acidic conditions with subsequent loss of biological activity. The isolation of such a compound was a formidable obstacle requiring a departure from the more
classical approach of hydrolysis or chemical derivatization followed by isolation of the lipophilic product. The necessity that pure
substance be isolated with retention of biological activity required some basic research in modern separation techniques to
develop a suitably mild isolation strategy.

The isolation procedure that we have developed involves ex-
tracting the leaf tissue with boiling, deionized water, precipitation of phenolic components with lead acetate ($PbOAc_2 \cdot 3H_2O$) and

extraction of the clarified aqueous layer with n-butanol. The n-butanol is then evaporated to dryness and the residue redissolved in deionized H_2O and the sample deionized through a cation exchange column (Dowex 50W X-8, H^+) using deionized water as the eluent. The column effluent is then reduced in volume and passed through a column packed with Sephadex G-15 (Pharmacia, Inc.) and eluted with deionized water. Several fractions are obtained with the activity confined to the fraction eluting at the void volume. HPLC analysis of this fraction revealed several components (Figure 1) so a preparative scale medium resolution reversed phase column was employed to fractionate the mixture further. Two active feeding fractions were obtained and each was analyzed by HPLC. The first fraction corresponded to the peak at 9.3 minutes and the second fraction corresponded to the peaks between 9.7 and 10.6 minutes. Final purification of the peak eluting at 9.3 minutes was accomplished by semi-preparative reversed phase HPLC. No attempts have been made thus far to isolate the several peaks eluting between 9.7 and 10.6 minutes.

The 400.13 MHz ^{1}H-nmr (10) was obtained and is presented in Figure 2. Although this spectrum is complex, certain structural features are discernable. The loss of signals in the region of 40-55 ppm upon the addition of D_2O suggests the presence of a glycoside. Other structural features include aromaticity (<10% of the total integrated proton area), olefinic protons and a methylene, methyl pattern typical of a highly substituted fused ring system. The ^{13}C-nmr (100.58 MHz) (10) is somewhat more informative and the salient features are presented in Table II. Further, it should be noted that molecular weight estimation from the 54 observed carbon signals suggests a molecular weight of 1,000-1,200 a.m.u.

Table II. ^{13}C-nmr of Feeding Stimulant (100.58 MHz)

Absorbance(ppm)	Structural moiety
197	benzylic ketone
164	enol ether
130 (7 signals)	aromaticity (6), enol ether (1)
100 (3 signals)	anomeric carbons of sugar residues
80-60	carbanols of a carbohydrate
<60	methylene and methyl of a steroid

We had earlier observed that the biologically active fractions after HPLC readily hydrolyzed to give a chloroform soluble product. 100 mg of the combined fractions (HPLC peaks 9.3-10.6)

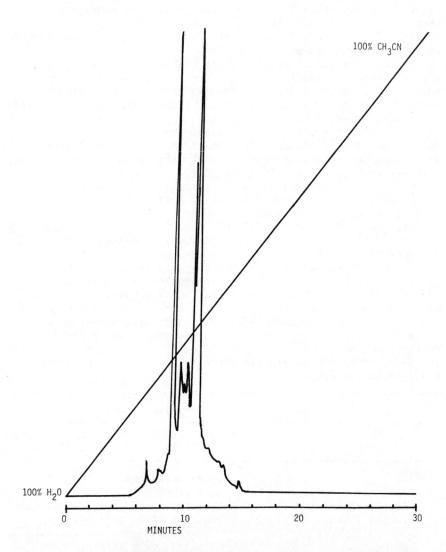

100% CH₃CN

100% H₂O

0   10   20   30
                         MINUTES

Figure 1. Analytical HPLC of the phagostimulants. Conditions: Partisil 10 ODS-2 (4.6 x 250 mm) column; mobile phase, 100% H₂O to 100% CH₃CN, 2 mL/min; sample, 10 μL 10 mg/mL; ambient temperature; detector, SP8310 (Spectra-Physics), 254 nm, 0.16 AUFS, 10 MVFS.

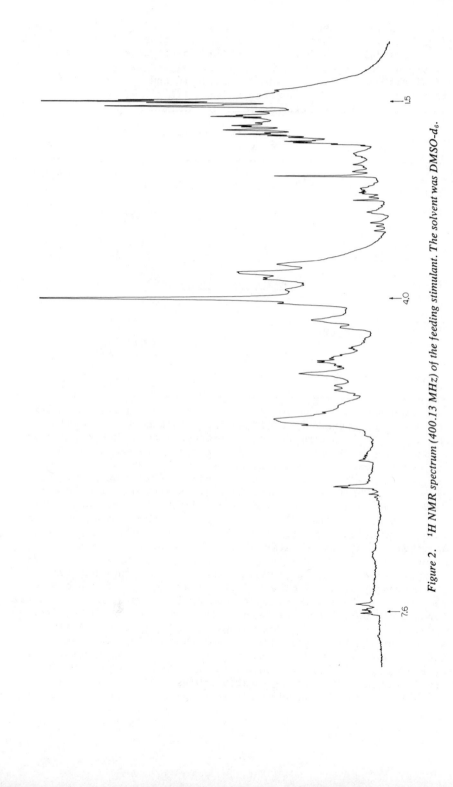

Figure 2. ¹H NMR spectrum (400.13 MHz) of the feeding stimulant. The solvent was DMSO-d₆.

were hydrolyzed under acidic conditions according to the procedure
described by M. E. Wall (11) in an attempt to see if the aglycones
of the several biologically active components were the same and to
isolate a readily identifiable product. A single product after
hydrolysis was obtained! The pertinent IR and UV data before and
after hydrolysis are presented in Table III.

Table III. IR and UV Data Before and After
Acid Hydrolysis

Feeding Factors Before Hydrolysis

UV(EtOH)   IR(Thin Film)

λmax 245nm   νmax 1690 cm^{-1}
      280nm   3400 cm^{-1}(br)

Hydrolysis Product

UV(EtOH)   IR(CHCl$_3$)

λmax 245nm   νmax 1690, 1715 &
      280nm   1725 cm^{-1}
      340nm   2820 cm^{-1}
                                                       3400 cm^{-1}(br)

Upon examination of the 400.13 MHz ^{1}H-nmr certain spectral similarities with nicandrenone, a withanolide isolated from Nicandra
physaloides (4), became obvious. A comparison of these two spectra
is presented in Figure 3. The dotted lines are drawn to line up
regions of equivalent chemical shifts in the two spectra. Numerous
similarities are seen, however two major differences should be
discussed. The first of these is that the protons at C-16 and
C-18 are shifted 0.5 ppm downfield while the C-15 proton remains
unchanged as does the splitting pattern of the aromatic multiplet.
This observation suggests that C-12 is a carbonyl. The other
major difference is the loss of the large multiplet centered
around 3 ppm in nicandrenone. This multiplet consists primarily
of hydroxyl and allylic protons. Deuterium exchange has removed
the hydroxyl signals in the hydrolysis product and the broad hump
remaining may be due to no more than 2 allylic proton. To account
for these observations, we propose the modifications of nicandrenone shown in Figure 4. Beginning with ring A, the enol-ether
allows for the hydrolytic lability of the feeding stimulant, the
assignment of C-1 at 164.5 ppm in the ^{13}C-nmr, assignment of C-2
as the seventh "aromatic" carbon in the 130 ppm region of ^{13}C-nmr,
the appearance of a ketone after hydrolysis (IR, 1715 cm^{-1}), and a
convenient point of attachment for the glycosyl residues. In the

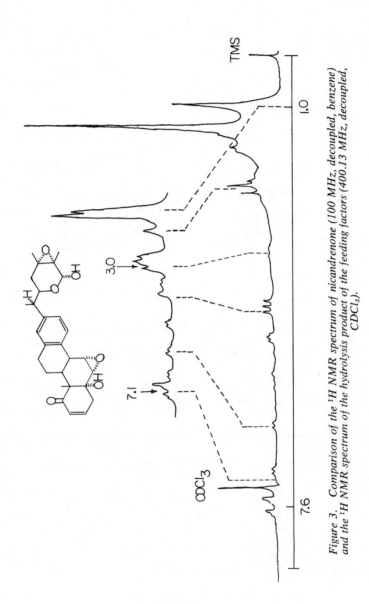

Figure 3. Comparison of the ^1H NMR spectrum of nicandrenone (100 MHz, decoupled, benzene) and the ^1H NMR spectrum of the hydrolysis product of the feeding factors (400.13 MHz, decoupled, $CDCl_3$).

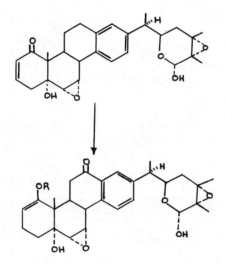

*Figure 4.   Comparison of nicandrenone structure (top) to the proposed structure
of the feeding stimulant (bottom). R = glycosyl.*

C ring, C-12 is modified to a benzylic ketone as suggested by ^{13}C-nmr (197 ppm) and IR (1690 cm^{-1}). The substitution pattern of the aromatic D ring is unchanged, however the presence of the 12-oxo carbon will shift the protons at C-18 and C-16 pattern downfield in the ^{1}H-nmr. The C-21 methyl of the feeding stimulant is assigned on the basis of the ^{13}C-nmr (13.8 ppm) and the ^{1}H-nmr (1.5 ppm, d) which compares favorably to that of nicandrenone (14 ppm and 1.25 ppm, d respectively). The structure of the E ring is proposed primarily to account for the appearance of an aldehyde upon acid hydrolysis of the feeding stimulant. To be sure, the structure proposed in Figure 4 is only tentative and experiments involving 2 dimensional J spectroscopy, double irradiation, and fast atom bombardment mass spectroscopy are currently underway to confirm this structure.

The isolation of the ovipositional factor for adult gravid female hornworms (M. sexta) is still on-going research and is not as far along as the two previously discussed topics. Nevertheless several interesting chemical and biological points have been revealed and are worthy of discussion.

The isolation procedure that we have developed involves an initial hot water extraction of the dried leaves of either horse-nettle (Solanum carolinense), tomato (L.esculentum), potato (S. tuberosum) or tobacco (Nicotiana spp.). Differential solubility utilizing a n-butanol-water partition followed by decreasing the polarity of the n-butanol layer with the addition of diethyl ether to a 50% ether-butanol mix and back extraction with water is an effective cleanup procedure. The biologically active aqueous layer has been analyzed by multiple thin layer chromatographic procedures and a relatively clean ovipositionally active fraction has been obtained. The larval feeding stimulant is also resolved from the ovipositional stimulant at this stage and therefore proves that the oviposition and feeding factors are two different compounds. HPLC analysis of the ovipositional fractions obtained from the various members of the Solanaceae shows that each of these fractions are 90% pure. However, each have slightly different retention times and are therefore different. Chemical analysis of the ovipositional factors obtained from these different sources suggests that it is a nitrogen containing phenolic glycoside (12, 13). The results of the chemical and spectral evidences are summarized in Table IV.

Further, the phagostimulative fraction was analyzed via HPLC and were chromatographically similar to the phagostimulative substances described earlier. The chemical and spectral data obtained from the phagostimulative fractions of these various members of the Solanaceae also suggests a glycosidic steroid.

Although the ovipositional stimulant and the phagostimulant are necessary to illicit their respective responses, they may not be sufficient to account for host-plant selection. In the case of Nicandra physaloides, a member of the Solanaceae not preferred as a host plant by M. sexta, we have shown that aqueous extracts

contain the feeding factor and hornworm larvae will feed on these
extracts once nicandrenone has been removed. Further, we have

Table IV. Oviposition Factor - Preliminary Characterization

Functionality	Chemical Tests	Spectral Data
Phenol (Yes)	Positive FeCl$_3$ & K$_3$Fe(CN)$_6$, Vanillin, Ammonia Vapor, Bromine water, FeCl$_3$, Precipitation with Pb(OAc)$_2 \cdot$3H$_2$O	IR: 3400–3200 cm^{-1}, 1650–1590 cm^{-1} UV: λmax 320, 255 nm, λmax(base) 328, 260 nm NMR: ^{1}H 6.9–8.0 ppm ^{13}C 55–77 ppm
Carbohydrate (Yes)	Positive Molisch, Phenol/H$_2$SO$_4$	NMR: ^{1}H 3.4–4.2 ppm ^{13}C 55–77 ppm
Amine (Yes)	Positive Ninhydrin Positive Dragendorff	IR: 3400–3200 cm^{-1}
Carbonyl (No)	Negative 2,4 DNPH	IR: No peak 1800–1700 cm^{-1} NMR: ^{13}C No peak 150–200 ppm

recently received a variety of N. physaloides (var. albiflora)
which has been shown to lack nicandrenone (14) and hornworm larvae
will feed upon this variety. Obviously, nicandrenone is a phago-
deterrent and its presence deters hornworm larvae in spite of the
presence of the feeding stimulant.

Jermy (15) has emphasized the importance of allomones in the
host plant selection process. Although the ovipositional and
phagostimulative kairomones do not appear to be sufficient to
account for host specificity by M. sexta in the host plant selection process, the mere avoidance of allomones does not appear to
be sufficient either. Rather, the presence of a detectable
allomone is sufficient to account for non-selection of a potential
host plant. For example, given a choice between L. esculentum
and any other suitable host plant M. sexta moths select L. excu-
lentum (16). No allomones are involved! To account for this
preference, the presence of volatile orientation factor(s) may be
involved. In fact, Morgan and Lyon (5) isolated amyl salicylate
from the host plant Datura stromonium as an orientation factor for
gravid female moths. We have also shown that an orientation
factor is present in the steam distillate of L. esculentum leaves.

It is interesting to note that the feeding stimulant is the
glycoside of a withanolide, a known class of allomones unique to

the Solanaceae (17). This observation tends to support the view
that specialization in host plant selection is the result of the
evolved ability of an insect species to tolerate or detoxify toxic
factors and in some cases utilize these compounds as kairomones.
Although the ovipositional factor is ubiquitous within M. sexta's
host range, it has not been determined whether or not these compounds are unique to the Solanaceae, if they occur in non-preferred members of the Solanaceae as well, or if they represent
another example of an allomone turned kairomone.

Another insect-hostplant system under investigation in our
laboratories is that of the oligophagous Mexican bean beetle
(Epilachna varivestis, Mulsant) and its predominant hosts in the
continental United States - snap beans (Phaseolus vulgaris), lima
beans (P. lunatum), and soybeans (Glycine max) - all of which are
members of the family Leguminosae. Our interest in the bean
beetle focuses on a frequently noted fraility of this insect pest-
its sensitivity to climatic stress. In this study our interest
goes beyond the role of secondary plant metabolites in modulating
insect behavior during the host plant selection process. Rather,
we are examining the role of major plant metabolites in insect
physiology once the selection process has been completed.

The supposed area of origin for the Mexican bean beetle (MBB)
is the high plateau region of Central America (18) - an area
characterized by daily rains and moderate temperatures during the
growing season. This is presumably the climate to which this
insect adapted in its evolutionary past. Early investigators
noted that populations of this beetle were limited to irrigated
Phaseolus fields in the southwestern United States (19).

As the MBB spread into the southern United States, field observations confirmed the sensitivity of this insect to high
temperatures and low humidities (20, 21). Qualitative laboratory
studies successfully substantiated the conclusions of the earlier
field investigators (22).

All of this early work was concerned with survival on preferred hosts in the genus Phaseolus. The introduction and rapid
expansion of soybean production coincided with the MBB's spread
to the southeastern United States, and this important agricultural
crop unfortunately became a secondary host for this insect.
Although soybeans are only a secondary host (i.e. given a choice,
the insect prefers Phaseolus species), the potential for adaptation of the MBB when presented large monocultures of Glycine
(essentially a no-choice situation) has not been overlooked by
entomologists. Several studies in the Department of Entomology
at NCSU have demonstrated an exaggerated sensitivity of this
insect to hot, dry conditions when feeding on soybean hosts. (23,
24, 25) Of particular pertinence to our work in the Department of
Chemistry were K. G. Wilson's findings (26) that transpirational
water loss in MBB larvae is dependent on both rearing temperature
and hostplant. Those larvae reared on soy hosts displayed increased transpiration, especially at elevated temperatures. Our

laboratories are currently involved in testing the hypothesis that these differences are related to the cuticle chemistry of this insect.

Cuticular permeability in insects is correlated with a number of factors including growth stage of the insect, surface to volume ratios, the degree of tanning in the cuticle, conditioning temperatures, and cuticular lipids (27, 28, 29). The cuticle of an insect is a complex matrix of chitin (a N-acetyl glucose amine polymer in the form of overlapping fibrils), a chitin-protein complex known as the exocuticle, a thin lipoprotein matrix (cuticulin) and an outermost layer of wax - the epicuticular lipid (Figure 5). The production and composition of this structure is regulated by the epidermal cells. Entomologists and chemists concerned with the structure and function of insect cuticle have demonstrated that the epicuticular wax layer of an insect is of principle importance in providing a barrier to desiccation in insects. (30, 31, 32) In particular, several studies relating insect lipids to rearing temperature (28, 33, 34) indicated that greater saturation of the fatty acids either in their free form or as part of a larger molecule (wax esters, glycerides, etc.) was correlated with those insects reared at higher temperatures. Few studies have attempted to correlate dietary lipid composition with the insect's cuticular lipids however. This relationship was one which intrigued us, especially since R. F. Wilson (35) of the Department of Crop Science had performed analyses of soy leaf lipids which showed that a high proportion of these fats were unsaturated. The culmination of our interests in bean beetle climate stress, the role of cuticular lipids in insect water balance, and the possibility that this insect's sensitivity to such stresses might be related to the lipid composition of its host led us to undertake a detailed analysis of Mexican bean beetle cuticular lipids as they relate to hostplant, life stage, and rearing temperature. In addition, analyses of plant foliar lipids have also been carried out.

Studies of changes as they depended on life stage (larvae, pupae, and adults) are being carried out on field insects reared on lima, snap, and soybean hosts. The experiments involving rearing temperature dependent changes are underway using larvae reared on limas or soybeans. Temperatures of rearing were 23, 27, and 32°C, which correspond to the values used in Wilson's water loss experiments. Gravimetric, thin layer, and gas chromatographic methodologies are employed in these analyses. Thin layer densitometry (36, 37, 38) of the whole lipid extract has given us composition by lipid classes. These analyses indicated that a majority of the lipid found in larval cuticle (65-70%) is composed of fatty acid containing substances. Major classes include hydro-carbon (the major fraction of adult cuticular lipid), wax esters (predominant class of larvae), free fatty acids, triglycerides, and sterols. Gas chromatographic analysis by individual class was deemed impossible due to time and quantity considerations, there-

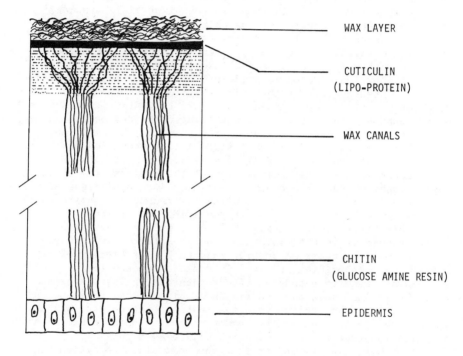

Figure 5.   The cuticle of an insect.

fore the analysis was carried out on hydrolyzed samples which gave
hydrocarbon, fatty acid, and alcohol fractions. These analyses
are based in part on methodologies employed in previous cuticular
lipid analyses (39, 40 and references therein).

Although these analyses are still underway in our laboratories, some preliminary information is available. Gas chromatography of the lima bean leaf lipids, in comparison with the soy
analysis carried out by Dr. R. F. Wilson shows that both contain
a high proportion of unsaturated fats (Table V). Both of these
analyses were carried out using packed column GC methodologies.
Capillary column analysis of the larval lipids, which provided
considerably more resolution than in the plant lipid studies,
indicate a rather complicated composition (Table VI). The degree
of unsaturation is higher in soy reared larvae, a potentially
positive correlation with the differences in cuticular permeability. In particular, the lima fed insects have a significantly
higher proportion of stearic acid in comparison with soy reared
beetles. Stearic acid is the major fraction in both cases.
Comparison with the leaf lipid data shows that stearic acid is a
minor component in both plants, but it nevertheless occurs in
approximately the same ratio as in the insect's cuticle. Any
correlation of plant lipid and insect lipid composition must,
however, await completion of all analyses. Of pertinence will be
comparisons of the lipid composition of the cuticle from those
insects reared under the controlled temperature conditions as well
as a re-analysis of plant lipid using the recently acquired capillary system with its much greater resolving capabilities.

It has become increasingly obvious to us that regulation of
water balance in insects is one of the keys to the success of
these animals in their adaptation to such a wide variety of habitats. The complexities involved in undertaking studies of such
systems may require expansion of our interdisciplinary approach
so as to include such diverse fields as biophysics, ecology, and
cell biology. The subject is a fascinating one, perhaps all the
more so due to its complexity, and promises to open up new areas
in which interdisciplinary research will play a role.

It is beginning to appear that the host plant selection process is indeed highly complex and involves the subtle interplay
of orientation factors, ovipositional factors, feeding factors and
allomonal factors as well as visual and tactile stimuli, environmental factors and plant nutrients. It has been the goal of our
research efforts to provide a clearer understanding of the dynamic
equilibrium that exists between an insect and its host plants.
Fortunately, the last decade has brought forth the technologies
necessary to isolate and identify complex bio-molecules and subsequently analyze biological systems in a manner that was previously impossible. Such holistic investigations promise to yield
methods of controlling insect pests on valuable crop plants. These
methods may then be integrated into a truly bio-rational approach
to pest management.

Table V. A Comparison of Plant Leaf Fatty Acids

Fatty Acid	Glycine max (35) (Soybeans - Bragg)	Phaseolus lunatum (Lima beans-Henderson)
12:0 lauric	- - -	1.3 ± 0.2
14:0 myristic	- - -	1.6 ± 0.2
16:0 palmitic	14.5 ± 0.7	11.0 ± 0.2
16:1 palmitoleic	2.8 ± 0.3	3.1 ± 0.3
16:2 hexadecadienoic	trace	0.4 ± 0.2
18:0 stearic	3.4 ± 0.5	6.4 ± 0.3
18:1 oleic	2.3 ± 0.3	2.3 ± 0.1
18:2 linoleic	13.1 ± 1.1	10.1 ± 0.4
18:3 linolenic	63.9 ± 1.6	63.7 ± 1.1
20:0 arachidic	- - -	0.4 ± 0.1
22:0 behenic	- - -	0.6 ± 0.1
22:1 erucic	- - -	0.3 ± 0.1
Total Unsaturated	82.1%	79.9%

Analysis done on a Carbowax 20M column,
2 meters x 1/4" O.D., 30 ml/min, N_2, T_{col}
120 → 220 at 5°/min. Perkin Elmer Sigma
3B with FID.

Table VI.   Comparison of Larvae Cuticle Fatty Acids

Fatty Acid	G. max Reared Larvae	P. lunatum Reared Larvae
12:0 lauric	trace	trace
14:0 myristic	5.97 ± 0.20	1.44 ± 0.00
16:0 palmitic	1.40 ± 0.03	0.58 ± 0.09
16:1 palmitoleic	trace	trace
16:2 hexadecadienoic	- - -	- - -
18:0 stearic	12.15 ± 0.65	32.19 ± 2.17
18:1 oleic	7.58 ± 1.00	7.48 ± 0.52
18:2 linoleic	10.71 ± 0.16	19.57 ± 1.44
18:3 linolenic	8.70 ± 0.30	8.36 ± 0.19
20:0 arachidic	8.56 ± 0.71	4.63 ± 0.31
20:1 eicosenoic	3.07 ± 0.10	0.67 ± 0.11
20:2 eicosadienoic	1.80 ± 0.21	0.47 ± 0.03
22:0 behenic	1.91 ± 0.28	1.90 ± 0.27
22:1 erucate	2.66 ± 0.35	0.31 ± 0.04
22:2 docosodienoic	2.83 ± 0.28	1.15 ± 0.24
24:0 lignoceric	0.65 ± 0.08	0.99 ± 0.09
24:1 nervonic	1.91 ± 0.54	0.26 ± 0.00
24:2 tetradodecadionoic	2.57 ± 0.05	1.26 ± 0.01
26:0 hexadodecanoic	1.52 ± 0.26	0.82 ± 0.00
% Unsaturated	56.53	48.16

Analyses performed on a 10 meter SP2100 capillary column, 50:1 split ratio, linear gas velocity 28.5 cm/sec, T_{col} 120 → 220°C at 5°/min, Perkin Elmer Sigma 3B w/FID interfaced to Spectra Physics SP4000 Data System.

Acknowledgements

The authors wish to gratefully acknowledge the financial support of this project by the Frasch Foundation and the United States Department of Agriculture (SEA/CRGO), NSF Grant #DEB 7822738.

Literature Cited

1. Schoonhoven, Louis M. "Offprints from Semiochemicals: Their Role in Pest Control"; Nordlund, Donald A., Ed.; John Wiley and Sons, Inc.:New York, 1981; p. 31.
2. Hedin, Paul A.; Maxwell, Fowden G.; Jenkins, Johnie N. "Proceedings of the Summer Institute on Biological Control of Plant Insects and Diseases"; Maxwell, Fowden G.; Harris, F.A., Ed.; University Press of Mississippi:Jackson, 1974; p. 494.
3. Blum, M.S., Chapter _ in this book.
4. Bates, R.B.; Eckert, D.J. J. Amer. Chem. Soc., 1972, 94, 8258.
5. Morgan, A.C.; Lyon, S.C. J. Econ. Ent. 1928, 21, 189-191.
6. Yamamoto, R.T.; Fraenkel, G. Ann. Ent. Soc. 1960a, 53, 499-503.
7. Williams, W.G.; Kennedy, George G.; Yamamoto, Robert T.; Thacker, J.D.; Bordner, Jon Science 1980, 207, 888-889.
8. Kennedy, G.G.; Yamamoto, R.T.; Dimock, M.B.; Williams, W.G.; Bordner, J. J. Chem. Ecol. 1981, 7, 707-716.
9. Yamamoto, R.T. J. Insect Physiol. 1974, 20, 641-650.
10. Ellis, P.D.; Inners, R.R. South Carolina Magnetic Resonance Laboratory, Department of Chemistry, University of South Carolina, Columbia, S.C. NSF Grant No. CHE 78-18723.
11. Wall, Monroe E.; Krider, Merle M.; Rothman, Edward S.; Eddy, C.Roland J. Biol. Chem. 1952, 198, 533-543.
12. Coco, Wm. J. M.S. Thesis, North Carolina State University, Raleigh, N.C. 1981.
13. Fraenkel, G.; Nayer, J.; Nalbandov, O.; Yamamoto, R.T. Proc. Intern. Congr. Entomol. 11th Vienna 3, 1960, p. 122.
14. Tishbee, Arye; Kirson, Isaac J. Chromatog. 1980, 195, 425-430.
15. Jermy, T. Entomol. Exp. Appl. 1966, 9, 1-12.
16. Yamamoto, R.T.; Fraenkel, G. Proc. Intern. Congr. Entomol. 11th Vienna 3. 1960, p. 127.
17. Kirson, Isaac; Glotter, Erwin J. Nat. Prod. 1981, 44, 633-647.
18. Landis, B.J.; Plummer, C.C. J. Agric. Res. 1935, 50, 989-1001.
19. Sweetman, H.L. Ecology 1929, 10, 228-244.
20. Howard, N.J. Florida St. Pl. Bd. Quart. Bull. 1921, 6, 15-24.
21. Eddy, C.O.; McAlister, L.C., Jr. S.C. Agric. Exp. Sta. Bull. 1927, #236. 33 pp.
22. Sweetman, H.L.; Fermald, H.T. Mass. Agric. Sta. Report 1980, #261. 25 pp.

23. Kitayama, K.; Stinner, R.E.; Rabb, R.L. Environ. Ent. 1979, 8, 458-64.
24. Sprenkel, R.K.; Rabb, R.L. Environ. Ent., in press.
25. Wilson, K.G.; Stinner, R.E.; Rabb, R.L. Environ. Ent. 1982, 11, 121-26.
26. Wilson, K.G. PhD Thesis, North Carolina State University, Raleigh, N.C., 1982, 86 pp.
27. Baker, John F.; Baker, Antje Ecol. Ent. 1980, 5, 309-14.
28. Munson, Sam C. Econ. Ent. 1953, 46, 657-66.
29. Gilby, A.R. "Advances in Insect Physiology" Vol. 15 Academic Press:New York, 1980; pp 1-30.
30. Wigglesworth, V.B. Ann. Rev. Entomol. 1957, 2, 37-54.
31. Armold, M.T.; Regnier, F.E. J. Insect Physiol. 1975, 21, 1827-33.
32. Jackson, L.L.; Baker, G.L. Lipids 1970, 5, 239-46.
33. Cherry, Lois M. Ent. Exp. Appl. 1959, 2, 68-76.
34. Hadley, Neil F. Insect Biochem. 1977, 7, 277-283.
35. Wilson, R.F.; personal communication.
36. Downing, Donald T. "TLC:Quantitative Environmental and Clinical Applications"; Touchstone, J.C.; and Rogers, Dexter, eds.; Wiley-Interscience:New York, 1980, pp 495-516.
37. Getz, Melvin. Ibid. pp 99-113.
38. Privett, Orville S.; Dougherty, Kathryn A.; Erdahl, Warren L. "Quantitative Thin Layer Chromatography"; Touchstone, Joseph C. ed.; Wiley-Interscience:New York, 1973; pp 57-78.
39. Baker, J.E. Insect Biochem. 1978, 8, 287-92.
40. Warthen, J.D.; Uebel, E.C.; Lusby, W.R.; Adler, V.E. Insect Biochem. 1981, 11, 467-72.

RECEIVED September 27, 1982

Detoxication, Deactivation, and Utilization of Plant Compounds by Insects

MURRAY S. BLUM

University of Georgia, Department of Entomology, Athens, GA 30602

While the roles of most plant natural products are poorly understood, it is at least well established that many of these compounds can function as feeding deterrents because they are either unpalatable or produce deleterious effects for a variety of animals. These toxic plants, endowed with such compounds, generally constitute a considerable challenge to animals that ingest their tissues, and are, in a real sense, "forbidden fruits" for the host of herbivores with which they share their living space. However, notwithstanding the pronounced toxicological attributes of these natural products, a multitude of herbivorous insects utilize toxic plant sources as food, usually selecting these plants as preferred hosts. These invertebrates demonstrate that the term toxic is relative at best, since these plants provide them with both a home and a resource that can be readily utilized to support both growth and development. Their predilection for attacking "so called" toxic plants has eventuated in a lifestyle that has enabled these insects to exploit a multitude of plant species that are "off limits" for most herbivores.

Some insect herbivores store these toxic compounds in their bodies during different developmental stages. Insects in at least six orders sequester a variety of plant natural products (1) including alkaloids (2), cardenolides (3), nitrophenanthrenes (4), mustard oils (5), and cannabinoids (6). On the other hand, various insect species feeding on the same species of plants do not appear to sequester these compounds (1). This diverse behavior emphasizes how inadequately we understand the physiological bases for sequestration (7) or the evolutionary factors that promote the retention of these compounds in the bodies of selected herbivores. In short, while it is one thing to demonstrate that a certain species of phytophagous insect may store specific natural products in its body tissues, it is quite something else to interpret this sequestrative ability in either detoxicative or metabolic terms. Indeed, it is now clear that the sequestration of a compound can represent an end product of a

0097-6156/83/0208-0265$06.00/0

concatenation of biochemical and physiological events that
reflects a species' idiosyncratic processing of the allelochemic.
Analyses of the interplay of factors that characterize the fates
of natural products ingested by selected insects indicate that
physiological variety is the spice of herbivorous life.

The demonstration that eclectic mechanisms have been evolved
by insects for coping with the potentially toxic concomitants of
their ingested nutrients necessitates careful analyses of the
processing of each of these compounds by each adapted herbivore.
Furthermore, it is important to realize that in itself,
sequestration is nothing more than an end product of a series of
reactions that may reflect selective absorption, metabolism of
specific compounds, and excretion of selected allelochemics (8).
In the present review, these diverse processing strategies will
be explored in order to illustrate the various ways in which a
multitude of herbivores accommodate potential plant toxins. Two
lepidopterous species will be used as models which, hopefully,
will emphasize both the elegance and complexity identified with
insects as processors of plant-derived compounds.

Processing of Plant Natural Products by Herbivores

The intrusion of a plant allelochemic in the gut of an
adapted herbivore triggers a series of reactions that may result
in the compound being excreted or absorbed and sequestered, with
or without being metabolized. The metabolism of an absorbed
compound may constitute detoxication or it may represent a means
of producing a compound that can be more readily sequestered than
the nonderived allelochemic, toxicity notwithstanding. Since
little information is available on the toxicities of plant
allelochemics or their metabolites to adapted herbivores, it is
virtually impossible to interpret the detoxicative significance
of these metabolic transformations. Superimposed on this
informational hiatus is a lack of data on the ecological
correlates of storing these allelochemics vis-à-vis pathogens
or predators. Thus, while the fates of specific plant natural
products can be explored after their ingestion and metabolism by
herbivores, their potential roles, as for example, possible
emetics or toxins for predators, is at best terra incognita.

Excretion. The lack of detectable plant allelochemics in the
bodies of a host of insect species that had developed on plants
containing potentially toxic natural products (1) may reflect the
rapid excretion of these compounds subsequent to ingestion.
However, the demonstration that selected herbivores eliminate,
during pupal or adult development, compounds sequestered by larval
stages (9), militates against drawing conclusions about the
sequestrative abilities of larvae based solely on analyses of
adults. Thus, while it seems reasonable to assume that
the lack of ingested allelochemics in the body of a herbivore

probably reflects excretion (or metabolism) of these compounds, without supportive experimental data it is best to be cautious in interpreting this information.

Nicotine is rapidly excreted by three insect species that normally feed on Nicotiana tabacum (10,11). Manduca sexta, Trichoplusia ni, and Heliothis virescens eliminate the unchanged alkaloid after ingestion, and no evidence was obtained to indicate that nicotine was subjected to any metabolic transformations. M. sexta absorbed small amounts of nicotine, but this compound could only be detected in the blood for a relatively short period of time.

Larvae of the lymantriid, Eloria noyesi, an obligatory feeder on Erythoxylum coca, show very little tendency to absorb cocaine, the major alkaloid present in the leaves. Almost all of the ingested alkaloid is excreted unchanged, only traces being detectable in the blood (12). Although this species rapidly excretes virtually all of the cocaine which it encounters in its gut, traces of this tropane alkaloid are present in adults, indicating that small amounts of this compound, absorbed by larvae, are retained during development.

The presence of high levels of alkaloids such as nicotine and cocaine in the guts of larval insects may be highly adaptive for these invertebrates in terms of potential predators. Molested larvae, such as those of E. noyesi, discharge enteric fluids from the mouth when disturbed (12), an act which can expose predators to high levels of alkaloids. Since these compounds are excellent repellents for a variety of invertebrates, their value as deterrents for predatory animals may be considerable.

Metabolism. Plant allelochemics ingested by herbivores may be metabolically altered in the digestive tract or after absorption across the gut into the hemolymph. In some cases metabolism of these compounds results in compounds known to be considerably less toxic than the original natural products and thus constitutes true detoxication. On the other hand, the toxicological significance of many of these metabolites is unknown for either their producers or for potential predators. While these allelochemical derivatives may be much more amenable to sequestration than their parent compounds, it will not prove surprising if they are sometimes true detoxication products.

Nicotine (I) is metabolized to continine (II) by various insects, including species adapted to feed on alkaloid-containing leaves and those that are not. Continine, which is virtually nontoxic to insects, is the primary nicotine metabolite produced by some coleopterous and orthopterous species that feed on tobacco; other minor metabolites are produced as well (11). Two species of cockroaches and the housefly, Musca domestica, also convert nicotine to continine, although these insects do not normally feed on nicotine-fortified plants (11).

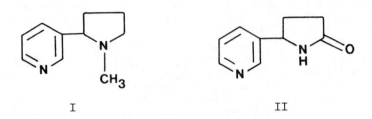

I   II

Sinigrin (III), a glycoside produced by many species of cruciferous plants, can be converted to a highly toxic aglycone, allylisothiocyanate (IV), after hydrolysis by adapted and nonadapted insects (5, 13, 14). On the other hand, the mustard oil is the preferred storage form of larvae of two _Pieris_ species, P. _brassicae_ and P. _rapae_, which had developed on food plants containing sinigrin (5). In the case of these pierids it must be assumed that the mustard oil can be sequestered more efficiently than sinigrin. The _Pieris_ species obviously have evolved mechanisms for avoiding the well-established toxicity of the former.

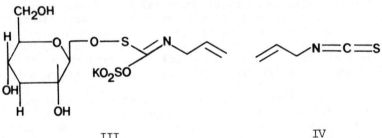

III   IV

Larvae of an arctiid, _Seirarctia echo_, have evolved a novel strategy for coping with the toxic effects of methylazoxymethanol (V), the aglycone of cycasin (VI), a constituent in the cycad leaves upon which they feed. When larvae of S. _echo_ are fed the aglycone, they convert it to the β-glycoside cycasin which is then sequestered (15). When cycad leaves containing an azoxyglucoside different from cycasin were ingested, the "foreign" glycoside was first converted to cycasin in the gut, following which it was sequestered. The ability of the larvae to selectively absorb cycasin, which is much less toxic than its aglycone, is correlated with the limited distribution of β-glucosidase. This enzyme, which can either synthesize cycasin from methylazoxy-methanol or hydrolyze it to the latter, is limited in its distribution to the gut, thus insuring that absorbed glycoside will not be converted to the toxic aglycone.

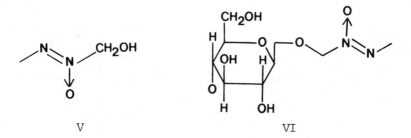

V   VI

Several pyrrolizidine alkaloids are sequestered by lepidopterous and orthopterous species after metabolic conversion from ingested alkaloids. Larvae of the arctiid Tyria jacobaeae primarily sequester the alkaloid seneciphylline (VII), although this compound is found in the leaves of the ragwort, Senecio jacobaeae, as the N-oxide (VIII) (2). In the case of T. jacobaeae, the liphophilic free compound, seneciphylline, is the preferred sequestration form of the alkaloid, the water-soluble N-oxide apparently being unsuited for ready storage. Conversely,

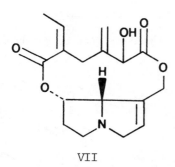

VII

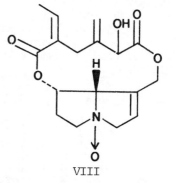

VIII

the pyrgomorphid grasshopper Zonocerus variegatus appears to convert a large proportion of an ingested pyrrolizidine alkaloid, monocrotaline (IX), to its N-oxide (X) before sequestering the alkaloids (16). Whereas the free alkaloid and the N-oxide are present in relatively similar amounts in the plant (Crotalaria sp.), Z. variegatus sequesters the alkaloid and N-oxide in a 1:3 ratio. Assuming that this sequestration pattern does not simply reflect selective sequestration of the N-oxide, it appears that this pyrgomorphid emphasizes the storage of polar alkaloids, some of which are oxidatively derived.

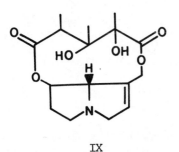

IX

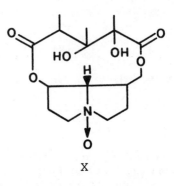

X

Notwithstanding the ability of larvae of _Arctia caja_ to develop on plants containing allelochemics as disparate as alkaloids and cardenolides (1), they cannot mature on marijuana (_Cannabis sativa_) plants rich in Δ^1-tetrahydrocannibinol (THC) (XI). On the other hand, these arctiids can develop on marijuana strains rich in a less toxic cannabinoid, cannabidiol (CBD) (XII), although developmental time is somewhat prolonged (6). Although larvae reared on the CBD-rich strain of _C. sativa_ contain only trace amounts of this cannabinoid, considerably more is sequestered if they are transferred to a THC-rich strain of marijuana. These larvae, which successfully completed development, sequestered substantial amounts of both THC and CBD, demonstrating a well-developed ability to metabolize cannabinoids.

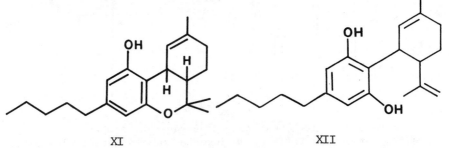

XI   XII

Cardenolides appear to be metabolized by a variety of species, possibly as a mechanism for converting these steroids into compounds that can be efficiently sequestered. The milkweed bug, _Oncopeltus fasciatus_, metabolizes (hydroxylates?) the nonpolar cardenolide digitoxin to more polar compounds that are subsequently sequestered in the dorsolateral space fluid (17, 18). Larvae of another cardenolide-adapted insect, the monarch butterfly, _Danaus plexippus_, also convert these steroids into compounds that are readily sequestered. For example, uscharidin, which contains a carbonyl group at C-3'(*) of the sugar moiety, is converted to calactin and calotropin (hydroxyl group at C-3') (XIII) by larvae after which these steroids are stored in specific tissues (19). The propensity of insects feeding on cardenolide-rich plants to selectively sequester these compounds probably reflects the evolution of specific metabolic pathways for producing cardenolide derivatives that are amenable to facile sequestration.

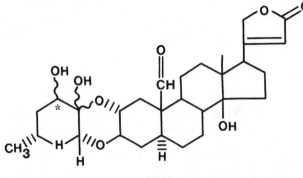

XIII

Selective Sequestration. Sequestration of plant natural products is an idiosyncratic phenomenon, and there is little indication that the profiles of insect-stored allelochemics in any way mirror those of their plant food sources. In general, each insect species treats ingested plant allelochemics distinctively, so that a compound excreted in toto by one species may constitute the main sequestration product of another. Unfortunately, while these sequestrative idiosyncrasies are quite evident, their significance is not.

The pyrgomorphid grasshopper Poekilocerus bufonius sequesters calactin and calotropin, two of the six cardenolides present in its ascelepiadaceous food source, and these steroids are present in the eggs as well (20). Another insect that feeds on a cardenolide-rich food source, the ctenuchid moth Syntomeida epilais, sequesters oleandrin, the main steroid present in leaves of its host plant, oleander (21). On the other hand, a variety of aphids and coccids feeding on oleander do not sequester oleandrin(1).

The sequestrative propensities of insects feeding on plants fortified with pyrrolizidine alkaloids are equally unpredictable. For example, larvae of the arctiid Amphicallia bellatrix sequester crispatine and trichodesmine, whereas the main alkaloid present in the leaves is crosemperine (22). Another arctiid, Tyria jacobaeae, concentrates senecionine in its tissues in spite of the fact that this compound is a trace constituent in the leaves of its food plant, Senecio sp. (23). On the other hand, jacobine, jacozine, and jacoline, three major alkaloids in the leaves of the plant, constitute minor storage constituents in adults of T. jacobaeae.

The ability of species to synthesize their own defensive allomones may have important ecological correlates vis-à-vis their tendency to sequester plant allelochemics. Chrysomelid beetles in the genus Chrysolina sequester hypericin, a naphthodianthrone present in their food plants, Hypericum spp. (24). On the other hand, many species in this genus synthesize cardenolides de novo, but those that feed on Hypericum do not (25). It appears that there has been no selection pressure for the biosynthesis of cardenolides in Chrysolina species that sequester hypericin, a known mammalian toxin, in their body tissues.

Insect Sequestrators and Plant Allelochemics -- Two Selected Case Studies

While it is firmly established that the plant natural products ingested by larval endopterygote insects may be ultimately sequestered in adult tissues, almost no information is available on the detailed fates of these compounds during development. What happens to these allelochemics during larval, pupal, and adult molts, especially when extensive tissue reorganization occurs? Are these compounds irrevocably sequestered in adult tissues or are they exchanged between different tissues, or for that matter excreted? Is sequestration sometimes based on a nonenergy-intensive physical process, as has been demonstrated for the cardenolides that are stored in the dorsolateral space fluid of the milkweed bug, Oncopeltus fasciatus (18)? These questions are illustrative of the lacunae in our knowledge of how adapted insects process the potential toxic compounds that are in their preferred food plants.

Some insights into how insects physiologically manipulate absorbed allelochemics have been provided recently by studying the fates of cardenolides in two unrelated lepidopterous species that develop on milkweeds fortified with these steroids. Analyses of cardenolide processing in the arctiid, Cycnia inopinatus, and the monarch, Danaus plexippus, demonstrate that each species has evolved distinctive physiological mechanisms for manipulating the medley of allelochemics that are internally omnipresent (26).

Cycnia inopinatus. Larvae of this arctiid develop on Asclepias humistrata, a milkweed species that contains very high levels of cardenolides. The hemolymph of the larvae sequesters and maintains these steroids at very high levels, thus insuring that internal concentrations of allelochemics will always be substantial. Significantly, the blood stores a predominance of polar cardenolides which appear to be retained in this fluid medium throughout metamorphic development (26). Elimination of cardenolides during larval life occurs primarily by means of the exuviae, which are shed at each larval molt. About 25% of the total cardenolide load of the larvae are excreted as exuvial constituents. Indeed, cardenolide loss as larval exuvial constituents corresponds to a concentration two to three times higher than those in the different body regions of adults of C. inopinatus. Pupal exuviae contain less cardenolides than those of the larvae, the concentration corresponding to about 50% of that in the different adult body regions.

The cardenolides in the larval hemolymph are about at an equivalent concentration with those distributed in the other larval tissues, negligible amounts of these steroids being present in the gut (26). In the adult, cardenolides are equally distributed between the three main body regions and the wings, with females containing about twice as much of these steroids as males. This female bias in cardenolide concentration appears to be directly attributable to the presence of substantial amounts of these steroids in the eggs.

Larvae of C. inopinatus emerge as major sequestrators of cardenolides primarily because their hemolymph, which is present in a relatively large volume, effectively sequesters high concentrations of polar cardenolides (26). Cardenolide excretion largely reflects loss of these steroids as components of the larval exuviae, the concentration of these compounds becoming relatively stable after pupal ecdysis. These steroids are ubiquitously distributed in the adult moth, having been derived primarily from the rich cardenolide pool in the larval blood.

Danaus plexippus. The mechanism of cardenolide processing by the monarch, D. plexippus, shares some common denominators with C. inopinatus but is mainly characterized by differences. Unlike larvae of C. inopinatus, those of the monarch have a large volume of gut fluid which contains polar cardenolides (26). Monarch larvae sequester far more efficiently from plants with low cardenolide concentrations than from those with high concentrations; in some cases their sequestrative efficiency is not great enough to store the high cardenolide concentrations in some plant species. In contrast, C. inopinatus larvae are effective sequestrators even when cardenolide concentrations in the host plant are inordinately high, such as are found in Asclepias humistrata, a milkweed species upon which both lepidopterans develop.

For monarch larvae, the large volume of gut fluid appears to constitute the evolutionary breakthrough that made a sequestrative lifestyle possible on cardenolide-rich milkweeds. This cardenolide-fortified fluid, which may exceed one-third of its total liquid volume, is withdrawn at pupation to become part of the hemolymph pool, being stored primarily under the wings (26). Subsequently, the wing scales, along with the hemolymph, become the richest source of cardenolides in the body, the steroids from the gut fluid having been exploited as an allelochemic source for the wings and blood. Gut fluid, which diminishes drastically during the prepupal period, reappears when the molting fluid is secreted, and increases in volume as ecdysis approaches. During pupal development the gut fluid again diminishes as it is converted to hemolymph only to emerge again in the pharate adult. This cardenolide-rich liquid is then again converted to hemolymph during adult development, so that very little is lost at emergence when the adult evacuates fluids from the gut.

Cardenolides in the gut fluid constitute the major steroidal pool which is manipulated eclectically in order to channel the cardenolides to different sites during development. Larval exuviae are rich in cardenolides and these discarded tissues constitute an excretory form for these compounds (26), as is the case for C. inopinatus larvae. Cardenolides are also concentrated in the prepupal molting fluid, and presumably part of this steroidal pool is eliminated as the exuvia discarded at larval-pupal ecdysis. In a two-day old pupa, cardenolides are present in low concentrations in the gut fluid but are much more concentrated in the hemolymph. However, before wing expansion the cardenolide concentration in the teneral adult is at its lowest imaginal level, rapidly climbing after this time so that it reaches the highest level encountered in any life stage.

The presence of high concentrations of cardenolides in the blood of adults demonstrates that a large percentage of these steroids is not locked in tissue sinks but is freely circulating in the body. Whether these blood-borne steroids exchange with those sequestered in tissues is unknown, but their presence as major constituents in adult hemolymph demonstrates that they are in a dynamic state, even in a stage which no longer ingests them (26). These results contrast with those obtained for C. inopinatus, and emphasize the importance of regarding each species and its allelochemic-fortified host plant as a unique evolutionary case.

Literature Cited

1. Rothschild, M. Symp. Roy. Ent. Soc. London 1972, 6, 59.
2. Aplin, R. T.; Rothschild, M. in "Toxins of Plant and Animal Origin"; de Vries, A.; Kochva, K., Eds., Gordon and Breach: New York, 1972; pp 579-595.
3. Reichstein, T.; von Euw, J.; Parsons, J. A.; Rothschild, M. Science 1968, 161, 861.
4. von Euw, J.; Reichstein, T.; Rothschild, M.; Israel, J. Chem. 1968, 6, 659.
5. Aplin, R. T.; Ward, R. d'Arcy; Rothschild, M. J. Ent., Ser. A. 1975, 50, 73.
6. Rothschild, M.; Rowan, M. G.; Fairburn, J. W. Nature 1977, 266, 650.
7. Duffey, S. S. Ann. Rev. Ent. 1980, 25, 447.
8. Blum, M. S. "Chemical Defenses of Arthropods"; Academic Press: New York, 1981.

9. Rothschild, M.; von Euw, J.; Reichstein, T.; Smith, D. A. S.; Pierre, J. Proc. Roy. Soc. London, Ser. B. 1975, 190, 1.

10. Self, L. S.; Guthrie, F. E.; Hodgson, E. Nature 1964, 204, 300.

11. Self, L. S.; Guthrie, F. E.; Hodgson, E. J. Insect Physiol. 1964, 10, 907.

12. Blum, M. S.; Rivier, L.; Plowman, T. Phytochem. 1981, 20, 2499.

13. Erickson, J. M.; Feeney, P. Ecology 1974, 55, 103.

14. Blau, P. A.; Feeney, P.; Contardo, L.; Robson, D. S. Science 1978, 200, 1296.

15. Teas, H. J. Biochem. Biophys. Res. Comm. 1967, 26, 686.

16. Bernays, E.; Edgar, J. A.; Rothschild, M. J. Zool. 1977, 182, 85.

17. Duffey, S. S.; Scudder, G. G. E. J. Insect Physiol. 1972, 18, 63.

18. Duffey, S. S.; Blum, M. S.; Isman, M. B.; Scudder, G. G. E. J. Insect Physiol. 1978, 24, 639.

19. Seiber, J. N.; Tuskes, P. M.; Brower, L. P.; Nelson, C. J. J. Chem. Ecol. 1980, 6, 321.

20. von Euw, J.; Fishelson, L.; Parsons, J. A.; Reichstein, T.; Rothschild, M. Nature 1967, 214, 35.

21. Rothschild, M.; von Euw, J.; Reichstein, T. Proc. Roy. Soc. London, Ser. B. 1973, 183, 227.

22. Rothschild, M.; Aplin, R. T. in "Chemical Releasers in Insects"; Tahori, A. S., Ed., Gordon and Breach: New York, 1971, pp 177-182.

23. Aplin, R. T.; Benn, M. H.; Rothschild, M. Nature, 1968, 219, 747.

24. Rees, C. J. C. Ent. Exp. Appl. 1969, 12, 565.

25. Pasteels, J. M.; Daloze, D. Science 1977, 197, 70.

26. Nishio, S. "The Fates and Adaptive Signficance of Cardenolides Sequestered by Larvae of Danaus plexippus (L.) and Cycnia inopinatus (Hy. Edwards)"; Ph. D. Thesis, University of Georgia, 1980.

RECEIVED August 23, 1982

ROLES OF PLANT CONSTITUENTS

L-Canavanine and L-Canaline: Protective Allelochemicals of Certain Leguminous Plants

GERALD A. ROSENTHAL

University of Kentucky, T. H. Morgan School of Biological Sciences and the
Graduate Center for Toxicology, Lexington, KY 40506

L-Canavanine and L-canaline are non-protein
amino acids of certain leguminous plants, that
function as protective allelochemicals. L-Canavanine is incorporated into de novo synthesized
proteins in place of arginine; there is suggestive evidence that formation of such anomalous
proteins figures significantly in canavanine's
adverse biological effects. Canavanine, however,
does not appear to inhibit overall protein
synthesis. Thus, an important basis for canavanine's antimetabolic properties resides in the
sustained production of biologically aberrant
proteins.
    L-Canaline is an ineffective antimetabolite
of L-ornithine since it has little ability to
antagonize ornithine-dependent reactions. On the
other hand, it forms a covalently bound Schiff-base
complex with the pyridoxal phosphate moiety of
B_6-containing enzymes. As such it is a potent
inhibitor of many decarboxylases and aminotrans-
ferases that utilize this vitamin.

L-Canavanine, 2-amino-4-guanidinooxybutyric acid, is a
structural analogue of L-arginine in which the terminal methylene
group is replaced with oxygen. This creates the guanidinooxy
group having a pK_a value of 7.01 as compared to 12.48 for arginine's guanidino group (1).
    Arginase (EC 3.5.3.1) mediates the hydrolytic cleavage of L-
canavanine to produce L-canaline and urea. L-Canaline, 2-amino-
4-aminooxybutyric acid, bears the same structural analogy to L-
ornithine as canavanine does to arginine. The aminooxy group
of canaline with its pK_a value of 3.96 differs markedly from the
δ-amino function of ornithine (pK_a = 10.76).
    These secondary plant metabolites are two of nearly 250
non-protein amino acids currently known to be proliferated by

0097-6156/83/0208-0279$06.00/0

higher plants (2). Canavanine and canaline are constituents of
a much more limited assemblage of non-protein amino acids, distinctive by virtue of their intrinsic toxicity and potent antimetabolic properties (3-6). They are markedly growth-inhibitory
to insects, and can be insecticidal as well (4,7,8,9).

Canavanine Toxicity and Aberrant Protein Formation

      Canavanine is a potent arginine antagonist that functions
in virtually all of the reactions for which arginine is the
preferred substrate (10). It is readily activated and amino-
acylated by arginyl-tRNA synthetase and subsequently incorporated
into the nascent polypeptide chain. This property is not unique
to canavanine but rather is observed with many non-protein amino
acids bearing structural analogy to a protein amino acid counterpart (11). Increasingly, it has come to be accepted that an
important, perhaps even the principal basis for the antimetabolic
properties of non-protein amino acids structurally akin to a
protein amino acid, resides in the formation of anomalous
proteins.

      Isolation of alkaline phosphatase from Escherichia coli in
which 85% of the proline residues were replaced by 3,4-dehydro-
proline affected the heat lability and ultraviolet spectrum of
the protein but the important criteria of catalytic function
such as the K_m and V_{max} were unaltered (12). Massive replacement
of methionine by selenomethionine in the β-galactosidase of E.
coli also failed to influence the catalytic activity. Canavanine
facilely replaced arginine in the alkaline phosphatase of this
bacterium; at least 13 and perhaps 20 to 22 arginyl residues were
substituted. This replacement by canavanine caused subunit
accumulation since the altered subunits did not dimerize to yield
the active enzyme (13). Nevertheless, these workers stated:
"There was also formed, however, a significant amount of enzy-
matically active protein in which most arginine residues had been
replaced by canavanine." An earlier study in which either 7-
azatryptophan or tryptazan replaced tryptophan resulted in active
protein comparable to the native enzyme (14).

      More recently, canavanine was incorporated into the pro-
collagen of cultured embryonic tendon cells of the chicken (15).
Canavanyl procollagen was secreted less rapidly and accumulated
intracellularly as compared to the native molecule. It is very
significant that no change in such critical processes as hydroxylation of proline to hydroxyproline, triple helix formation in
collagen, basic glycosylation, or the functional parameters of
the resulting collagen was reported (15). Canavanine is also
placed into pro-opiomelanocortin, a glycoprotein progenitor of
adrenocorticotropin and β-lipotropin. While canavanine affected
prohormone conversion to biologically active components, the
question of impaired biological function of the assembled macromolecule was not addressed (16).

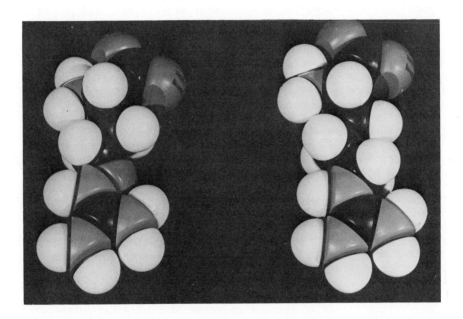

CPK space-filling model of L-*arginine (right) and* L-*canavanine (left).*

As mentioned earlier, canavanine is less basic than arginine, a property that can affect R-group interactions and disrupt tertiary and/or quaternary interactions essential for the uniquely correct three dimensional configuration of a particular macromolecule. Canavanine is incorporated into the proteins of organisms such as insects that do not normally synthesize this compound. All female insects produce vitellogenin—an extraovarial protein secreted by the fat body into the hemolymph and used in the manufacture of vitellin, a major storage protein of the oöcyte (17). Excised fat bodies of the migratory locust, Locusta migratoria migratorioides maintained by in vitro culture techniques retain their ability to produce and secrete vitellogenin at a rate comparable to in vivo activity (18). This system established the formation of canavanyl vitellogenin and permitted demonstration of its altered physicochemical but not immunological properties. Thus, canavanine can be incorporated into de novo synthesized protein which can have a significant effect on the resulting physicochemical properties. The germane question of the biochemical and biological ramifications of formation of an aberrant canavanyl protein has not been properly addressed nor resolved.

Canavanine and Protein Synthesis

Experiments conducted with the tobacco hornworm, Manduca sexta and the aquatic plant, Lemna minor are consistent in finding that canavanine does not affect whole organism ability to incorporate [^{3}H]leucine into trichloroacetic acid-insoluble materials (see Table I). Such determinations evaluate the balance between reactions fostering protein synthesis and those responsible for the turnover and degradation of proteins. If the treatment time is reduced to 30 min, so as to accentuate synthetic reactions over those of catabolism, the result is the same.
Similar evaluations employing an in vitro reticulocyte assay system failed to provide evidence of diminished polypeptide formation (unpub. obser.). Thus, a consistent pattern emerges in these three systems: canavanine does not impede protein synthesis, including aberrant, canavanyl proteins.

Canavanine and Nucleic Acid Metabolism

Studies of canavanine interaction with the tobacco hornworm and L. minor also revealed the marked ability of canavanine to inhibit whole organism incorporation of [^{3}H]thymidine and [^{3}H]-uridine into trichloroacetic acid-precipitated materials. When canavanine is provided simultaneously with the appropriate radiolabeled precursor, ample evidence for curtailed nucleic acid metabolism emerges but protein synthesis is unaffected (Table I, exp. I). In experiment II, canavanine is allowed to assimilate

into the metabolic reactions of the rapidly growing larvae for
24 hrs (at this time, at least 3.5% of the administered canavanine
can be recovered from de novo-synthesized proteins). In experiment II relative to I, there is approximately a 3-fold increase
in canavanine-mediated inhibition of both DNA and RNA metabolism;
a result possibly reflecting the formation of errant proteins
(Table I). Even under this extremely deleterious situation, however, protein synthesis is not altered.

Table I

The Effect of L-Canavanine on Macromolecular Synthesis

In experiment A, newly ecdysed fifth stadium tobacco
hornworm larvae received 5 µCi of labeled compound and 1 mg
canavanine/g fresh body weight. Three hrs later, the treated
larvae were collected and processed. In experiment B, canavanine was injected first and 24 hrs later, the labeled compound
was administered. All larvae were collected and processed 3
hrs later (from the work of Rosenthal and Dahlman).

| Injected compound | Radioactivity in the trichloracetic-acid insoluble fraction[1] | |
	Experiment A	Experiment B
	(% of the control)	
[^{3}H]leucine	96 ± 1	103 ± 4
[^{3}H]thymidine	85 ± 6	57 ± 1
[^{3}H]uridine	76 ± 4	34 ± 5

[1]Each value is the mean of 3 determinations involving 3
larvae per determination ± SE.

Canavanine and Vitellogenin Production

The previously described test system involving in vitro
analysis of fat body vitellogenin production and secretion is
apparently distinctive in providing evidence for canavanine-
mediated curtailment of protein synthesis. At least, canavanine
attenuates the amount of [^{3}H]leucine-labeled protein secreted
by the fat body (Fig. 1). As shown in figure 1, the effect of
arginine depletion itself is debilitating and it becomes increasingly deleterious with time—quite apart from any effect
canavanine may elicit. With increased treatment time, canavanine-

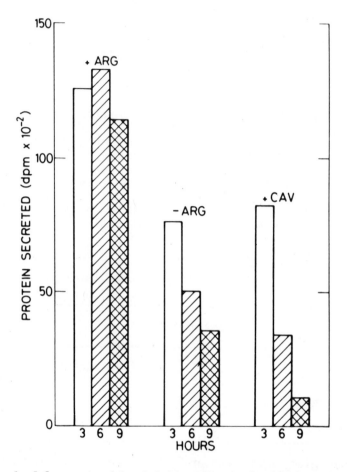

Figure 1. Influence of arginine depletion and canavanine treatment on protein secretion by cultured fat bodies of Locusta migratoria migratoroides. *The fat body samples were maintained in standard incubation medium containing 5 μCi of [³H] leucine and supplemented with 5 mM arginine or 10 mM canavanine or lacking arginine. After 3 and 6 h the fat bodies were transferred to appropriate fresh radioactive medium. The time on the abscissa denotes the hour at the beginning of which the samples were removed for determining protein production. The data presented are typical of several determinations that gave essentially the same results. (Reproduced, with permission, from Ref. 19. Copyright 1981, National Academy of Sciences.)*

exposed fat bodies exhibit greater inhibition in the secretion
of radioactive proteins than their arginine-denied counterpart.
It is possible that what is actually occurring is the ability of
canavanine to block the uptake and utilization of arginine
critical to protein synthesis. Consistent with this assertion,
addition of 5 mM arginine and 10 mM canavanine to the arginine-
depleted fat body reduced the extent of canavanine-mediated
inhibition significantly since protein secretion fell to only
55% of the control level (19). At lower canavanine concentrations,
supplemental arginine was even more effective in blocking the
adverse effect of canavanine. This point is being evaluated
further at this time.

Insectan Canavanyl Protein Production

The female bruchid beetle, Caryedes brasiliensis deposits
her eggs as a cluster on the outside of the fruit of the Central
American legume, Dioclea megacarpa. The seed of this plant is
distinctive in its containment of sufficient canavanine to
account for as much as 13% of the seed dry weight (20) and 95%
of the nitrogen allocated to free amino acid formation (21). If
canavanyl protein production lacks adverse biological effect, it
is reasonable to expect that this seed predator would produce
canavanine-containing proteins. In contrast, if an adaptive
advantage accrues to the developing larva by avoiding the production of canavanyl proteins, then one would expect acquisition
of the ability to avoid errant protein formation.

Examination of this question with the tobacco hornworm, an
insect known to be canavanine-sensitive (this insect normally
feeds on canavanine-free plants) revealed that it readily incorpor-
ates [^{14}C]canavanine into its newly synthesized proteins. Caryedes
brasiliensis, however, very effectively avoids the production of
such radiolabeled proteins. When the arginyl-tRNA synthetase
activity of these insects was compared, tobacco hornworm larvae
readily activated canavanine while the larvae of the bruchid
beetle possess an arginyl-tRNA synthetase with a marked ability
to discriminate between arginine and its structural analogue (22).

Recent study of the arginine activating system of the bruchid beetle disclosed that injection of 7,500 nCi of L-[guani-
dinooxy-^{14}C]canavanine resulted in the incorporation of only
0.95 nCi of radioactive canavanine into newly produced proteins.
This compares with 347 nCi of incorporated L-[guanidino-^{14}C]
arginine from 2,850 nCi of injected material (the difference in
the amount of injected radioactive amino acid reflects the
internal pool size of each compound. In this way, the initial specific activity of both compounds was identical). Nearly
300 times more radioactive canavanine was incorporated into the
proteins of the tobacco hornworm, a canavanine-sensitive organism.
I am presently examining the question of whether bruchid beetle
acquisition of an activating system able to discriminate between

canavanine and arginine, provides broad spectrum resistance to other arginine analogues capable of being incorporated into proteins.

Elucidation of whether or not canavanine affects protein synthesis is a point of considerable significance in our understanding of non-protein amino acid toxicity. All of the available evidence support the ability of arginyl-tRNA synthetase to activate canavanine. This observation is reported consistently for organisms in which this arginine analogue is a foreign compound. On the other hand, canavanine-producing organisms have an arginyl-tRNA synthetase that does not form canavanyl-tRNA^{Arg} and thus maintains a high fidelity in the manufacture of essential proteins. While this point proves nothing, it is consistent with a link between canavanyl protein production and its intrinsic toxicity.

If canavanine placement into proteins is a significant basis for its toxicity and if it lacks the capacity to deter protein synthesis—assertions which are both reasonable and consistent with the available evidence; then these properties would be mutually complementary and reinforcing. It would foster a marked antimetabolic effect in species naive to canavanine. Higher plant allelochemical production of a compound that produces anomalous proteins while scrupulously avoiding diminution in the formation of these aberrant macromolecules certainly represents a subtly deceitful and highly effective form of chemical defense.

Canaline Toxicity

L-Canaline shares with canavanine an appreciable potential for eliciting adverse biological effects. Evaluations with L. minor of some 55 naturally occurring and synthetic amino acids indicated that canavanine and canaline were amongst the most toxic; canaline exhibited slightly greater growth-inhibition than canavanine (23). These toxic non-protein amino acids interact to curtail additively the proliferation of this aquatic plant (24).

Little is known of canaline toxicity in whole animals or plants. Canaline-fed tobacco hornworm larvae grew poorly, exhibited much more deformity, and succumbed in larger numbers than the controls (8). This ornithine analogue is neurotoxic to the adult moth where it induces almost continuous motor activity. Axonal conduction of action potentials was unaffected but the postsyntaptic potential of flight muscle fibers was prolonged. Central nervous system functions were affected but its exact mode of action remains unknown (7).

Canaline and Pyridoxal Phosphate Interaction

Canaline reacts facilely with the aldehyde group of free pyridoxal phosphate to yield a covalently linked complex having the following postulated structure at pH 3.0.   NMR data of the complex are consistent with this structure and assert to a Schiff-base linkage between canaline and pyridoxal phosphate (25).

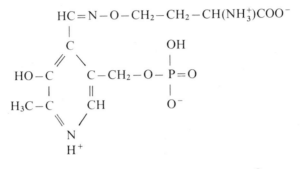

L-Canaline-pyridoxal phosphate complex

The canaline-pyridoxal phosphate complex lacks discernible toxicity as evaluated by the L. minor bioassay, even at a concentration of 10^{-5}M where canaline virtually stops plant growth.

Canaline reacts with the pyridoxal phosphate moiety of enzymes possessing this vitamin; B_6-containing decarboxylases and transaminases typically are inhibited strongly by this non-protein amino acid (25, 26, 27).   When canaline reacts with free pyridoxal phosphate, distinctive changes in the absorbance spectrum of this vitamin result; similar spectral shifts also occur with B_6-containing enzymes treated with canaline (27, 28).   Such alterations in absorbance correlate closely with canaline-mediated inhibition of enzyme activity (Table II).

Table II

L-Tyrosine Decarboxylase Activity Determinations

L-Tyrosine decarboxylase (2 mg/ml, 0.8 ml) in 100 mM sodium acetate buffer (pH 5.5) was treated with canaline as indicated for 30 min. A unit of tyrosine decarboxylase activity is that amount of enzyme forming 1 µmol CO_2/min. See original for additional details.

Enzyme form	Treatment	Activity
		(units/mg)
Holoenzyme	none	0.221
Holoenzyme	0.1 mM canaline	0.178
	0.25 mM canaline	0.116
	0.5 mM canaline	0.071
	1.0 mM canaline	0.023
Apoenzyme	none	0.017
Apoenzyme	5.0 mM pyridoxal phosphate	0.218
Apoenzyme	1.0 mM canaline	0.022

Source: Reproduced with permission from Ref. 25.

Canaline as an Ornithine Antagonist

Several lines of investigation assert to the inability of canaline to function as an effective ornithine antagonist. Ornithine interaction with canaline has been evaluated with the ornithine carbamoyltransferase (EC 2.1.3.3) of human liver. Neither canaline nor ornithine inhibited this enzyme when the other member of this set served as the carbamoyl group recipient (29). The ornithine antagonist, 2,4-diaminobutyric acid drastically reduced urea production in the rat; this reflected curtailment of the ornithine carbamoyltransferase-mediated conversion of ornithine to citrulline. Yet, canaline had no such effect on urea formation in this mammal (30).

Canaline is a potent inhibitor of all seven pyridoxal phosphate-containing enzymes studied by Rahiala et al. (27) but it lacks adverse effects on three ornithine-utilizing enzymes lacking a B_6 cofactor. Finally, in jack bean, Canavalia ensiformis, ornithine carbamoyltransferase can form O-ureido-L-homoserine from canaline and carbamoyl phosphate as it does citrulline from ornithine and carbamoyl phosphate. Nevertheless, neither compound inhibited formation of the reaction products (31). It is evident, therefore, that it is the binding of canaline to the pyridoxal phosphate moiety of the enzyme rather than effective competition with ornithine for the active site that is responsible for the antimetabolic properties of canaline.

Acknowledgment

The author gratefully acknowledges the support of the National Institutes of Health (AM-17322), the National Science Foundation (PCM-78-20167) and the Kentucky Research Foundation for studies by the author described in this communication.

Literature Cited

1. Boyar, A., Marsh, R.E. J. Am. Chem. Soc. 1982, 50, 1995.
2. Rosenthal, G.A. "Plant Nonprotein Amino and Imino Acids. Their Biological, Biochemical, and Toxicological Properties"; Academic Press, New York, 1982.
3. Fowden, L.; Lea, P.J.; and Bell, E.A. Adv. Enzymol. 1979, 50, 117.
4. Rosenthal, G.A.; Bell, E.A. In: "Herbivores: Their Interaction with Secondary Plant Metabolites"; (G.A. Rosenthal and D.H. Janzen, eds.), Academic Press, New York, 1979; p. 353.
5. Bell, E.A. Prog. Phytochem. 1980, 7, 171.
6. Fowden, L. Perspect. Exptl. Biol. 1976, 2, 263.
7. Kammer, A.E.; Dahlman, D.L.; Rosenthal, G.A. J. Exptl. Biol. 1978, 75, 123.
8. Rosenthal, G.A.; Dahlman, D.L. Comp. Biochem. Physiol. 1975, 52A, 105.
9. Rehr, S.S.; Bell, E.A.; Janzen, D.H.; Feeny, P.P. Biochem. Syst. Ecol. 1973, 1, 63.
10. Rosenthal, G.A. Quant. Rev. Biol. 1977, 52, 155.
11. Fowden, L; Lea, P.J. In: "Herbivores: Their Interaction with Secondary Plant Metabolites"; (G.A. Rosenthal; D.H. Janzen, eds.), Academic Press, New York, 1979, p. 135.
12. Fowden, L; Lewis, D; Tristram, H. Adv. Enzymol. 1967, 29, 89.
13. Attias, J; Schlesinger, M.J.; Schlesinger, S. J. Biol. Chem. 1969, 244, 3810.
14. Schlesinger, S. J. Biol. Chem. 1968, 243, 3877.
15. Crine, P.; Lemieux, E. J. Biol. Chem. 1982, 257, 832.
16. Schein, J.; Harsch, M.; Cywinski, A.; Rosenbloom, J. Arch. Biochem. Biophys. 1982, 203, 572.
17. Hagedorn, H.H.; Kunkel, J.G. Annu. Rev. Entomol. 1979, 24, 475.
18. Harry, P; Pines, M; Applebaum, S.W. Comp. Biochem. Physiol. 1979, 63B, 287.
19. Pines, M.; Rosenthal, G.A.; Applebaum, S.W. Proc. Natl. Acad. Sci. 1981, 78, 5480.
20. Rosenthal, G.A.; Janzen, D.H.; Dahlman, D.L. Science 1977, 196, 658.
21. Rosenthal, G.A. Biochem. Syst. Ecol. 1977, 5, 219.
22. Rosenthal, G.A.; Dahlman, D.L.; Janzen, D.H. Science 1976, 192, 256.

23. Gulati, D.K.; Chambers, C.L.; Rosenthal, G.A.; Sabharwal, P.S. Envir. Exptl. Bot. 1981, 21, 225.
24. Rosenthal, G.A.; Gulati, D.K.; Sabharwal, P.S. Plant Physiol. 1976, 57, 493.
25. Rosenthal, G.A. Eur. J. Biochem. 1981, 114, 301.
26. Rahiala, E.L.; Kekomäki, M.; Janne, J.; Raina, A.; Raiha, N.C.R. Biochim. Biophys. Acta 1971, 227, 337.
27. Katunuma, N.; Okada, M.; Matsuzawa, T.; Otsuka, Y. J. Biochem. (Tokyo) 1965, 57, 445.
28. Beeler, T.; Churchich, J.E. J. Biol. Chem. 1976, 251, 5267.
29. Natelson, S.; Koller, A.; Tseng, H.-Y.; Dods, R.F. Clin. Chem. 1977, 23, 960.
30. Kekomäki, M.; Rahiala, E.-L.; Raiha, N.C.R. Am. Med. Exp. Fenn. 1969, 47, 33.
31. O'Neal, T.D. Plant Physiol. 1975, 55, 975.

RECEIVED August 23, 1982

Cytotoxic and Insecticidal Chemicals of Desert Plants

ELOY RODRIGUEZ

University of California, Phytochemical Laboratory, Department of Ecology and Biology, Irvine, CA 92717

A diverse group of natural chemicals are produced by arid land plants. These natural products, which are stored in glandular hairs (trichomes), include terpenoids, alkaloids, phenolics and amines. Some of these compounds coat the leaf and stem surfaces and prevent water loss through the cuticle and probably protect the plant from excessive damage by radiation. Another important ecological role of secondary metabolites is that of defense against phytophagous insects and pathogens. Recent phytochemical investigations in our laboratory indicate that chromenes, prenylated quinones and sesquiterpenes esterified with phenolic acids, are excellent repellents and in some cases are cytotoxic and inhibit larval growth and development. The chemistry and role of these cytotoxins and insecticides in desert plants of Baja California and Chihuahua is reviewed.

Deserts cover approximately one-seventh of the earth's land surface, with the North American deserts populated by a diversity of bizarre plant life forms ranging from cacti to creosote bushes to the boojum trees of Baja California, Mexico. An outstanding characteristic of a majority of plants that dominate the desert landscape is their enormous photosynthetic capacity to produce an array of secondary metabolites. Many of these natural products are essential to the everyday survival of plants exposed to the harsh, hot and dry environment of the desert. These natural products, which include terpenoids, alkaloids, phenolics, amines and tannins are produced in large quantities by specialized glandular hairs called trichomes and coat leaf, stem and flower surfaces. This thin layer of secondary chemicals are believed to prevent water loss (antidessicants) through the cuticle and in some cases prevent excessive cell damage by blocking ultraviolet

0097-6156/83/0208-0291$06.00/0

radiation. More important, these biologically active constituents
also repel and in some cases kill phytophagous insects and plant
pathogens. It is the desert allelochemicals (constituents which
have a deleterious effect on herbivorous insects and pathogens)
which are of considerable interest to phytochemists concerned
with controlling insect pests of crop plants. With the current
agricultural development of desert hydrocarbon crop plants, such
as guayule (Parthenium argentatum) and jojoba (Simmondsia
chinensis), it is important that a line of natural insecticides
and repellents be developed against insects especially adapted
to plants cultivated in marginal arid lands (1).

     In this communication, we present our latest phytochemical
findings on prenylated quinones and chromenes that are cytotoxic
and inhibit larval growth and development. Many of these
bioactive chemicals are unique to desert plants and have probably
been selected against desert invertebrates that feed on plants.

Desert Plants and Their Secondary Metabolites

     As previously noted, desert plants are no different from
temperate plants in producing a wide-range of secondary
metabolites (2). Dominant plant species such as Larrea tridentata
(Zygophyllaceae) may contain up to 15% of the dry weight in
lignans, flavonoids, chromenes, triterpenes and volatile
terpenes (3). Important desert families include the Asteraceae,
Fabaceae, Agavaceae, Burseraceae, Euphorbiaceae and Fonquieriaceae
which synthesize diverse products such as sesquiterpene lactones,
methylated flavonoids, polyacetylenes, steroids, saponins,
alkaloids, cyanogenic glycosides, aromatic terpenes and amines
(refer to Figure 1 and 2 for representative structures). These
natural products are not only used by native peoples for
medicines, food and natural antibiotics, but many of these
compounds are very effective against a host of insects and
pathogens (4). In our studies we have concentrated on prenylated
quinones and chromenes that are present in desert species of the
Asteraceae and Hydrophyllaceae. In the ensuing paragraphs we
summarize the chemistry and biological effects of a selected group
of quinones and chromenes that are cytotoxic and insecticidal
to milkweed bugs (Oncopeltus) and mealworm beetles (Tenebrio).

Prenylated Quinones

     Quinonoid compounds are quite common throughout the plant
kingdom, but only in a few cases have they been reported in
glandular hairs (trichomes) of desert herbaceous plants. Many of
these quinonoid compounds are extremely active and are the major
cause of allergic reactions in humans. For example, Primula
obconica (Primulaceae), an ornamental plant common to Europe,
produces a quinone which is stored in small non-capitate
trichomes which release their contents when touched. These

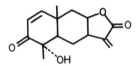

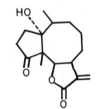

Flourensia Acid
Sesquiterpene
Flourensia cernua

Farinosin
Sesquiterpene Lactone
Encelia farinosa

Coronopilin
Sesquiterpene Lactone
Ambrosia dumosa

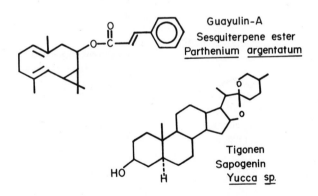

Guayulin-A
Sesquiterpene ester
Parthenium argentatum

Tigonen
Sapogenin
Yucca sp.

Figure 1. Terpenoid constituents of desert plants.

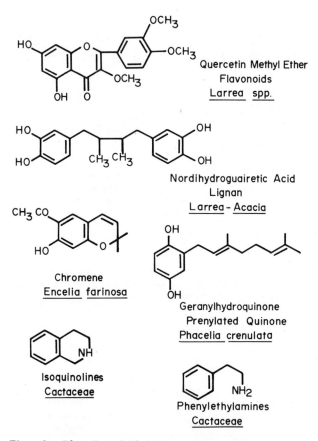

Figure 2.   *Phenolic and alkaloid constituents of desert plants.*

secreted chemicals are very effective against aphids that c rawl
on the leaf. These chemicals, like the secretion of Solanum
tuberosum (Solanaceae), hardens on mouth parts and tarsi of
insects and inhibits movement and feeding (5-6).

Recently we have isolated a series of prenylated quinones
and quinonoid compounds from the viscid capitate-glandular
trichomes of Phacelia (Hydrophyllaceae) that are cytotoxic,
allergenic and insecticidal. Most of the species noted to have
the prenylated quinones are restricted to the Sonoran desert of
Baja California, Mexico and the Mojave desert of the Southwest
(7-10). Phacelia crenulata, an annual which is common along
desert roadsides in the southwestern United States and northern
Mexico, is responsible for a dermatitis similar to that of
poison oak, but is is also been demonstrated to inhibit larval
formation. The effects are similar to those noted for milkbugs
(Oncopeltus) treated with the precocenes, compounds well-noted
for their anti-juvenile activity (11). The principal
constituents are geranylhydroquinone (I) and in lesser amounts
geranylbenzoquinone (II)(8). Both compounds are new natural
products to higher plants, but have been reported in marine
urochordates (12). Compound (I) has been synthesized as a
drug and found to have cancer preventive properties in
experimental animals (13). Another species, P. ixodes from Baja
California, Mexico contains numerous quinones in the trichomes
(10). The compounds have been identified as geranylhydroquine
(I), 3-geranyl-2,5-dihydroxyphenyl acetate (III),
geranylbenzoquinone (II), 2-geranyl-6-hydroxy-4-methoxyphenyl
acetate (IV), 2-geranyl-4-hydroxyphenyl acetate (V) and
6-hydroxy-2-methyl-2,4-methyl-3-pentenyl)-chromene (VI).
Compounds (I) through (IV) were assayed for their cytotoxic
and allergenic potential on guinea pigs, which are effective
indicators of contact allergenicity in humans (10). The same
compounds were also tested on Tenebrio sp. (mealworm beetles),
an experimental insect used to test the potential of insecticides.
Geranyl-benzoquinone proved to be a very potent elicitor of
allergic skin reactions as well as a potent insecticide (10).
A topical application of 100μg of geranylbenzoquinone on pupae
of Tenebrio caused severe abnormalities and death. Although
the exact mechanism of toxicity for (II) is not known, its
similarity to the juvenile hormones suggests that (II) might
be a powerful alkylator of enzymes regulating juvenile hormone
synthesis. Compound (VI) was not as potent as (II), but previous
reports of (VI) in Cordia alliodora (Boraginaceae) indicate that
(VI) is also an effective insecticide (14).

Two species of Phacelia, P. minor and P. parryi, widespread
throughout the semi-arid mountains of southern California
contained as the major constituent geranylgeranylhydroquinone
(VII). A minor constituent, 2-(-1-oxofarnesyl)-hydroquinone
(VIII) was also present and previously reported in a brown alga
(15) and Wigardia kunthii of the family Hydrophyllaceae (16).

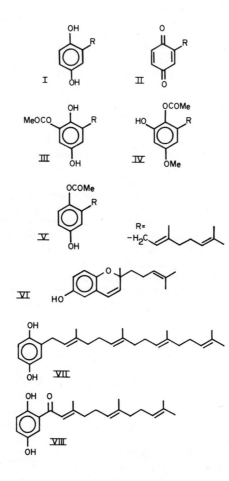

Geranylgeranylhydroquinone (VII) was applied (100μg) to fifth
instars of Oncopeltus (milkbugs) and found to stunt growth and
wing development.

It is apparent from our investigations of prenylated
quinones from desert annuals, that these compounds are not only
potent skin sensitizers but also potential insecticides. The
toxicity of the prenylated quinones is in part due to their
lipophilicity which permits the compounds to penetrate the
cuticle and alkylate important metabolic enzymes. Another
possible explanation for their antijuvenile action is that the
prenylated quinones are transformed into a quinone methide
species which mimic the juvenile hormones (JH) and possibly
interfere with the production of JH or deactivate the hormones.

Chromenes in the Asteraceae

In a recent literature survey of chromenes and benzofurans
in flowering plants, we have documented that approximately 90%
of the 200 compounds isolated are present in the sunflower
family (Asteraceae). These naturally occurring chemicals have
recently received considerable attention because of their potent
cytotoxic and insecticidal activity. Chemotaxonomically, not all
tribes of the Asteraceae seem to produce chromenes or benzofurans,
with the major tribes capable of synthesizing chromenes identified
as the Eupatorieae, Heliantheae, Inuleae, Senecioneae and the
Astereae (17).

The chromenes and benzofurans are rather simple compounds
built from acetate and isoprene metabolites. Heterocyclic ring
formation gives rise to 2,2-dimethyl chromene or 2-isoprophenyl
benzofurans. The majority of known chromenes and benzofurans
exhibit a methyl ketone moiety at a position para to the oxygen
of the heterocyclic ring. Constituents esterified with phenolic
acids or lacking methyl ketones are rare.

The chromenes are well-known for their cytotoxic and
antijuvenile activity in insects (18-19). The prococenes, simple
chromenes first isolated from Ageratum (Asteraceae), have been
shown to act directly on the corpus allatum by direct cytotoxic
destruction of the parenchymal cells (20). The precocenes, when
applied externally to the second instar larvae of the milkweed
bug (Oncopeltus fasciatus) cause the nymphs to molt to normal
third and fourth instar larvae, and then to precocious adults.
Recent biochemical studies have suggested that the precocenes
undergo oxidative activation with the corpus alluatum and form
reactive epoxides that alkylate nucleophilic substrates. The
reactive precocene intermediate is suggested to be the quinone
methide (20), a similar reactive species that we have proposed
for the prenylated quinones of Phacelia.

Although considerable information has been gathered on
cytotoxic and antijuvenile action of the precocenes, little
information is available on the distribution of chromenes and

benzofurans in desert plants and their possible role as
phototoxic agents and feeding deterrents. With this in mind, we
have begun a detailed phytochemical investigation of chromenes
and benzofurans in dominant desert shrubs of the Sonoran and
Chihuahuan deserts that are cytotoxic, phototoxic and
insecticidal.

Chromenes and Benzofurans in Desert Shrubs

Using HPLC, we have surveyed in detail taxa of Encelia,
Flourensia, and Geraea. All of these genera were found to
produce large quantities of chromenes and benzofurans (21). In
the genus Encelia, chromenes can comprise up to 10% of the dry
weight material. The compounds are more common in leaves, but
have been detected in the stem, flowering heads and in some cases
in seeds.

A detailed quantitative study of the chromenes has been
conducted on Encelia farinosa (brittlebush), a desert shrub common
throughout Baja California, Sonora and southern California. The
major chromene was identified as encecalin (IX), but also present
was the benzofuran euparin (X) which is primarily produced in the
stems (22). Application of encecalin to the first instar of
larvae of the milkweed bug proved to be moderately insecticidal.
Encecalin, euparin, and 7-hydroxyencecalin (XI) were dissolved
in methanol and were applied at concentrations 5 mg-100μg. Each
petri dish contained 25-30 first instar larvae of O. fasculatus.
Concentrations of 1.2 mg/petri dish (or higher) of encecalin
were lethal to the larvae within a period of three days. Lesser
concentrations of encecalin showed no effects, while euparin and
7-hydroxyencecalin was not toxic (23). Encecalin does not
compare in toxicity with the precocenes, since 44 μg of precocene
II have been reported to induce precocious metamorphosis in
milkweed bug larvae. The presence of a methylketone moiety in
encecalin instead of a methoxy substituent probably results in a
loss of antijuvenile activity noted for the precocenes. On the
other hand, 7-hydroxyencecalin was less active than encecalin and
this could be due to a more rapid detoxification of phenolic
compounds exhibiting free hydroxyl groups rather than methoxyl
groups (23).

As a feeding deterrent, encecalin was more active. Fifth
instars of Heliothis zea (Lepidoptera) were exposed to artificial
diets containing varying amounts of encecalin. At concentrations
of 0.35%, H. zea starved to death (24). It should be noted that
encecalin is present in higher amounts in the leaves, therefore
suggesting that encecalin and other less cytotoxic chromenes and
benzofurans are feeding deterrents. It should be added that most
desert phytophagous insects either chew or suck plant parts and
therefore are likely to be repelled before they consume the
chromenes. Topical applications are less likely, but one could
speculate that the accidental rupturing of glandular hairs could

IX

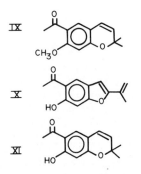

X

XI

XII

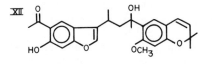

XIII

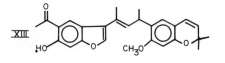

XIV

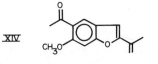

XV

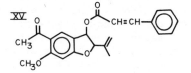

result in the deposition of active cytotoxins on an insect
cuticle.

Analysis of the stems of Encelia ventorum from Baja
California showed the presence of a number of benzopyran
and benzofuran derivatives and two stereo isomers of a
novel euparin-encecalin (XII) dimer (25). A closely related
euparin-encecalin dimer (XIII), was previously isolated from E.
farinosa (26). Although the dimer is not photoactive,
preliminary studies indicate that it is a feeding deterrent to
Heliothis zea.

The benzofuran 6-methoxyeuparin (XIV) and the two chromenes
encecalin and 7-hydroxyencecalin from species of Encelia
(Asteraceae) from Baja California have been shown to be phototoxic
to several bacteria and yeast in long wave UV light (23).
Compound (XIV) was the most activity against Pseudomonas
fluorescens, an organism that is not affected by the potent
photosensitizer 8-methoxypsoralen (27). The three compounds were
active against Saccharomyces cerevisiae and Candida albicans.

Preliminary experiments with human erythrocytes with this
new class of photosensitizers rules out the membrane as a target,
since the chromenes seem to behave like the photosensitizing
furanocoumarins by interacting with nucleic acids or intracellular
molecules in light. Further experimentation is needed to clearly
understand their mode of action.

Numerous chromenes and benzofurans have also been isolated
from Flourensia, a genus that is dominant in the desert of
Chihuahua, Mexico. A benzofuran (XV) with a cinnamic acid moiety
from F. dentata and F. ilicifolia has recently been shown to be
extremely toxic to milk weed bugs and highly phototoxic (28).
Approximately 30 chromenes and benzofurans have been isolated
from other desert sunflowers and we are currently screening them
for insecticidal and phototoxic activity.

Concluding Remarks

Desert plants are remarkable phytochemical factories. In
this chapter, we have covered only two classes of compounds;
chromenes and prenylated quinones that are allergenic, cytotoxic
and insecticidal. Many other desert plants produce resins that
are complex mixtures of sesquiterpenoids, chromenes, flavonoids
and quinones. These mixtures might not be of interest to
phytochemists, but to the plant, secondary metabolites are
essential for survival and reproduction. Like the chromenes and
quinones, many natural constituents of arid land plants play a
dual role. In some cases, the natural products excreted on the
leaf or secreted by trichomes are functioning as antidessicants
(prevent water loss through the cuticle), protecting the leaf
from harmful radiation and, most important, keeping the plant
healthy against phytophagous insects and pathogens. The chromenes
and benzofurans are chemicals that are antifeedants and exhibit

antijuvenile activity when applied topically. The chromenes are primarily restricted to members of the Asteraceae and are more widespread in desert sunflower species than previously thought. In combination with sesquiterpene lactones, the chromenes and benzofurans are another group of defensive compounds that desert invertebrates have to detoxify. Indeed, the success of many desert sunflower species is in part due to their diverse secondary chemistry. Prenylated quinones, on the other hand seem to be restricted to the Hydrophyllaceae and tropical trees. The quinones are extremely active chemicals that exhibit insecticidal activity at concentrations lower than many chromenes. The prenylated quinones of Phacelia are also potent cytotoxins and allergens.

Acknowledgements

This research has been supported by NIH Grant AI 18398, NSF PCM-8209100 and the Focused Research Program at UCI. I am greatly indebted to my research associates and colleagues cited in the references, namely, Dr. G.H.N. Towers (UBC, Canada), Dr. Gary Reynolds (LSU), Dr. Peter Proksch (UCI), Charles Wisdom (UCI), Manuel Aregullin (UCI) and Margareta Proksch (UCI).

Literature Cited

1. Rodriguez, E. "Aspects of American Hispanic and Indian Involvement in Biomedical Research." Martinez, J.V. and Marinez, D.I. Eds. SACNAS, Bethesda, Maryland. 1981, p. 244.
2. Campos-Lopez, E.; Roman-Alemany, A. J. Agric. Food Chem. 1980, 28, 171-183.
3. Mabry, T.J. and Ulubelen, A. J. Agric. Food Chem. 1980, 28, 188-196.
4. Rosenthal, G.A.; Janzen, D.H. "Herbivores: Their Interaction with Secondary Plant Metabolites."; Academic Press: New York, 1979, p. 604.
5. Kelsey, R.; Reynolds, G.; Rodriguez, E. "Biology and Chemistry of Plant Trichomes." Healey, P., Mabry, T. and Rodriguez, E. Eds. Plenum Press: New York, in press.
6. Gibson, R.W. Ann. Appl. Biol. 1976, 68, 113-117.
7. Rodriguez, E. Rev. Latinoamer. de Quimica 1978, 9, 125-131.
8. Reynolds, G.: Rodriguez, E. Phytochem. 1979, 18, 1567-68.
9. Reynolds, G.; Epstein, W.L.; Terry, D.; Rodriguez, E. J. Contact Dermatitis 1980, 6, 272-274.
10. Reynolds, G.; Rodriguez, E. Planta Medica 1981, 43, 187-193.
11. Bowers, W.S. "The Juvenile Hormones" Gilbert, L.I. Ed. Plenum: New York, 1976, pp. 394-408.
12. Fenical, W. "4th Proceeding of Food Drugs from the Sea." Marine Technological Society: Washington D.C., 1974.
13. Rudali, P.G.; Menetrier, L. Therapie 1967, 22, 895-899.

14. Jurd, L.; Manners, G.D. J. Agric. Food Chem. 1980, 28, 183-188.
15. Ochi, M.; Kotsuki, H.; Inoue, S.; Taniguchi, M.; Tokoroyama, T. Chem. Lett. 1979, 831-833.
16. Gomez, F.; Quijano, L.; Calderon, J.S.; Rios, T. Phytochem. 1980, 19, 2202-2203.
17. Proksch, P.; Rodriguez, E. Phytochem. 1982, in press.
18. Bowers, W.S.; Ohta, T.; Cleere, J.S.; Marsella, P.A. Science 1979, 193, 542-548.
19. Bowers, W.S. Am. Zool. 1981, 21, 737.
20. Bowers, W.S. Science 1982, 217, 647-648.
21. Proksch, P.; Rodriguez, E. J. of Chroma. 1982, 240, 543-546.
22. Wisdom, C.; Rodriguez, E. Biochem. System. and Ecology, 1982, 10, 43-48.
23. Proksch, P.; Proksch, M.; Towers, G.H.N.; Rodriguez, E. J. of Natural Products, 1982, in press.
24. Wisdom, C. Ph.D. Thesis, University of California, Irvine, CA, 1982.
25. Proksch, P., Aregullin, M. and Rodriguez, E. Planta Med., 1982, in press.
26. Steelink, C.; Marshall, G.P. J. Org. Chem.: 1979, 44, 1429-1433.
27. Towers, G.H.N.; Graham, E.A.; Spenser, I.D.; Abramowski, Z. Planta Med. 1981, 41, 136-138.
28. Aregullin, M.; personal communication.

RECEIVED September 16, 1982

Role of Lipids in Plant Resistance to Insects

DAVID S. SEIGLER

University of Illinois, Department of Botany, Urbana, IL 61801

In the coevolutionary interactions of plants and animals, lipids play a major role. They function as ecomones (pheromones, allomones and kairomones) and have been classified by their function. Host plant resistance is partially dependent on these chemical constituents. Lipids may be subdivided into two types. Volatile lipids are generally involved in long distance interactions whereas non-volatile lipids are generally involved after the insect has contacted the host plant. Several examples of each are reviewed. Utilization of these compounds to promote increased host plant resistance could be accomplished by selection of plants rich in allomones, lacking kairomones for a particular pest or those with inducible systems of defense. Another approach is to isolate the defensive compounds of one plant and apply them to crop plants. Trap crops could also be used to lure insects away from other crops.

Despite the fact that a majority of insects are phytophagous (i.e. they eat plants), the world around us is still green (1,2). Of approximately 1,000,000 known insect species, only a few thousand are "pests" and of these only about 500 cause appreciable damage (3). Although many factors are involved in maintaining this balance between plants and insects, plant secondary compounds are generally conceded to play a major role.

Coevolution of Plants and Insects

In order to understand fully the importance of these chemical factors, it is necessary to consider the processes which are probably responsible for their diversification. A mechanism of coevolution of insects and plants was set forth eloquently by Erlich and Raven (4). According to their hypothesis, angiosperms produced a series of chemical compounds which were not directly related to their basic (or primary) metabolic pathways, but which were otherwise not harmful to the plants growth and development. In practice, these compounds may play other roles such as interactions within plants as primary compounds (5-8) or in interactions

0097-6156/83/0208-0303$07.25/0

with other organisms (plant-plant, plant-fungus, plant-bacterium, etc.). Functions such as these almost certainly predated the roles of these secondary compounds in plant-insect interactions, and it is clear that repellancy and attraction cannot be the sole raison d'être of these substances (9,10).

Fortuitously, some of the compounds may have reduced the palatability of the plants in which they were produced and/or in some manner reduced the fitness of the insects which normally ate the plant, and then these plants entered a new adaptive zone. Evolutionary radiation of the plants might then have followed and eventually what began as a chance mutation, might ultimately have characterized an entire family or group of families (4).

In a similar fashion, if a recombinant or mutation occurred in a population of insects that enabled individuals to feed on some previously protected group of plants, selection would carry that line into a new adaptive zone (4). Here the new group would be able to diversify with little competition from other phytophagous insects. All in all, the diversity of plants would tend to augment the diversity of insects and the diversity of phytophagous insects would tend to enhance the diversity of plants.

The situation is more complex than that proposed as no insect nor any plant evolves with regard to only one other organism (11, 12). Any change which occurs must not modify unfavorably the organism's overall fitness with regard to other organisms, physical and environmental factors and the organisms own developmental sequence and physiological processes. The success or failure of an organism is rarely ascribable to any single factor.

As an insect became more highly adapted to a particular food source, recognition of the food plant by visual, chemical or other means would be strongly favored.

The secondary compounds involved in these coevolutionary processes have been called "allelochemics" (13) and "ecomones" 14,15). Three major classes have been defined: pheromones, which are intra-specific chemical messengers, and allomones (compounds which are deleterious to the receiving organism) and kairomones (compounds which are beneficial to the receiving organism) (13,16,17). Blum (15) suggests that the term "ecomone" be used for all external chemical messengers as was proposed by Florkin (14) and that terms other than pheromone (intraspecific messengers) and allomone (interspecific messengers) be abandoned as many individual compounds may at the same time serve in different roles among various organisms. By explicitly stating the function of the allomone in the particular case analyzed, its idiosyncratic role can be stressed without obscuring any other functions that it may possess. The potential use of pheromones for control of insect pests has recently been reviewed (18). Allomones are those substances involved in interspecific communication which give an adaptive advantage to the producing organism and kairomones are compounds which give an adaptive advantage to the receiving organism. Kairomones are often allomones to which

insects have become coadapted (19), but may have other origins
(20).

For example, gossypol, a terpenoid substance, is an allomone
that limits herbivory by several lepidopteran species on cotton
(Gossypium spp., Malvaceae), whereas it is a kairomone for the
boll weevil, Anthonomus grandis and acts as a feeding stimulant
for this insect (3,21,22).

The means by which many insects select a suitable host plant
is by being attracted by plant secondary compounds that also serve
as allomones. In other cases the insect may avoid the presence of
a toxic compound by the correlated presence of other materials
which may be repellant.

Some compounds such as gymnemic acid from Gymnema sylvestre
(Asclepiadaceae) (which depresses the preceived sweetness of
sugars) are known to distort the taste of others (23) and may play
a role in "disguising" the presence of kairomones or nutritional
substances.

It has been suggested that insects were primitively polyphagous herbivores (24,25) although others (9) have suggested that
many primitive insects were saprophagus feeders. In either case,
in the process of coevolution, there has been a tendency toward
specialization (2,26) and today, most insects are monophagous or
oligophagous. Many apparently polyphagous groups are comprised of
closely related subtaxa which are each monophagous (1). Factors
other than plant chemistry are also sometimes important in the
evolution of biochemically specialized insects (27).

The process of host selection by an insect is complex and
involves five major steps: host habitat finding, host finding,
host recognition, host acceptance, and host suitability (25).
Each step in the process of host-plant selection may be mediated
by plant components. Both the secondary chemistry and nutritional
value play a major role in the suitability of the host.

About 85% of all insects are holometabolous, and the food
plants of larval and adult stages often differ (28). In most holometabolous insects host recognition may have been predetermined
for the larva by the ovipositing female (1). In some cases it has
been demonstrated that the larva prefers to eat the food upon
which they are initially fed (induction), even if it's not the   \
appropriate host (29).

Mechanisms of Host Plant Resistance

For an insect to successfully utilize its host plant, it must
complete each of the five aforementioned steps. Any circumstances
which preclude this will convey an advantage to the host plant.
All components must be present at the proper time and in adequate
amounts. Plant resistance may result from disruption of the
normal sequence of events or reduced presence of kairomones or
enhancement of allomones (25). This resistance is under genetic
control but may vary with environmental conditions; factors
involved in ecological resistance have been reviewed (25).

Three principal types of genetic resistance have been proposed: preference or non-preference, antibiosis, and tolerance (30).

A number of resistance factors influence behavioral processes and hence determine an insect's preference or non-preference for a particular plant. Hosts which do not contain the proper kairomonal compounds are often totally rejected as food plants and by ovipositing females. Dethier (29) noted, however, that plants are almost never neutral, but are almost always either attractive or repellant. As previously observed, the ovipositional choice of the female imago and the food choice of the larvae usually coincide (25).

The presence of kairomones is usually involved in the selection of a food plant. Resistant plants usually lack or have too little of the normal kairomones, the kairomones are inhibited or blocked by antagonistic compounds or only allomones are present (25,31).

Adverse physiological effects (antibiosis) resulting from ingestion of a plant by an insect may range from mild to lethal; the principal symptoms are death of the larvae in the first few stadia, abnormal growth rates, abnormal conversion of ingested and/or digested food, failure to pupate, failure of adult emergence from the pupae, malformed or subsized adults, failure to concentrate food reserves followed by unsuccessful hibernation, decreased fecundity, reduced fertility, and restlessness or other irregular behavior (25). These antibiotic effects may be caused by several factors, one of which is the presence of toxic metabolites (see reference 25 for a more complete list). Antibiosis is often the most evident mechanism of resistance.

Morphological or structural plant features which impair normal feeding or oviposition by insects or contribute to the action of these mortality factors are often grouped as "phenetic resistance" (25).

Certain plants can repair injury or produce an adequate yield despite supporting an insect population at a level capable of damaging a more susceptible host. Several factors are important in tolerance (see reference 25).

The role of other plants in the community and of other trophic levels in resistance has been emphasized (11,12).

Insect Resistance in Crop Plants

Resistance to insects has been successfully selected and introduced into important cultivars of a number of plant species including potatoes, wheat, corn, grapes, alfalfa, barley, beans sorghum, rice and sugar cane (25). This resistance is usually due to a combination of factors and only in a few cases is a single chemical factor identifiable as responsible for resistance (32).

In general, decreased pest resistance has occurred in the

domestication of crop plants. This factor and the fact that many
crops are "apparent" or "predictable" has increased the levels of
herbivory on them (33).
     There are a number of possibilities for using plant secondary
chemistry to control herbivory in crop plants. One possibility is
to select for insect resistant lines and though it has been done
in only a few cases, select for specific allomones. There are,
however, some potential problems with this approach. There is a
cost for the production of the secondary compounds which may be
useful for defense (33). Insect resistant soybean cultivars produce lower yields of seeds and accumulate nitrogen at a slower
rate than insect susceptible varieties in the absence of herbivores
(34). Conversely, varieties of crop plants selected for high yield
are often more susceptible to insects, pathogens, and weeds (35).
     Several trophic levels must be considered. Breeding plants
with greater allomone content in some cases causes specialist
herbivores to accumulate higher levels of these compounds and
discourages parasites that normally control herbivore levels (36).
The presence of secondary compounds may also alter the usefulness
of the crop plant to man or his domestic animals. Lines of cotton
with high gossypol content have increased insect resistance with
regard to a number of insects, but have reduced value as food
materials for livestock.
     In the plant, part of the metabolic cost of producing and
maintaining pools of secondary compounds may be reduced by using
compounds which contain only carbon, hydrogen and oxygen (which
are rarely limiting), by recycling the compounds or by use of
inducible systems of defense.
     Inducible systems of defense (phytoalexins) are widespread in
plants and are effective against many types of fungi and bacteria
(37,38). Similar systems have been demonstrated in a few cases
with insects and are probably common in nature (see for example
reference 39). Although inducible systems of insect resistance
would seem to be efficient and effective, no system is foolproof.
The larvae of Epilachna tredecimnotata cut a circular trench in
Cucurbita leaves and prevent mobilization of the deterrent substances to the area which is then consumed (40).
     It should be possible to create "trap crops", preferentially
attract the insects (via kairomones), destroy them and thus
protect the desired crop (25).
     Another approach may be to isolate the allomones from one
plant and apply them to the surface of another plant and thus
protect it (25,31). The practicality of this approach depends on
many factors including residual toxicity, cost, and stability of
the compounds involved.
     Several lists of plant secondary compounds which are involved
in plant-insect interactions have been compiled (1,9,13,17,31,41-
46). In addition, lists of plants with insecticidal properties
have been published (47,48). Many of these compounds are physi-
cally located on the outside of the plant, either in the epidermal

layer, in trichomes or glands or on the outer surface. Others are
located in flowers, seeds and fruits (1,33,44,49-51).

Table 1. Principal Classes of Chemical Plant Factors (Allelochemics) and the Corresponding Behavioral or Physiological Effect
on Insects.

Allelochemic Factors	Behavioral or Physiological Effects
Allomones	Give adaptive advantage to the producing organism
Repellents	Orient insects away from plant
Locomotor excitants	Start or speed movement
Suppressants	Inhibit biting or piercing
Deterrents	Prevent maintenance of feeding or oviposition
Antibiotics	Disrupt normal growth and development of larvae; reduce longevity and fecundity of adults
Antixenotics	Disrupt normal host reduction behavior
Kairomones	Give adaptive advantage to the receiving organism
Attractants	Orient insects toward host plant
Arrestants	Slow or stop movement
Feeding or oviposition excitants	Elicit biting, piercing, or oviposition; promote continuation of feeding

Source: Reproduced with permission from Ref. 134.

A system of classification (based largely on the previously
proposed systems of Dethier et al. (52), and Beck (53), and
Whittaker and Feeney (16)) of the major types of chemical factors
involved in plant-insect interactions has been proposed (25)
(Table 1). Unfortunately, it is often difficult to judge from
literature data, into which class of allomone or kairomone a
particular compound should be placed (see also reference 15).
Compounds which are referred to as "antifeedants" in the literature are involved in the inhibition of biting and piercing and
inhibition maintenance of feeding. These compounds generally do
not kill the insect but cause starvation because they prevent
feeding (54).
     Many of the plant secondary compounds involved in plant-
insect relations are lipids. This group is not of a single biosynthetic origin but is comprised generally of compounds soluble
in non-polar solvents. Among the common groups of lipids found in
plants are fatty acids and their derivatives (hydrocarbons, aldehydes, alcohols, esters, glycerides, acetylenic compounds, waxes,
etc.), phenylpropanoids (flavonoid aglycones, lignans, coumarins

etc.), and terpenoid derived compounds (monoterpenes, sesquiterpenes, diterpenes, triterpenes, steroids, tetraterpenoids or
carotenoids). Although many alkaloids are lipids, they are not
discussed in this review. Lipids are often localized on the surface of the palnt or produced by special glands or trichomes (50).
Other compounds of this type serve as energy reserves within the
plant and are especially important to both plants and herbivores.
Lipids may be subdivided into two general groups: volatile and
non-volatile lipids. There is considerable overlap between the
two groups.

In the following sections, recent examples from the literature
have been cited. They are by no means exhaustive. Many previously
described examples are cited in earlier works, especially that of
Hedin (41).

Volatile Lipids

Volatile compounds are often involved in long distance
attraction and are especially important as attractants and repellents (as defined by Kogan, 25). One major class of volatile
materials, essential oils, is comprised of complex mixtures of
terpenes, phenylpropanoid derived compounds and a number of esters,
alcohols, aldehydes, ketones, acids, and hydrocarbons. The constituent compounds are mostly of low to medium molecular weight
and generally not highly oxygenated. Some of the biological
properties of these compounds have been reviewed (17,41,46,55,56).

Many of the monoterpenes found in essential oils of plants
also occur as pheromonal substances in insects (45,57-60) and are
often involved in plant-insect interactions. Some compounds found
both in plants and insects are the monoterpenes citronellal, citronellol, geraniol, myrcene, citral, β-phellandrene, limonene,
2-terpinolene, α-pinene, β-pinene, 1,8-cineole, and verbenone.

Oil of citronella (the essential oil of Andropogon nardus,
Poaceae or Gramineae) has long been used as a mosquito repellent.
This oil is mostly composed of geraniol with lesser amounts of
citronellol, citronellal, and borneol. Other essential oils have
also been used to repel insects (49). A number of monoterpenes
and methyl esters of fatty acids were evaluated for their repellent and attractant properties toward Ips, Dentroctonus, and
Hylurgops species. Although activity was observed in the laboratory, none of the compounds tested appeared active in field tests
(61). Bay leaves (Laurus nobilis, Lauraceae) and cineole,
geraniol, and piperidine (which occur in bay leaves) possess
repellent properties toward cockroaches (62). Essential oils are
involved in the feeding response of several Papilio species to
members of the Apiaceae (Umbelliferae) (9). The aphid Cavariella
aegopodii which lives on members of the Apiaceae in summer, is
attracted to the monoterpene carvone but less so in the presence
of linalool which has a repellent effect (63). (+)-3-Thujone and
(-)-3-isothujone, which make up most (80-90%) of the leaf essential

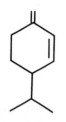 

β-phellandrene   limonene   2-terpinolene   α-pinene

 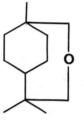

β-pinene   1,8-cineole   verbenone

oil of <u>Thuja plicata</u>, western red cedar, are feeding deterrents to
the white pine weevil (<u>Pissodes strobi</u>) (<u>64</u>).

Essential oils are known to inhibit microbial activity in
ruminants and disrupt digestive processes (<u>65</u>).

Some volatile iridoid monoterpenes with biological activity
are also found in essential oils and in insect pheromonal and
defensive substances. Eisner (<u>66</u>) found that 17 species of insects
were repelled by the iridoid monoterpene nepetalactone. Lacewings
(<u>Chrysopa septempunctata</u>) are attracted by the leaves and fruits
of <u>Actinidia polygama</u> (Actinidiaceae) which contain a series of
volatile iridoid monoterpenes (<u>67</u>).

Many sesquiterpenes which are not highly oxygenated are also
found in essential oils. Several of these are reported to possess
activity. α-Farnesene from apples is an attractant and oviposition
stimulant for the codling moth (<u>68</u>) and farnesol has been demonstrated to be an active feeding deterrent to gypsy moth larvae
(<u>Lymontria dispar</u>) (<u>69</u>).

A number of volatile phenylpropanoid compounds have pronounced biological properties and are also found in essential

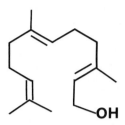

farnesol

oils (70). The toxicity of phenylpropanoid compounds in seed oils
has been reviewed (55). Several simple, volatile phenylpropanoid
compounds such as eugenol and myristicin are known to be allelochemics (17). Myristicin, isolated from parsnip, Pastinaca sativa
(Apiaceae), has been shown to be an antifeedant compound (71).
The essential oil of Acorus calamus (Araceae) contains β-asarone.
This compound causes depression of development of the gonads and
ovaries of the insect Dysdercus koenigii (72). Although third and
fourth instar larvae moulted normally to the next instars, 5th

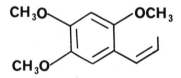

β-asarone

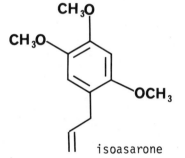

isoasarone

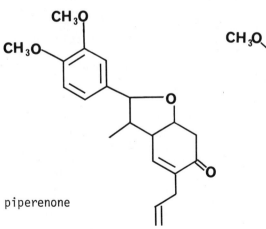

piperenone

myristicin

instar larvae moulted normal to adults but the ovaries were
irreversibily affected. In such adults, the ovaries remained
permanently immature.

Several phenylpropanoid compounds have pronounced antifeeding
activity. Isoasarone (from Piper futokadzura, Piperaceae) and
piperenone were highly active against Spodoptera litura (54).

Many short chain fatty acids, aldehydes, alcohols, esters,
ketones and hydrocarbons are produced by metabolism of fatty acids
(C_{16}-C_{18}). These compounds are common in essential oils and are
also found in insects.

Acetaldehyde, ethyl alcohol and ethyl acetate, which are
present in ripening figs are attractive to several Carpophilus
species (Coleoptera: Nitidulidae) (73). Several short chain
alcohols including ethanol, n-propanol, 2-propanol, isobutanol and
n-butanol serve as oviposition stimulants for the moth Ectomyelois
ceratoniae. This moth only oviposits on carob fruits which are
infested with a fungus of the genus Phomopsis which apparently is
responsible for producing the alcohols (74). Compounds such as
trans-2-hexen-1-ol, 1-hexanol, cis-3-hexen-1-ol and trans-2-hexenol
(and linalool) are involved in the olfactory orientation of the
Colorado beetle, Leptinotarsa decemlineata to the foliage of
potato (Solanum tuberosum, Solanaceae) (75). A similar complement
of compounds seem to be involved in the attraction of the alfalfa
seed chalcid (Bruchophagus roddi) to alfalfa (Medicago sativa,
Fabaceae) (76). Similar groups of compounds are probably involved
in many problems of attraction of insects over relatively long
distances.

Mustard oils which are found in some essential oils and are
probably the hydrolysis or breakdown products of glucosinolates,
are involved in host plant location of a number of groups of
insects (9) e.g. Pieris braccicae and P. rapae on plants in the
Brassicaceae (Cruciferae).

Essential oils are especially important in mutualistic relationships between plants and insects such as pollination and seed
and fruit dissemination (37,45,77-79). In these instances essential oil components serve as attractive substances for the plant.
These compounds vary widely in composition but contain most of the
chemical types described above. In some cases even volatile
amines and skatole are involved, as in the pollination of
Sauronatum guttatum (Araceae) (80).

Non-Volatile Lipids

While volatile lipids are often involved in attraction of the
insect to the plant from a distance, non-volatile lipids are
frequently involved in biting stimuli, continued feeding or pre-
vention of feeding and the disruption of normal growth (25). Many
types of compounds that are lipophilic in nature grade into more
polar groups, especially in series which are extensively oxygenated.
Among these are oxygenated sesquiterpenes, diterpenes, triterpenes,

tetraterpenes (carotenoids), phenylpropanoid compounds (simple, and oxygenated phenylpropanoids, coumarins, flavonoids, lignans, etc.), but also glycerides, fatty acids and their derivatives.

A number of lipids are pigments in plants. The most important group is probably carotenoids. The role of carotenoids in plant-insect interactions has been reviewed (81,82).

Several terpenoid compounds serve as juvenile hormones (or mimics) and ecydsones respectively (83) in insects.

The biological activity and toxicity of terpenoids to herbivores has been discussed (56,84) and representatives of each major type of terpene are known to be active. Well known examples are sesquiterpene lactones, pyrethrins, and several classes of diterpenes and triterpenes.

Several non-volatile iridoid monoterpenes occur as glycosides and have been observed to have biological activity. For example, xylomolin, from the unripe fruits of Xylocarpus moluccensis Roem. (Meliaceae) has antifeedant activity against Spodoptera exempta an African armyworm at 100 ppm (85) and crotepoxide from Croton macrostachys (Euphorbiaceae) possesses antifeedant activity against Spodoptera exempta (85).

A number of sesquiterpenes have been demonstrated to have pronounced biological activity (84); among the non-volatile compounds the sesquiterpene lactones are best known (86) but other oxygenated sesquiterpenes are also known to be active. For example, the role of gossypol, a dimeric sesquiterpene and structurally related compounds has been investigated (21,22). The oxygenated sesquiterpenes, shiromodiol monoacetate and diacetate, from Parabenzoin trilobum (=Lindera triloba Blume) possess potent antifeeding activity toward Spodoptera litura larvae (85).

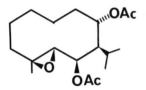

shiromodiol

helenalin

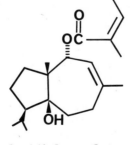

lasidiol angelate

Leaf cutter ants, abundant from Texas to Argentina are polyphagous
herbivores, but will not attack several plants. The ant, Atta
cephalotes, for example, does not feed on Lasianthaea fruticosa
(Asteraceae). The active repellent substance has been demonstrated
to be lasidiol angelate (87).
    Polygodial, ungandensidial, and warburganal from the bark of
Warburgia stuhlmanii and W. ugandensis (Cannellaceae) have also
proven to be highly active against Spodoptera exempta. Warburganal
is a highly active antifeedant (0.1 ppm against S. exempta) (85).

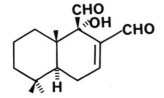

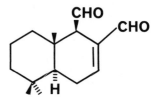

                warburganal   polygodial

    The iridoid monoterpenes, catalpal and catalposide occur in
nectar in flowers of Catalpa speciosa (Bignoniaceae) and are toxic
to many non-coadapted insects which attempt to rob nectar. The
bees which normally pollinate this plant are relatively insensitive
to the effects of these compounds (88).
    A number of diterpenes are known to be active against herbi-
vores (84). The diterpenes abietic, dehydroabietic, 12-methoxy-
abietic, sandaracopimaric, and isopimaric acid serve as feeding
deterrents for the larch sawfly, Pristiphora erichsonii in single
needles from new shoots of tamarack (Larix laricina) (133). The
larvae of this insect do eat tufted needles on short shoots of the
same trees.
    Varieties of sunflower (Helianthus annums, Asteraceae) that
are resistant to attack by larvae of the sunflower moth (Homeosoma
electellum) contain high concentrations of trachyloban-19-oic acid
and (-)-16-kauren-19-oic acid in their florets (84,132).
    Ryania speciosa (Flacourtiaceae) and several related species
are unique in that their insecticidal activity was discovered as
a part of a search for new insecticides (89). The toxic principle
is a diterpene esterified to pyrrole-2-carboxylic acid. This
insecticide proved to be somewhat selective in its activity (89).
    A series of complex oxygenated diterpenes is known to be
antifeedant. Several of these occur in the Lamiaceae (Labiatae)
and Verbenaceae, but are found in other families as well (e.g. in

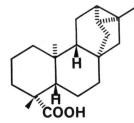

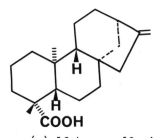

trachyloban-19-oic acid   (-)-16-kauren-19-oic acid

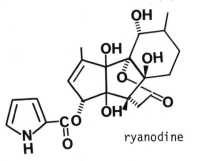

ryanodine

the Asteraceae). Clerodendrin A & B from verbenaceous plants
possess potent antifeeding activity against some insects (e.g.
Spodoptera litura), but less against others (Calospilos miranda).
Several genera of plants from this family contain similar compounds: for example Clerodendron, Caryopteris, Callicarpa (54).
Ajugarins, from Ajuga remota (Lamiaceae or Labiatae) have ent-
clerodane diterpenoid structure. They have antifeedant activity
against Spodoptera exempta at 100 ppm (85). Ent kaurenoid diterpenes such as inflexin from Isodon inflexus (Lamiaceae) and iso-
domedin from I. shikokianus var. intermedius had antifeedant
activity against the African armyworm (Spodoptera exempta). They
are also highly cytotoxic (LD$_{50}$ 5.4 and 4.0 µg/ml respectively)
(85). Two diterpenes, cinnzeylanine and cinnzeylanol from
Cinnamomum zeylanicum (Lauraceae) have been demonstrated to possess
insecticidal activity (90). At 2 to 4 ppm these compounds inhibited larval ecdysis in Bombyx mori.
   Many triterpenes also have antiherbivore activity. In
general, those which are highly oxygenated seem to be more
active in this regard (84). The role of cardiac glycosides,
insects and their predators has been reviewed (91-94). A number
of metabolically altered triterpenes from the Rutaceae, Meliaceae
and Simaroubaceae are antifeedants. Extracts of neem tree seeds
(Azadirachta indica, Meliaceae) were shown to be repellent to a
number of insects when applied to various crop plants at low
concentrations. The probable active compound is tetranortriter-
pene, azadirachtin (95). This compound from the leaves and fruits

of A. indica and Melia azedarach (Meliaceae) gives 100% feeding
inhibition at 40 μg/liter against Schistocerca gregaria (desert
locust) (85). Harrisonin from Harrisonia abyssinica (Simaroubaceae)
had antifeedant activity against Spodoptera exempta at 20 ppm (85).
Both harrisonin and azadarichtin are limonoid compounds.

Several simple non-volatile phenylpropanoids are known to be
allelochemics. Among these are p-coumaric acid, cis and trans-
caffeic acid (17), chlorogenic acid, and a caffeyl derivative of
an aldaric acid (97).

Coumarin (0.1%), ferulic acid (0.1%) and p-coumaric acid (5%)
were shown to be toxic to the larvae of the bruchid beetle,
Callosobruchus maculatus (96).

Furocoumarins such as isopimpinellin, bergapten, and kokusagin
have antifeedant activity against Spodoptera litura (54). A number
of similar compounds from umbelliferous plants have been demonstrated to be active antifeedants against Spodoptera litura,
Periplaneta americana, Musca domestica, Blattela germanica, and
Stylopyga rhombifolia (98).

Xanthotoxin, a linear furocoumarin occurs in many plants of
the Apiaceae. This compound is not appreciably toxic to the
larvae of Papilio polyxenes which normally feed on umbelliferous
plants. Angelicin, an angular furanocoumarin, which is found in
only a few relatively advanced tribes of the Apiaceae, reduces
growth rate and fecundity in this insect (100). A few insects can
utilize plants with angular furocoumarins, however, and these
authors suggest that the pathway leading to angular attachment of
the furan ring may have been favored in the Apiaceae by specialized
herbivores that had adapted to feeding on linear furocoumarins.

The toxicity of furocoumarins to mammals has been reviewed
(99).

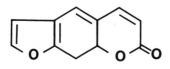   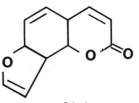

          xanthotoxin   angelicin

Flavonoid aglycones and especially highly methylated ones are
often found as resinous exudates on plants (101).

Many flavonoids are known to be phytoalexins, antiviral
agents, and to serve as antiinflammatory and antitumor compounds.
Several isoflavones have estrogenic activity in mammals (70).
(-)-Vestitol and sativan, isoflavans from Lotus species, are
phytoalexins. 3R(-)-Vestitol from the resistant pasture legume
Lotus pedunculatus, has been demonstrated to be a feeding deterrent

to larvae of <u>Costelytra zealandica</u> (Coleoptera: Scarabaeidae), a
serious agricultural pest in New Zealand (<u>102</u>). Pastures containing perennial rye grass (<u>Lolium perenne</u>) and as little as 20%
<u>Lotus pedunculatus</u> were relatively resistant to attack by this
insect.
  The lipophilic material found on the surface of <u>Larrea</u>
species (Zygophyllaceae) is comprised of several methylated flavonoid aglycones and lignans such as nordihydroguaiaretic acid.
This resinous material was shown to act as an antiherbivore substance and appeared to reduce digestibility of the plant for
several herbivores (<u>103</u>).
  Although somewhat less known than other groups of phenylpropanoid compounds, lignans are widely distributed among higher
plants (<u>70</u>). Several lignans are known to have antitumor activity
and many are cytotoxic. Lignans are also known to be active in
plant-insect relationships, for example sesamin which occurs in
<u>Sesamum indicum</u> (Pedaliaceae) seed oil as well as in other plants
was isolated from <u>Magnolia kobus</u> (Magnoliaceae) and demonstrated
to be a growth inhibiting substance for <u>Bombyx mori</u> (<u>104</u>).

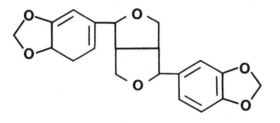

sesamin

  2-Methyl, 2-hydroxymethyl-, and 2-formylanthraquinones in the
heart wood of teak (<u>Tectona grandis</u>, Verbenaceae) are effective in
inhibiting termite activity (<u>105</u>). Several naphthoquinone derivatives including lapachol are thought to impart marine borer
resistance to woods, <u>e.g.</u> that of <u>Tabebuia guayacan</u> (Bignoniaceae)
(<u>108</u>).
  Often related insect species are not all sensitive to a
particular allomone. For example, juglone is a feeding deterrent
to the smaller European elm bark beetle (<u>Scolytus multistriatus</u>)
but not to the closely related hickory bark beetle (<u>Scolytus
quadrispinosus</u>) (<u>106,107</u>).
  Most insects have a dietary requirement for polyunsaturated
fatty acids, usually linolenic acids, but the exact requirements
appear to differ from species to species and few have been studied
thoroughly (<u>109</u>). Moreover, the presence of certain fatty acids

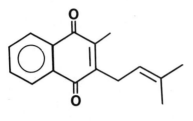

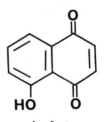

lapachol   juglone

in the diet may inhibit growth and cause mortality in insects that
do not require them.

General physiological roles for fatty acids in cellular lipids
are caloric storage, membrane fluidity, and prostaglandin precursors. The first of these mainly involved the formation and
hydrolysis of triacylglycerols, transport and activation of non-
esterified fatty acids, and other steps leading to energy
conversion (110). The second role primarily involves activation
and incorporation into 1- and 2- positions of different phospholipids which form a major part of membranes. The third role is
linked to the requirement for certain unsaturated fatty acids in
the diets of most animals (110).

Incorporation of different fatty acids into lipids depends on
the relative abundance of their CoA derivatives and their acyl-
transferase K_m values. The synthetic enzymes which form membrane
phospholipids may select the acid by molecular features not in
accord with the optimal physiological properties of the products
(110), resulting in the formation of membranes which do not function adequately.

When trans-fatty acids are fed to rats with adequate amounts
of essential fatty acids, they have little effect on growth,
longevity, or reproduction, but when fed as the sole source of
lipids they exaggerate the symptoms of essential fatty acid
deficiency (111). An effect on the metabolism of long chain polyunsaturated fatty acids was noted however.

Fatty acids are known to be feeding stimulants for certain
insects. Linolenic acid (as a free fatty acid) in mulberry leaves
stimulates potent feeding activity in Bombyx mori, the silkworm.
Linoleate and laurate also have activity whereas myristate,
palmitate, stearate, elaidate, oleate and vaccenate have none.
Oleate promotes growth but does not promote feeding activity. A
synergistic effect of β-sitosterol has been observed with lino-
lenate, linoleate, vaccenate and laurate. Linoleic and linolenic
acids are phagostimulants for the fire ant Solenopsis saevissima
var. richteri (112). Palmitate and stearate produces a positive
feeding response in Dermestes maculata, and valeric acid in Trogo-
derma granarium. Fatty acids (C_5-C_{11}) are repellent to Tribolium
castaneum.

The antifungal properties of fatty acids and several of their
derivatives such as amides and methyl esters have been reviewed
(113) as have their antimicrobial activities (114). Monolaurin
(the ester of lauric acid and glycerol) is the most potent lipid
derivative tested to date with regard to antibiotic activity.

The effects of selected fatty acid (C_{10}-C_{12}) methyl esters on
the pink bollworm (Pectinophora gossypiella), bollworm (Heliothis
zea) and tobacco budworm (Heliothis virescens) were determined,
and a number of cyclopropyl, olefinic and acetylenic methyl esters
were also tested (115). Methyl (Z,Z)-deca-2,8-diene-4,6-diynoate
(matricaria ester) was lethal at low concentrations to all three
insects. This last ester was isolated from Conyza canadensis but
is found in vegetative matter of many plants of the Asteraceae.
It was toxic to the pink bollworm at 0.005% and to the bollworm
and the tobacco budworm at 0.15% in artificial diets. Esters
(C_{10}-C_{12}) were also toxic to the insects as sprays. Matricaria
ester was also shown to be a potent insect antifeedant compound to
these insects (115).

A number of mammalian and fungal enzyme systems, whole cells
and entire animals have been shown to respond differently to fatty
acids which vary in number and position of unsaturation, geometrical isomers, or cyclopropyl ring positions (110,116).

The salts of fatty acids (not naturally occurring) have long
been known to have insecticidal properties. The most effective
potassium salts center around oleate in the monounsaturated and
saturated series, although potassium caprate (C_{10}) was especially
active against Choristoneura occidentalis (Western spruce budworm)
and Acleris gloverana (Western blackheaded budworm) (117).

A number of lipid materials were shown to be toxic to the
larvae of the bruchid beetle, Callosobruchus maculatus (96).
Among these were a cyclopropane fatty acid from Sterculia foetida
(Sterculiaceae) seed oil (0.1%) and cyanolipids from Koelreuteria

$$CH_3\left(CH_2\right)_7 \overset{\displaystyle \triangle}{C}=C\left(CH_2\right)_7COOH \qquad \text{sterculic acid}$$

paniculata, Sapindus drummondii, and Ungnadia speciosa (Sapinda-
ceae). A pentane extract of the seed of tung (Aleurites fordii,
Euphorbiaceae) contained a compound which was shown to be strongly
repellent to Anthonomis grandis, the boll weevil. Some fractions
were also active to the striped and spotted cucumber beetles, the
codling moth and the redbanded leafroller (95). The active compounds were α-eleostearic acid and erythro-9,10-dihydroxy-1-
octadecanyl acetate (118).

Fatty acids and various esters are also involved in mutualistic relationships between certain types of bees and several
plant species (123). These compounds serve as food reserve for
the pollinating species. For example, several species of Krameria

produce free β-acetoxy-fatty acids which are collected by the genus Centris.

Seeds of the violet, Viola odorata are disseminated by the ant Aphaenogaster rudis. Elaiosomes (appendages attached to the outside of the seed coat) often contain high concentrations of lipids and are associated with attraction of the ants. 1,2-Diolein, a diglyceride, is largely responsible for this attraction (124).

$$CH_3(CH_2)_3CH \overset{E}{=} CH\ CH \overset{E}{=} CH\ CH \overset{Z}{=} CH(CH_2)_7COOH$$

α-eleostearic acid

$$CH_3(CH_2)_7\underset{\underset{HO}{|}}{CH}\ \underset{\underset{OH}{|}}{CH}\ (CH_2)_7CH_2OH$$

erythro-9,10-dihydroxy-1-octadecanoate

Amide derivatives of unsaturated aliphatic fatty acids (C_{10}-C_{18}) are especially common in members of the Asteraceae, Piperaceae, and Rutaceae. Among these, the isobutyl amides have pronounced insecticidal activity (119). Most of these compounds have not proven useful as commercial insecticides because of their irritating properties to mammals and their instability.

A series of structurally similar compounds derived by acetate extension of phenylpropanoid precursors is also found in the Piperaceae. Pipercide from Piper nigrum (Piperaceae) was insecticidal, but the mixture of amides from this plant was significantly more toxic and it appears that co-occurring compounds exert synergistic effects (120).

Many non-volatile lipids in plants are also deived from fatty acids. A number of these compounds are known to possess biological activity.

The wild tomato, Lycopersicon hirsutum f. glabratum is covered with trichomes which contain 2-tridecanone. The level of this compound is much lower in the domesticated tomato, L. esculentum. This exudate proved to be toxic to Manduca sexta (tobacco hornworm) and to Heliothis zea (121). The density of glandular trichomes, which secrete 2-tridecanone, was influenced

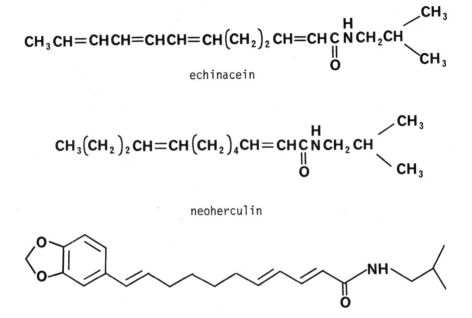

$$CH_3 CH\!=\!CHCH\!=\!CHCH\!=\!CH(CH_2)_2 CH\!=\!CHC\overset{H}{\underset{O}{\overset{\|}{N}}}CH_2CH\overset{CH_3}{\underset{CH_3}{<}}$$

echinacein

$$CH_3(CH_2)_2 CH\!=\!CH(CH_2)_4 CH\!=\!CHC\overset{H}{\underset{O}{\overset{\|}{N}}}CH_2 CH\overset{CH_3}{\underset{CH_3}{<}}$$

neoherculin

pipercide

by an interaction between day length and light intensity. The toxic compound was significantly more abundant on foliage of plants grown under long day regimes (122). This finding is of considerable importance as there is a possibility of introducing this resistance into acceptable tomato cultivars.

Most acetylenic compounds in plants are derived from meta-bolically altered fatty acids. These often are active in plant-insect relationships. 8-cis-Dihydromatricaria acid is also found in the defensive secretion of the soldier beetle (Chauliognathus lecontei) (125), and has subsequently been shown to have anti-feedant properties against Phidippus spp. (jumping spiders) (126). As previously mentioned matricaria ester has antifeedant properties to the pink bollworm, bollworm and tobacco budworm (115).

Several nematicidal acetylenic compounds have been isolated. Most are from the Asteraceae (127). Recently isolated are (8R,9R, 10S)-9,10-epoxyheptadec-16-ene-4,6-diyne-8-ol and other compounds from Cirsium japonicum (128,129). 1-Phenylhepta-1,3,5-triyne and 2-phenyl-5-(1'-propynyl)-thiopene, from Coreopsis lanceolata and cis-dehydromatricaria ester from Solidago altissima have been shown to be fly ovicidal substances (130,131).

$$CH_3CH_2CH_2C\equiv C\ C\equiv C\ CH\ \underset{\underset{OH}{|}}{CH}-\overset{\overset{O}{\diagup\diagdown}}{CH}(CH_2)_5\ CH=CH_2$$

(8R,9R,10S)-9,10-epoxyheptadec-16-ene-4,6-diyne-8-ol

Literature Cited

1. Kogan, M. <u>Proc. XV Int. Cong. Ent.</u> 1976, 211-226.
2. Feeny, P. P. <u>In</u> "Coevolution of Animals and Plants" (L. E. Gilbert and P. H. Raven, eds.), 3-19. Univ. of Texas Press, Austin, 1975.
3. Waiss, A. C. Jr., Chan, B. G. and Elliger, C. A. <u>In</u> "Host Plant Resistance to Pests" (P. A. Hedin, ed.) 115-128. Symposium Series No. 62, Amer. Chem. Soc., Washington, 1977.
4. Ehrlich, P. R. and Raven, P. H. <u>Evolution</u> 1964, <u>18</u>, 568-608.
5. Seigler, D. S. <u>Biochem. Syst & Ecol</u>. 1977, 5, 195-199.
6. Seigler, D. S. and Price, P. W. <u>Amer. Nat</u>. 1976, <u>110</u>, 101-104.
7. Jones, D. A. <u>Amer. Nat</u>. 1979, <u>113</u>, 445-451.
8. Swain, T. <u>Nova Acta Leopoldina, Suppl. 7</u> 1976, 411-421.
9. Jones, O. T. and Coaker, T. H. <u>Ent. Exp. and Appl</u>. 1978, <u>24</u>, 272-284.
10. Jermy, T. <u>Symp. Biol. Hung</u>. 1976, <u>16</u>, 109-113.
11. Atsatt, P. R. and O'Dowd, D. J. <u>Science</u> 1976, <u>193</u>, 24-29.
12. Price, P. W., Bouton, C. E., Gross, P., McPheron, B. A., Thompson, J. N. and Weis, A. E. <u>Ann. Rev. Ecol. Syst</u>. 1980, <u>11</u>, 41-65.
13. Whittaker, R. H. and Feeny, P. P. <u>Science</u> 1971, <u>171</u>, 757-770.
14. Florkin, M. <u>Bull. Cl. Sci. Acad. Roy. Belg</u>. 1965, <u>51</u> 239-256.
15. Blum, M. S. <u>In</u> "Animals and Environmental Fitness" (R. Gilles, ed.). Pergamon, Oxford, 1980, pp. 207-222.
16. Brown, W. L., Jr., Eisner, T. and Whittaker, R. H. <u>BioScience</u> 1970, <u>20</u>, 21-22.

17. Beck, S. D. and Reese, J. C. <u>In</u> "Biochemical Interaction Between Plants and Insects" (J. W. Wallace and R. L. Mansell, eds.) 41-92. Plenum, New York, 1976.
18. Silverstein, R. M. <u>Science</u> 1981, <u>213</u>, 1326-1332.
19. Blum, M. S. <u>In</u> "Chemical Control of Insect Behavior" (H. H. Shorey and J. J. McKelvy, Jr., eds.) 149-167. New York, 1977.
20. Weldon, P. J. <u>J. Chem. Ecol</u>. 1980, <u>6</u>, 719-724.
21. Stipanovic, R. D., Bell, A. A. and Lukefahr, M. J. <u>In</u> "Host Plant Resistance to Pests" (P. A. Hedin, ed.) 197-214. Symposium Series No. 62, Amer. Chem. Soc., Washington, 1977.
22. Hedin, P. A., Thompson, A. L. and Gueldner, R. C. <u>In</u> "Biochemical Interaction Between Plants and Insects" (J. W. Wallace and R. L. Mansell, eds.) 271-350. Plenum, New York, 1976.
23. Eisner, T. and Halpern, B. P. <u>Science</u> 1971, <u>172</u>, 1362.
24. Dethier, V. G. <u>Evolution</u> 1954, <u>8</u>, 33-54.
25. Kogan, M. <u>In</u> "Introduction to Insect Pest Management" (R. L. Metcalf and W. H. Luckman, eds.) 103-146. Wiley, New York, 1975.
26. Gilbert, L. E. <u>In</u> "Coevolution of Animals and Plants" (L. E. Gilbert and P. H. Raven, eds.) 210-240. Univ. of Texas Press, Austin, 1975.
27. Smiley, J. <u>Science</u> 1978, <u>201</u>, 745-747.
28. Price, P. W. "Insect Ecology", Wiley, New York, 1975.
29. Dethier, V. G. <u>In</u> "Chemical Ecology" (E. Sondheimer and J. B. Simeone, eds.) 83-102. Academic Press, New York, 1970.
30. Painter, R. H. "Insect Resistance in Crop Plants", MacMillan, New York, 1951.
31. Chapman, R. F. <u>Bull. Ent. Res</u>. 1974, <u>64</u>, 339-363.
32. Maxwell, F. G. <u>In</u> "Chemical Control of Insect Behavior" (H. H. Shorey and J. J. McKelvy, Jr., eds.) 299-304. Wiley, New York, 1977.
33. Rhoades, D. F. <u>In</u> "Herbovores" (G. A. Rosenthal and D. H. Janzen, eds.) 3-54. Academic Press, New York, 1979.
34. Tester, C. F. <u>Phytochemistry</u> 1977, <u>16</u>, 1899-1901.
35. Pimentel, D. <u>Bull. Ent. Soc. Am</u>. 1976, <u>22</u>, 20-26.
36. Campbell, B. C. and Duffey, S. S. <u>Science</u> 1979, <u>205</u>, 700-702.
37. Harborne, J. B. <u>In</u> "Introduction to Ecological Biochemistry", 2nd Ed. Academic Press, London, 1982.
38. Gross, D. <u>Fort. Chem. Org. Naturst</u>. 1977, <u>34</u>, 187-247.
39. Haukioja, E. <u>Oikos</u> 1980, <u>35</u>, 202-213.
40. Carroll, C. R. and Hoffman, C. A. <u>Science</u> 1980, <u>209</u>, 414-416.
41. Hedin, P. A., Maxwell, F. G. and Jenkins, J. N. <u>In</u> "Proc. of Summer Inst. of Biol. Control of Plant Insects and Diseases" (F. G. Maxwell and F. A. Harris, eds.) 494-527. University of Mississippi Press, Jackson, 1974.
42. Hedin, P. A., Jenkins, J. N. and Maxwell, F. G. <u>In</u> "Host

Plant Resistance to Pests" (P. A. Hedin, ed.) 231-275.
Symposium Series No. 62, Amer. Chem. Soc., Washington, 1977.

43. Jacobson, M. and Crosby, D. G. In "Naturally Occurring
    Insecticides", Dekker, New York, 1971.
44. McKey, D. In "Herbivores" (G. A. Rosenthal and D. H. Janzen,
    eds.) 56-133. Academic Press, New York, 1979.
45. Rodriguez, E. and Levin, D. A. In "Biochemical Interaction
    Between Plants and Insects" (J. W. Wallace and R. L. Mansell,
    eds.) 214-270. Plenum, New York, 1976.
46. Schoonhoven, L. M. Rec. Adv. Phytochem. 1973, 6, 197-224.
47. Jacobson, M. "Insecticides from Plants", Agriculture Hand-
    book No. 154, USDA, Washington, D.C., 1958.
48. Jacobson, M. "Insecticides from Plants", Agriculture Hand-
    book No. 461, USDA, Washington, D.C., 1975.
49. Levin, D. Quart. Rev. Biol. 1973, 48, 3-15.
50. Levin, D. A. Ann. Rev. Ecol. Syst. 1976, 7, 121-159.
51. McKey, D. Amer. Nat. 1974, 108, 305-320.
52. Dethier, V. G., Brown, L. B. and Smith, C. N. J. Econ.
    Entom. 1960, 53, 134-136.
53. Beck. S. D. Ann. Rev. Entomol. 1965, 10, 207-232.
54. Munakata, K. In "Host Plant Resistance to Pests" (P. A.
    Hedin, ed.) 185-196. Symposium Series No. 62, Amer. Chem.
    Soc., Washington, 1977.
55. Seigler, D. S. In "Herbivores (G. H. Rosenthal and D. H.
    Janzen, eds.) 449-470. Academic Press, New York, 1979.
56. Herout, V. In "Progress in Phytochemistry", Vol. 2. (L.
    Reinhold and Y. Livschitz, eds.) 143-202. Interscience,
    London, 1970.
57. Blum, M. S. In "Biochemistry of Insects" (M. Rockstein, ed.)
    466-513. Academic Press, New York, 1978.
58. Nahrstedt, A. Planta Med. 1982, 44, 2-14.
59. Weatherston, J. and Percy, J. E. In "Chemicals Controlling
    Insect Behavior" (M. Beroza, ed.) 95-144. Academic Press,
    New York, 1970.
60. Duffey, S. S. Proc. XV Int. Cong. Ent. 1976, 15, 323-394.
61. Vasechko, G. I. Z. Angew. Entomol. 1978, 85, 66-76.
62. Verma, M. and Meloan, C. E. Amer. Lab. 1981, 64-69.
63. Chapman, R. F., Bernays, E. A. and Simpson, S. J. J. Chem.
    Ecol. 1981, 7, 881-888.
64. Alfaro, R. I., Pierce, H. D., Jr. Borden J. H. and
    Oehlschlager, A. C. J. Chem. Ecol. 1981, 7, 39-48.
65. Oh, H. K., Sakai, T., Jones, M. B. and Longhurst, W. M.
    Appl. Microbiol. 1967, 15, 777-784.
66. Eisner, T. Science 1964, 146, 1318-1320.
67. Sakan, T., Isoe, S. and Hyeon, S. B. In "Control of Insect
    Behavior by Natural Products" (D. L. Wood, R. M. Silverstein
    and M. Nakajima, eds.) 239-247. Academic Press, New York,
    1970.
68. Sutherland, O. R. W., Wearing, C. H. and Hutchins, R. F. N.
    J. Chem. Ecol. 1977, 3, 625-631.

69. Doskotch, R. W., Cheng, H. Y., Odell, T. M. and Girard, L. J. Chem. Ecol. 1980, 6, 845-851.
70. Mabry, T. J. and Ulubelen, A. J. Agr. Fd. Chem. 1980, 28, 188-196.
71. Lichtenstein, E. P. and Casida, J. E. J. Agr. Fd. Chem. 1963, 11, 410-415.
72. Saxena, B. P., Koul, O., Tikku, K. and Atal, C. K. Nature 1977, 270, 512-513.
73. Smilanick, J. M., Ehler, L. E. and Birch, M. C. J. Chem. Ecol. 1978, 4, 701-707.
74. Gothilf, S., Levy, E. C., Cooper, R. and Lavie, D. J. Chem. Ecol. 1975, 1, 457-464.
75. Visser, J. H., Van Straten, S. and Maarse, H. S. J. Chem. Ecol. 1979, 5, 13-25.
76. Buttery, R. G. and Kamm, J. A. J. Agr. Fd. Chem. 1980, 28, 978-981.
77. Yeo, P. F. In Roy. Ent. Soc. Sympos. No. 6, (H. F. Van Emden, Ed.) 51-57. Blackwell, London, 1973.
78. Faegri, K. and van der Pijl, "The Principles of Pollination Ecology", 3rd Ed. Pergamon, London, 1979.
79. Dodson, C. H. In "Coevolution of Animals and Plants" (L. E. Gilbert and P. H. Raven, eds.) 91-99. University of Texas Press, Austin, 1975.
80. Smith, B. N. and Meeuse, B. J. D. Plant Phys. 1966, 41, 343-347.
81. Rothschild, M. In "Biochemical Aspects of Plant and Animal Coevolution", (J. B. Harborne, ed.) 259-276. Academic Press, New York, 1978.
82. Rothschild, M. In "Coevolution of Animals and Plants", (L. E. Gilbert and P. H. Raven, eds.) 20-50. University of Texas Press, Austin, 1975.
83. Sláma, K. In "Herbivores" (G. A. Rosenthal and D. H. Janzen, eds.) 683-700. Academic Press, New York, 1979.
84. Mabry, T. J. and Gill, J. E. In "Herbivores" (G. A. Rosenthal and D. H. Janzen, eds.) 501-537. Academic Press, New York, 1979.
85. Kubo, I. and Nakanishi, K. In "Host Plant Resistance to Pests" (P. A. Hedin, ed.) 165-178. Symposium Series No. 62, Amer. Chem. Soc., Washington, 1977.
86. Burnett, W. C., Jr., Jones, S. B. Jr. and Mabry, T. J. In "Biochemical Aspects of Plant and Animal Coevolution, (J. B. Harborne, ed.) 233-257. Academic Press, New York, 1978.
87. Wiemer, D. F. and Ales, D. C. J. Org. Chem. 1981, 46, 5449-5450.
88. Stephenson, A. G. J. Chem. Ecol. in press.
89. Crosby, D. G. In "Naturally Occurring Insecticides", (M. Jacobson and D. G. Crosby, eds.) 177-239. Dekker, New York, 1971.
90. Isogai, A., Murakoshi, S., Suzuki, A. and Tamura, S. Agr. Biol. Chem. 1977, 41, 1770-1784.

91. Roeske, C. N., Seiber, J. N., Brower, L. P. and Moffitt, C. M. In "Biochemical Interaction Between Plants and Insects", (J. W. Wallace and R. L. Mansell, eds.) 93-163. Plenum, New York, 1976.

92. Nelson, C. J., Seiber, J. N. and Brower, L. P. J. Chem. Ecol. 1981, 7, 981-1009.

93. Seiber, J. N., Tuskes, P. M., Brower, L. P. and Nelson, C. J. J. Chem. Ecol. 1980, 6, 321-339.

94. Fink, L. S. and Brower, L. P. Nature 1981, 291, 67-70.

95. Jacobson, M., Reed, D. K., Crystal, M. M., Moreno, D. S. and Soderstrom, E. L. Ent. Exp. Appl. 1978, 24, 248-257.

96. Janzen, D. H., Juster, H. B. and Bell, E. A. Phytochemistry 1977, 16, 223-227.

97. Elliger, C. A., Wong, Y., Chan, B. G. and Waiss, A. C. Jr., J. Chem. Ecol. 1981, 7, 753-758.

98. Yajima, T. and Munakata, K. Agr. Biol. Chem. 1979, 43, 1701-1706.

99. Ivie, G. W. In "Effects of Poisonous Plants on Livestock" (R. F. Keeler, K. R. van Kampen and L. F. James, eds.) 475-485. Academic Press, New York, 1978.

100. Berenbaum, M. and Feeny, P. Science 1981, 212, 927-929.

101. Wollenweber, E. and Dietz, V. H. Phytochemistry 1981, 20, 869-932.

102. Russell, G. B., Sutherland, O. R. W., Hutchins, R. F. N. and Christmas, P. E. J. Chem. Ecol. 1978, 4, 571-579.

103. Rhoades, D. F. and Cates, R. G. In "Biochemical Interaction Between Plants and Insects" (J. W. Wallace and R. L. Mansell, eds.) 168-213. Plenum, New York, 1976.

104. Kamikado, T., Chang, C., Murakoshi, S., Sakurai, A. and Tamura, S. Agr. Biol. Chem. 1975, 39, 833-836.

105. Rudman, R. and Gay, F. J. Holzforschung 1961, 15, 117.

106. Gilbert, B. L. and Norris, D. M. J. Insect Physiol. 1968, 14, 1063-1068.

107. Gilbert, L. I. and King, D. S. Physiology Insecta 1973, 1, 249-370.

108. Manners, G. D., Jurd, L., Wong, R. and Palmer, K. Tetrahedron 1975, 31, 3019-3024.

109, McFarland, J. E. In "The Pharmacological Effects of Lipids" J. J. Kabara, ed.) 97-104. Amer. Oil Chem. Soc. Monograph No. 5, Champaign, IL, 1978.

110. Lands, W. E. M. In "Geometrical and Positional Fatty Acid Isomers" (E. A. Emken and H. J. Dutton, eds.) 181-212. Amer. Oil Chem. Soc. Monograph No. 6, Champaign, IL, 1979.

111. Alfin-Slater, R. B. and Aftergood, L. In "Geometrical and Positional Fatty Acid Isomers" (E. A. Emken and H. J. Dutton, eds.) 53-74. Amer. Oil Chem. Soc. Monograph No. 6, Champaign, IL, 1979.

112. Vinson, S. B., Thompson, J. L. and Green, H. B. J. Insect Physiol. 1967, 13, 1729-1736.

113. Gershon, H. and Shanks, L. In "The Pharmacological Effects

of Lipids" (J. J. Kabara, ed.) 51-62. Amer. Oil Chem. Soc. Monograph No. 5, Champaign, IL, 1978.

114. Kabara, J. J. In "The Pharmacological Effects of Lipids" (J. J. Kabara, ed.) 1-14. Amer. Oil Chem. Soc. Monograph No. 5, Champaign, IL, 1978.

115. Binder, R. G., Chan, B. G. and Elliger, C. A. Agr. Biol. Chem. 1979, 43, 2467-2471.

116. Holman, R. T. In "Geometrical and Positional Fatty Acid Isomers" (E. A. Emken and H. J. Dutton, eds.) 283-298. Amer. Oil Chem. Soc. Monograph No. 6, Champaign, IL, 1979.

117. Puritch, G. S. In "The Pharmacological Effect of Lipids" (J. J. Kabara, ed.) 105-112. Amer. Oil Chem. Soc. Monograph No. 5, Champaign, IL, 1978.

118. Jacobson, M., Crystal, M. M. and Warthen, J. D. Jr., J. Agr. Fd. Chem. 1981, 29, 591-593.

119. Jacobson, M. In "Naturally Occurring Insecticides" (M. Jacobson and D. G. Crosby, eds.) 139-176. Dekker, New York, 1971.

120. Miyakado, M., Nakayama, I. and Yoshioka, H. Agr. Biol. Chem. 1980, 44, 1701-1703.

121. Williams, W. G., Kennedy, G. G., Yamamoto, R. T., Thacker, J. D. and Bordner, J. Science 1980, 207, 888-889.

122. Kennedy, G. G., Yamamoto, R. T., Dimock, M. B., Williams, W. G. and Bordner, J. J. Chem. Ecol. 1981, 7, 707-716.

123. Simpson, B. B. and Neff, J. L. Ann. Missouri Bot. Gard. 1981, 69, 301-322.

124. Marshall, D. L., Beattie, A. J. and Bollenbacher, W. E. J. Chem. Ecol. 1979, 335-344.

125. Meinwald, J., Meinwald, Y. C., Chalmers, A. M. and Eisner, T. Science 1968, 160, 890-892.

126. Eisner, T., Hill, D., Goetz, M., Jain, S., Alsop, D., Camazine, S. and Meinwald, J. J. Chem. Ecol. 1981, 7, 1149-1159.

127. Bohlmann, F., Burckhardt, T. and Zdero, C. In "Naturally Occurring Acetylenes", Academic Press, London, 1973.

128. Kawazu, K., Nishii, Y. and Kakajima, S. Agr. Biol. Chem. 1980, 44, 903-906.

129. Sugiyama, K. and Yamashita, K. Agr. Biol. Chem. 1980, 44, 1983-1984.

130. Nakajima, S. and Kawazu, K. Agr. Biol. Chem. 1980, 44, 1529-1533.

131. Kawazu, K., Ariwa, M. and Kii, Y. Agr. Biol. Chem. 1977, 41, 223-224.

132. Elliger, C. A., Zinkel, D. F., Chan, G. B. and Waiss, A. C. Jr., Experientia 1976, 32, 1364-1366.

133. Ohigashi, H., Wagner, M. R., Matsumura, F. and Benjamin, D. M. J. Chem. Ecol. 1981, 7, 599-614.

134. Kogan, M. In Introduction to Insect Pest Management" (R. L. Metcalf and W. H. Luckman, eds.) 2nd Ed. Wiley, New York, 1982 in press.

RECEIVED August 23, 1982

Isolation of Phytoecdysones as Insect Ecdysis Inhibitors and Feeding Deterrents

ISAO KUBO and JAMES A. KLOCKE

University of California, Division of Entomology and Parasitology, College
of Natural Resources, Berkeley, CA 94720

Phytoecdysones, due to their effects on the
behavior and the development of certain species of
insects, appear to be components of multichemical
defensive strategies found in some insect-resistant
plant species.

Observations in nature obviate the fact that certain plant
species and cultivars are more resistant to insect attack[1-3]
than are others. *Ajuga remota* (Labiatae) is an example of this.
A survey of a Kenyan savannah following a locust attack revealed
that the only vegetation to survive the assault was *A. remota*[4].
In order to test for chemical factors involved in this observed resistance, extracts of *A. remota* foliage were incorporated into artificial diets optimized for several economically important pest insects (Fig. 1)[5]. Briefly, a methanolic extract
was dissolved in solvent and added to a non-nutritive filler
(α-cellulose), evaporated to dryness, and added to the components
of a meridic artificial diet, including solid nutrients (casein,
sucrose, wheat germ, Wesson salts), vitamins (C and B-complex),
and 4% agar. Newly-hatched larvae of the pink bollworm, *Pectinophora gossypiella* and of the fall armyworm, *Spodoptera frugiperda*
were placed singly on portions of the diet in plastic vials.
Additional bioassays were conducted with the silkworm, *Bombyx
mori*, by incorporating dissolved *A. remota* extracts directly into
dried mulberry powder (Nihon Nosan), evaporating the solvent to
dryness, and adding a 2% agar solution.
Analysis of the test insects fed the *A. remota* extracts revealed a developmental disruption in which the insects died in
the pharate condition following initiation of molting (apolysis),
but before completion of molting (ecdysis) (Fig. 2-4)[6]. Insect
molting cycle is initiated when the cuticular epithelium sepa-
rates from the overlying cuticle in the process of apolysis. The
molting cycle is terminated, upon the completion of cuticle synthesis, by hydrostatic expansion of the new cuticle during the
process of ecdysis. The *A. remota* extract apparently upset the

0097-6156/83/0208-0329$06.00/0

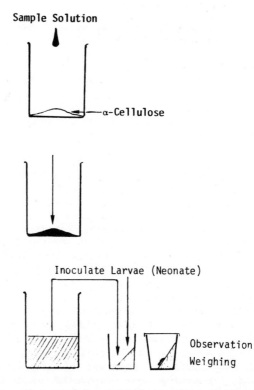

Figure 1.   Artificial diet feeding bioassay for lepidopterous larvae. The diet contained (1) Solid nutrients (casein, sugar, salt, wheat germ); (2) Buffered vitamin solution (B vitamins, vitamin C); and (3) 4% agar.

new head capsule

old head capsule

trunk exuviae

Figure 2.   A molting cycle failure of the silkworm, Bombyx mori, *caused by ingestion of the crude methanol extract of* Ajuga remota *root. The insect underwent normal apolysis, but failed to complete ecdysis. Thus, it could not remove its head capsule or its trunk exuviae. Magnification* × 11.

Figure 3.   Electron micrograph of a fall armyworm, Spodoptera frugiperda, *after ingestion of the crude methanol extract of* Ajuga remota *roots. This insect has three head capsules that mask its functional mouthparts. The insect eventually starved to death. Magnification* × 38.

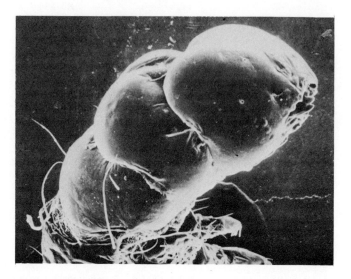

Figure 4.   Electron micrograph of a pink bollworm, Pectinophora gossypiella, *after ingestion of the crude methanol extract of* Ajuga remota *roots. This insect has three head capsules that mask its functional mouthparts. The insect eventually starved to death. Magnification × 113.*

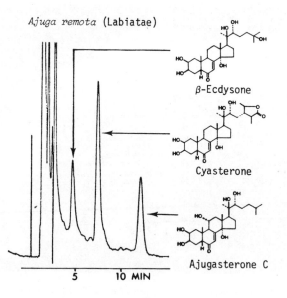

Figure 5. C_{18} reversed phase HPLC. Conditions: $H_2O–CH_3CN–MeOH$ (82:18:1.8 v/v); 1.5 mL/min; 254 nm.

temporal patterning such that the molting cycle failed due to an inhibition of ecdysis. Fig. 2 exemplifies this as the newly-molted *B. mori* larva died while encased by the old cuticular skin and head capsule (pharate condition). The effects of this pharate condition are to prevent feeding because of masking of the mouthparts by the head capsule and also to prevent locomotory and excretory functions because the whole body is trapped by the retention of the entire cuticular skin.

Figs. 3 and 4 are electron micrographs of a fall armyworm, and a pink bollworm, respectively. Both insects have three head capsules because they underwent two failed molting cycles before death. That is, even though feeding became impossible after the first inhibited ecdysis because the adhering second head capsule covered the mouthparts, these larvae could synthesize a third head capsule[7].

Chromatographic fractionations following this molting cycle failure bioassay resulted in the isolation of several bioactive compounds. Spectral identification (UV, IR, MS and NMR (^{1}H and ^{13}C)) resulted in ß-ecdysone, cyasterone, and ajugasterone C as the active principles from the *A. remota* extract (Fig. 5)[8]. We have used the molting cycle failure bioassay to detect phytoecdysones in other plant species observed to be resistant to insect attack, including *A. reptans*, *A. chameacistus*, *Podocarpus gracilior*, and others. *A. reptans* was found to contain ß-ecdysone and cyasterone, in addition to ajugalactone. *P. gracilior* was found to contain ponasterone A as the active principle in this bioassay[9].

In order to obtain sufficient quantities of these phytoecdysones for more detailed biological studies, droplet countercurrent chromatography (DCCC) was adapted[10]. DCCC is an especially efficient method for the preparative separation of polar compounds like the phytoecdysones. Thus with DCCC, while requiring only small volumes of solvent, more than 50 mg of each of the *Ajuga* phytoecdysones were rapidly and nondestructively separated and fully recovered[8] (Fig. 6) from each 500 mg injection.

A comparison of the five isolated phytoecdysones, tested in the artificial diet feeding bioassay with pink bollworm larvae, showed the importance of the phytoecdysone side chain in the structure/activity relationship (Table 1). Thus, ponasterone A showed the most potent ecdysis inhibitory (EI_{95}) as well as growth inhibitory (ED_{50}) activities, while ajugalactone was inactive as ecdysis inhibitor to concentrations as high as 1000 ppm. ß-ecdysone, cyasterone, and ajugasterone C were comparatively active in this bioassay (Table 1).

Ponasterone A was also the most potent of three phytoecdysones orally injected into fourth instar *B. mori* larvae (Table 2). All of the larvae treated with >5 µg ponasterone A were induced to initiate molting (apolysis), but were unable to complete this molt due to an inhibition of ecdysis. Thus, all of these ponasterone A-treated larvae died in the pharate condition.

Table 1. Effects of 5 phytoecdysones on growth and
          development of pink bollworm larvae.

Compound	Amount in diet (ppm)	Effect
ß-ecdysone	35	ED_{50}
	50	EI_{95}
Cyasterone	25	ED_{50}
	40	EI_{95}
Ponasterone A	1	ED_{50}
	2	EI_{95}
Ajugasterone C	14	ED_{50}
	45	EI_{95}
Ajugalactone	430	ED_{50}
	*	EI_{95}

Values are based on three or more replicates, each
of which consisted of 30 or more neonate pink bollworm assayed for 12 days. ED_{50} refers to the
effective dose for 50% growth inhibition, while
EI_{95} refers to the effective dose for 95% kill
due to ecdysis inhibition.

* No ecdysis inhibition was observed to
concentrations of 1000 ppm.

Table 2. Effects of ß—ecdysone as compared to cyasterone and ponasterone A when orally injected to 4th instar *B. mori*. 25 larvae were used per treatment.

Developmental stage in 4th instar (day)	µg phytoecdysone per os	Activity		
		ß—ecdysone	Cyasterone	Ponasterone A
1st	30	P; 74% EI	$I_{(P)}$; 10%EI	P; 100% EI
2nd	30	P; 67% EI	$I_{(P)}$; 13%EI	P; 100% EI
	20	P; 33% EI	$I_{(P)}$; 18%EI	P; 100% EI
	10	P; 10% EI	$I_{(P)}$; 16%EI	P; 100% EI
	5	I	--	P; 100% EI
	2.5	NE	--	--
3rd	20	P; 46% EI		--
	10	P; 15% EI		--
4th	30	P	--	--
	20	P	--	--
	10	P	--	--
	5	NE	--	--

P : Promotion of 100% of treated larvae to apolysis within 24—48 hr post-injection.

I : Delay of 100% of treated larvae to apolysis for 24—72 hr compared to control.

$I_{(P)}$: Delay of >82% of treated larvae to apolysis for >72 hr compared to control. (P) indicates <18% of treated larvae were promoted to apolysis within 24—48 hr postinjection. All of the latter group died through ecdysis inhibition (EI).

EI : Ecdysis inhibition resulting in death. All other larvae recovered.

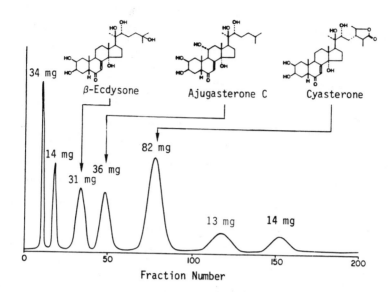

Figure 6. DCCC of the ethyl acetate extract of Ajuga remota (500 mg) with CHCl₃–MeOH–H₂O (13:7:4) by the ascending method.

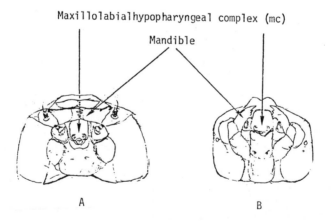

Figure 7. Schematic depicting (ventral side) dysfunctional mouth-parts in ecdysis-inhibited Bombyx mori larvae.

Left: Normally ecdysed 5th instar head capsule with fully closed mandibles.

Right: Ecdysis-inhibited 5th instar head capsule following artificial removal of 4th instar head capsule. The adhering 4th instar head capsule prevented full expansion of the 5th instar head capsule resulting in the forward position of the maxillolabial-hypopharyngeal complex (mc) such that the mandibles could not fully close.

Oral injections >10 µg of ß-ecdysone also resulted in a promoted apolysis, but depending upon concentration and the exact age of the treated fourth instar larvae, a variable number of the treated larvae were able to complete molting and essentially recover. Cyasterone injections >10 µg, in general, caused a delay (antiecdysone) of molting of the treated fourth instar larvae to the fifth instar. This 'antiecdysone' effect of cyasterone, at its most severe, resulted in prothetely (precocious development) of a small % of the treated larvae.

In order to examine anatomically the newly-synthesized head of fourth instar larvae undergoing molting cycle failure, the adhering exuvial head capsule was carefully removed with forceps. This procedure revealed the morphological disruption of the feeding apparatus (Fig. 7). The fifth instar head capsule was compressed by the adhering fourth instar head capsule, which resulted in the pushing forward of the maxillolabial-hypopharyngeal complex (mc) such that the mandibles could not fully close.

In order to study how injected (or ingested) phytoecdysone causes an inhibition in ecdysis, several biochemical parameters of larval molting fluid were analysed, including phenoloxidase activity, total protein, and total ascorbic acid (Table 3). Molting fluid may be implicated in the inhibition of ecdysis since either total removal of the molting fluid (unpublished) or a simple deletion of ascorbic acid from the molting fluid of *Spodoptera littoralis* (Navon, 1978)[11] (fed with no phytoecdysone) duplicated the phytoecdysone-induced inhibition of ecdysis. The antioxidant ascorbic acid has been hypothesized to control phenoloxidase hardening of the newly-synthesized cuticle before its hydrostatic expansion. In addition, ß-ecdysone is known to activate the enzyme which catalyses the synthesis of a phenoloxidase from its proenzyme[12].

Nevertheless, ascorbic acid levels in both the molting fluid and the hemolymph of phytoecdysone-treated larvae are comparable to those levels found in untreated larvae. In addition, phenoloxidase activity, as measured with catechol substrate, is actually less in phytoecdysone-induced larvae as compared to control larvae.

These negative results do not eliminate the involvement in ecdysis inhibition of other biochemical parameters in the molting fluid, i.e. chitinase, protease, but do seem to indicate that premature phenoloxidase-catalysed cuticle hardening is not the cause for phytoecdysone-induced failed ecdysis.

More detailed data with *B. mori* showed that the effects of ingested ß-ecdysone included, besides an inhibition of ecdysis, death without molting, death following completion of promoted molting, and an inhibition in growth with no effect on molting. These various effects are dependent upon the concentration of exogenous ß-ecdysone, the precise developmental stage of the treated larvae, and the duration of exposure.

Concentrations of ß-ecdysone >50 ppm in artificial diet in-

Table 3. Comparison of phenoloxidase activity, protein, and
          ascorbic acid levels in the molt fluid and ascorbic
          acid levels in the hemolymph approximately 36 hr
          following apolysis to the 5th instar. *B. mori* larvae
          were injected per os at 4th instar 2nd day with 10 µl
          30% aq/EtOH alone or with 10 µl 30% aq/EtOH + phyto-
          ecdysone. 25 larvae were used/treatment/parameter and
          analyzed.

Orally applied treatment	A_{470}/min/ µl molt fluid	µg Ascorbic acid/µl molt fluid	µg Ascorbic acid/µl hemolymph	µg Protein/ µl molt fluid
Control (solvent only)	0.031 ± 0.012	0.05	0.13	16.0
30 µg β-Ecdysone (in solvent)	0.008 ± 0.005	0.06	0.12	--
10 µg Ponasterone A (in solvent)	0.006 ± 0.003	--	--	15.0

duced premature molting in *B. mori* and resulted in 100% mortality. Much of this mortality, however, occurred not as a result of an inhibition of ecdysis, but actually after molting had been completed. In fact, while the lower concentrations of ß—ecdysone resulted in <100% mortality, the % of the total mortality occurring during the molting process (i.e., ecdysis inhibition) increases as the dose of ß—ecdysone is decreased (Table 4). Thus, death of *B. mori* by ingested ß—ecdysone cannot be entirely explained by an inhibition of ecdysis.

At 25 ppm ß—ecdysone, molting is delayed (Table 5) so that 88% of the exposed larvae die during the second instar (that is, before molting occurs). Doses > 6.25 ppm but < 25 ppm also result in a delay of molting to the third instar, such that 25 ppm seems to be the dietary concentration above which enhancement, and below which retardation, of molting takes place.

Total mortality and exposure time required to reach this mortality are dependent upon the concentration of the dietary ß—ecdysone. In addition, growth inhibition by dietary ß—ecdysone is also a concentration—dependent phenomenon.

Hypothetically, the larvae fed high concentrations of ecdysone (> 25 ppm) had a high titre of ecdysone built in the hemolymph, a titre which could not be metabolized or excreted rapidly enough to prevent hormonal imbalance resulting in molting promotion and death. Larvae fed lower concentrations (< 25 ppm) of ß—ecdysone grow more slowly than control and molt later than control. Possibly the ecdysone at these lower concentrations induced metabolism of both the exogenous and the endogenous ecdysone such that there was a delay in apolysis.

Hikino *et al.* (1975)[13] showed that the catabolic activity on ß—ecdysone of *B. mori* varied during the course of its growth and development. This is illustrated in Tables 2 and 6 in which it can be seen that the larvae are more sensitive to either ingested (Table 6) or injected (Table 2) ß-ecdysone during the earlier phase of the fourth instar.

An additional bioassay with a pest aphid species was conducted to test for a feeding deterrent effect of dietary phytoecdysones[14] (Fig. 8). One of several phytoecdysones was dissolved directly into an aqueous diet optimized for maximal aphid feeding. The control aqueous diet consisted of vitamins (C and B-complex), sucrose, amino acids, trace metals, salts, cholesterol, brought to pH 8.7 with K_3PO_4. The aqueous diet was placed into poly-ethylene vial caps and each of these caps were fitted into circular holes punched into plastic snap-on lids for polystyrene catsup cups (1 oz). Between 50-100 Biotype C greenbugs, *Schizaphis graminim*, an important pest on sorghum (and other economically important grains) in the midwestern U.S., were transferred from sorghum plants into each of the 1 oz catsup cups which were immediately fitted with the diet cap-containing snap-on lids. After 24 hrs at room temperature the no. of aphids feeding/total no. aphids was determined for each treatment.

Table 4. Effects of ß-ecdysone on the larval development
of second instar *Bombyx mori*[a]

Concen-	%			% of total mortality		
tration	Development to		Total	occurring as:		
in diet	3rd	4th	mortality	2nd	Ecdysis	3rd
(ppm)	instar	instar	%	instar	inhibition	instar
100	28	0	100	72	0	28
50	28	0	100	56	16	28
25	8	0	100	88	4	8
12.5	63	50	51	41	33	25
6.25	73	69	31	26	61	13
3.125	100	96	4	0	0	100
Control	100	96	4	0	0	100

a) 25 second instar first day larvae/treatment

Table 5. Effects of ß-ecdysone on the larval
developmental period of second instar *Bombyx mori*[a]

Concentration in diet (ppm)	Average Period (days) to				
	Molted 2nd Instar	Ecdysed 3rd Instar	Ecdysed 4th Instar		
100	3.0(7)[b,c]	4.1(7)[c]	–	↑	Molting
50	3.9(11)[c]	4.9(7)[c]	–		Promotion
25	8.0(2)[c]	9.5(2)[c]	–		Molting
12.5	6.4(19)[c]	6.9(15)[c]	14.3(10)		Delay
6.25	6.0(24)[c]	6.0(19)[c]	14.3(18)	↓	
3.125	5.0(25)	6.0(25)	13.7(24)		
Control	4.9(25)	5.9(25)	14.0(24)		

a) 25 second instar first day larvae/treatment.

b) Figures in parenthesis show the number of larvae
alive out of the original 25 second instar first
day larvae.

c) Significant difference from control at $P=0.001$
by Mann-Whitney U test.

Aqueous
Sample Solution

Akey Diet
Amino acids, Sucrose, B-vitamins
Vitamin C, Salts, Trace Metals
Cholesterol ajusted to pH 8.7

350 μl

 1.5 cm ID polyethylene vial cap

Parafilm

Catshup cup lid with inserted diet cap

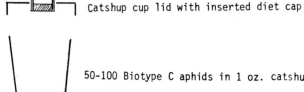

50-100 Biotype C aphids in 1 oz. catshup cup

Observation of Number Feeding after 24 h

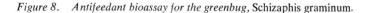

Figure 8.   Antifeedant bioassay for the greenbug, Schizaphis graminum.

Table 6. Effects of varying no. consecutive days feeding on
dietary ß-ecdysone by second instar *B. mori* larvae

Days of treatment with ß-ecdysone following molt to 2nd instar of *B. mori* larvae[a]	Concentration of dietary ß-ecdysone (ppm)				
	100	50	25	12.5	6.25
1st–5th days inclusive	0.20	0.22	0.30	0.48	0.84
1st day only	0.61	0.67	0.66	0.69	0.86
2nd day only	0.58	0.66	0.67	0.74	0.88
3rd day only	0.68	0.72	0.75	0.80	0.86
4th day only	0.83	0.78	0.86	0.93	0.90[b]

a) 25 second instar first day larvae/treatment

b) No. indicate weight ratio of treated larvae/control
    larvae on the fifth day of second instar.

☐   indicates treatment affecting inhibition of molting to
    the third instar.

☐   indicates treatment affecting promotion of molting to
    the third instar.

Appropriate controls, which consistently resulted in > 90% feeding, were then used to compare to each of the treatments in order
to determine ED_{50} values for each of the tested phytoecdysones.
ED_{50} is the effective dose for 50% feeding compared to control.

From Table 7 it can be seen that ajugasterone C is more than
10-fold more potent than ß-ecdysone, and more than 30-fold more
potent than cyasterone, as a feeding deterrent to *S. graminum*
when incorporated into the artifical diet of this behavioral bioassay.

Although ingested phytoecdysones do have a potent and unique
hormonal activity against susceptible species like the silkworm
and the pink bollworm, other insect species are unaffected by
dietary phytoecdysones. For example, *Heliothis* complex fed more
than 3000 ppm phytoecdysones in artificial diet show no obvious
morphological or developmental changes. However, other chemicals
besides phytoecdysones contained in 'resistant' plants like *A.
remota* and *P. gracilior* do have a variety of effects against
Heliothis and other insect species. *A. remota* extracts have
yielded 6 diterpenes causing antifeedant and insecticidal activity[1-3,15], while several insecticidal nagilactones coupled to
an antifeedant activity as well as two growth-inhibiting bisfla-
vones have been isolated from *P. gracilior* foliage[9].

Such multicomponent defensive strategies, as those elucidated
in *A. remota* and in *P. gracilior*, may be more the rule than the
exception in resistant plant cultivars. The elucidation of
these strategies, particularly the chemical aspects of them, is
important for an understanding of ecological and evolutionary
aspects of host plant resistance. In addition, the mechanistic
understanding of host plant resistance may have economical implications in that 'resistance' chemicals may be bred into crop
plants, or they may be extracted from one plant species and
applied directly to another economically important plant species,
or they may serve as leading structures in synthetic pesticide
research.

The role of phytoecdysones in this scheme is that of an
important component in a rather complex defensive strategy of
some plants. Their presence in plants probably serves a limited,
yet important, protective role. For example, cotton bolls bred
with several ppm of ponasterone A would very likely be resistant
to attack by pink bollworm.

In summary, then, whether acting alone or in conjunction with
other chemicals, the unique and potent physiological, biochemical, and morphological effects induced by the phytoecdysones confers an integral role for them in host plant resistance.

Table 7. Feeding deterrency of three phytoecdysones on greenbug, *Schizaphis graminum*[a]

Compounds	ED_{50} (ppm in diet)[b]
ß-ecdysone	650
Cyasterone	2000
Ajugasterone C	62

a) Biotype C of *S. graminum* from a mixed
population in a 24 h no-choice bioassay.

b) ED_{50} is the effective dose for 50% feeding
compared to control.

Acknowledgement

Insects were kindly supplied by the agencies of the USDA in Brownsville, Tx; Phoenix, Az; and Tifton, Ga. The authors thank J. DeBenedictis for his help with electron micrographs and D. Dreyer and K. Jones for their help with the aphid bioassay. Authentic samples of phytoecdysones were gifts from Professor T. Takemoto ;and Professor K. Nakanishi.

References

1. Kubo, I.; Lee, Y. W.; Balogh-Nair, V.; Nakanishi, K.; Chapya, A. *J. Chem. Soc. Chem. Commun.* 1976, 949.
2. Kubo, I.; Kido, M.; Fukuyama, Y. *J. Chem. Soc. Chem. Commun.*, 1980, 897.
3. Kubo, I.; Klocke, J. A.; Miura, I.; Fukuyama, Y. *J. Chem. Soc. Chem. Commun.*, 1982, 618.
4. Kubo, I. *Science Year 1982*, World Book-Childcraft International, Chicago, 1981, 126.
5. Chan, B. G.; Waiss, A. C. Jr.; Stanley, W. L.; Goodban, A. E. *J. Econ. Entomol.*, 1978, *71*, 366.
6. Kubo, I.; Klocke, J. A.; Asano, S. *Agric. Biol. Chem.*, 1981, *45*, 1925.
7. Kubo, I.; Klocke, J. A.; Asano, S., in preparation.
8. Kubo, I.; Ganjian, I.; Klocke, J. A., in preparation.
9. Kubo, I.; Klocke, J. A., in preparation.
10. Hostettmann, K. *Planeta Medica*, 1980, *39*, 1.
11. Navon, A. *J. Insect Physiol.*, 1978, *24*, 39.
12. Chapman, R. F. "The Insects, Structure and Function, American Elsevier Publishing Company, New York, 1971, 700.
13. Hikino, H.; Ohizumi, Y.; Takemoto, T. *J. Insect Physiol.*, 1975, *21*, 1953.
14. Dreyer, D. L.; Reese, J. C.; Jones, K. C. *J. Chem. Ecol.*, 1981, *21*, 273.
15. Kubo, I.; Nakanishi, K. *Host Plant Resistance to Pests*, ACS Symposium Series 62, American Chemical Society, Washington, D.C., 1977, 165.

RECEIVED September 24, 1982

Multiple Factors in Cotton Contributing to Resistance to the Tobacco Budworm, *Heliothis virescens* F.

P. A. HEDIN, J. N. JENKINS, D. H. COLLUM, W. H. WHITE, and
W. L. PARROTT

U.S. Department of Agriculture, Science and Education Administration,
Agricultural Research Southern Region, Boll Weevil Research Laboratory,
Chemistry Research Unit, Mississippi State, MI 39762

Cyanidin-3-β-glucoside has been shown to be an
important factor of resistance in cotton Gossypium
hirsutum L. leaves to the feeding of tobacco buworm
Heliothis virescens (Fab.) in the field. The
reported effectiveness of gossypol was confirmed, but
the condensed tannins (proanthocyanidins) in terminal
leaves were not correlated with resistance. Para-
doxically, these 3 compounds when incorporated in
laboratory diets are equally toxic to larvae. These
findings provide a potential basis for achieving
insect resistance in non-glanded cotton and other
crops infested by Heliothis.

Since the original development of glandless cotton by
McMichael (1), entomologists and plant breeders have noted that
the experimental glandless lines are generally susceptible to
certain phytophagous insects. Bottger et al. (2), reporting on
the relationship between the gossypol content of cotton plants
and insect resistance, noted that several insects fed on a
glandless line in preference to glanded lines. Jenkins et al.
(3) reported increased susceptibility of several glandless lines
to the bollworm (Heliothis zea Boddie). Lukefahr et al. (4)
showed that the growth of bollworm and tobacco budworm (Heliothis
virescens F.) larvae increased on diets of glandless cotton
lines compared with that on diets of the corresponding glanded
line. Lukefahr and Martin (5) incorporated 3 cotton pigments,
gossypol, quercetin, and rutin, into a standard bollworm diet
whereupon larval growth was decreased. They suggested that plant
breeders might select for cotton plants with higher pigment
(gossypol and quercetin) content as a mechanism of resistance.
It was observed that flower buds from certain wild and primitive cottons showed more insecticidal activity than could be
accounted for by gossypol, and the additional activity was
ascribed to "X" factors (6, 7). The "X" factors were identified

as sesquiterpenoid quinones, hemigossypols, and heliocides in a
series of investigations by this College Station, Texas group
(8-11). In summary, they isolated at least 15 terpenoid aldehydes and related compounds from glanded cotton plants, present in
varying amounts from different lines. They reported that toxicity (ED_{50}; concentration to reduce larval growth by 50%)
approached or equalled gossypol (0.05%) for a number of these
compounds when incorporated in diets fed to tobacco budworms.
Gossypol was generally found in considerably higher concentations
than the other terpenoid aldehydes in the lines investigated.
This group (12) also chromatographically identified catechin,
gallocatechin, quercetin, and condensed tannin from Verticillium
wilt-resistant young cotton leaves in higher concentrations than
from susceptible larger leaves.

Subsequently, a series of investigations by a group at the
Western Regional Research Center in California and associates
have shown condensed cotton tannin to be an antibiotic chemical
for the bollworm, tobacco budworm, and pink bollworm (13, 14,
15). Initially, they sought to investigate the "X" factor that
had been described by its extraction from freeze dehydrated
tissue with ethyl ether or acetone. They found that if freeze
dehydrated cotton powder was first extracted with hexane or a
similar non-polar solvent system, gossypol and the other terpenoid aldehydes were isolated in the extract. This extract when
incorporated in insect diets decreased larval growth as previously reported. The residual powder was then extracted with
methanol; this extract when incorporated in the larval diet
depressed growth severely. In fact, in a study where control
tobacco budworm larvae weighed 306 mg after 14 days, larvae fed
the hexane extract (from 6 g of powder added to 30 g of diet)
weighed 56.6 mg, those fed the acetone extract weighed 343 mg,
and those fed the comparable amount of the ethanol extract
from the methanolic extract weighed 1.1 g. The major antibiotic
compound in the methanol soluble fraction was subsequently
characterized as condensed tannin, which when hydrolyzed with HCl
in n-butanol yielded cyanidin and delphinidin. By osmotic
measurements, the average molecular weight was estimated to be
4850. In related work, Chan et al. (14) isolated the cycloprope-
noid fatty acids, the terpene aldehydes including gossypol, the
flavonoids, and the condensed tannins. When fed to tobacco budworm [also bollworm and pink bollworm (Pectinophora gossypiella
Saunders)] hatchling larvae in diets, the ED_{50} values (percent of
diet) were as follows: gossypol, 0.12; hemigossypolone, 0.08;
heliocide H_1, 0.12; heliocide H_2, 0.13; catechin, 0.13; quercetin, 0.05; condensed tannin, 0.15; methyl sterculate, 0.41; and
methyl malvalate, 0.49. Stipanovic (16) obtained similar ED_{50}
values for the tobacco budworm with gossypol, hemigossypolone,
and heliocides H_1 and H_2.

In another test, Chan et al. (14) analyzed the condensed tannin and gossypol (with analogs) content of 10 cotton plant parts.

In brief, the tannin content of various leaf tissues was fairly high while the gossypol content was relatively low. In the anthers, corolla, and calyx, the reverse situation existed. When the plant parts were fed to budworms, they had comparatively little toxicity (ED_{50} = 1.7%) despite the relatively high gossypol content. Except for the high toxicity of corolla tissue (0.15%) which was attributed to flavonoids, the various leaf tissues, (early feeding sites in the field) were otherwise the most toxic (0.27-0.50%).

Shaver and Parrott (17) reared bollworm and tobacco budworm larvae on a standard larval diet, and transferred them at 5 ages onto media containing 0-0.4% gossypol. The influence on development increased with larval age at the time of transfer. Recently Waiss and his co-workers (18) incorporated condensed tannin in diets fed to the tobacco budworm where the larvae initially were of different ages. The ED_{50} values were as follows: 1 day, 0.10; 3 day, 0.10-0.15; 5 day, 0.20-0.30; and 7 day, non-toxic. Thus, it is evident that larvae also become more tolerant to condensed tannin with age.

Shaver et al. (19) studied feeding and larval growth in the laboratory of the tobacco budworm on component parts of cotton flower buds. They found that most of the feeding of 2-4 day larvae occurred on the anthers after they penetrated into the interior of the bud through the petals. For perspective, it should be noted that females oviposit on the young terminals and young leaves, so that the hatchling larvae feed on these sites before migrating to the bud. When component parts of the flower bud were incorporated into larval diet, only the petals (corolla) inhibited growth of 3 day larvae. This inhibition occurred both on glanded and glandless lines, so it can not be attributed to the gossypol content alone. Later, Shaver et al. (20) fed 3-day-old tobacco budworm larvae laboratory diets into which ethyl ether extracts of flower buds were incorporated. They reported that lines with high gossypol and gossypol related compounds, as determined by the aniline test, reduced larval weight gains. Bud tannins were not analyzed, nor were they extracted and fed in this study.

Schuster et al. (21) identified cotton plant resistance to the two-spotted spider mite (*Tetranycus urtica* Koch) by mass screening seedlings. Later Schuster and Lane (22) were able to show that high tannin lines, particularly TX-1055, showed resistance to this arachnid and the bollworm.

Another trait of cotton affecting larval growth of tobacco budworm is pollen color which ranges from cream to yellow to even orange. Hanny et al. (23) fed 1st instar tobacco budworm larvae whole fresh anthers of 5 cream and yellow lines. In a number of tests conducted during 1977 and 1978, weight gains after 7 days were 13-15% less on the yellow pollen. Hanny (24) reported the following average analyses: gossypol, yellow 0.88%; cream 0.70%; condensed tannins, yellow 4.79%; cream 5.34%; and flavonoids,

yellow 0.56%, cream 0.54%. The isolation of 4 gossypetin glycosides (but not gossypin), 8 quercetin glycosides, and an anthocyanin were reported.

In preliminary work at this location (25), gossypol and tannin were negatively correlated with weight gains of tobacco budworm larvae fed in the field on plant terminals. This was based
on seasonal averages for a five cultivar test. In laboratory
tests, the ED_{50} values for several flavonoids and condensed cotton tannins were determined and found to be similar (0.05-0.15%)
to the values reported by Chan et al. (14) and Stipanovic (16).
The ED_{50} values for a number of cotton cultivars, hibiscus,
sorghum, and sinfoin, were found to range from 0.03-0.10%, indicative that the tannin could be biologically similar. Larval
feeding tests were also performed on a number of chromatographic
fractions obtained from solvent extracts of cotton terminal
tissue. Those containing gossypol, tannins, and flavonoids were
most toxic.

From the preceding, it appears that chemical resistance in
cotton to Heliothis insects is due to multiple factors.
Different lines, each with insect resistance, may possess different ratios of antibiotic compounds. Thus, it may be possible
to increase resistance by crossing lines where each contributes
genes for biosynthesis of different antibiotic compounds. The
tobacco budworm was selected for study in preference to the cotton bollworm because it is easier to rear and use in the laboratory, is more resistant to insecticides in the field, and it is
approximately as susceptible to cotton constituents incorporated
in laboratory diets (14). This present study was carried out to
identify and analyze for cotton constituents that were toxic in
laboratory feeding tests, and to determine whether there were
positive correlations of their content in leaves and/or other
tissue with field resistance. From this information, the generation of lines with multiple factors for resistance could be initiated.

Materials and Methods[1]

Agronomic and entomological practices. Plants of diverse
cotton lines were grown in field plots during the years
1978-1982 on the Plant Science Farm at Mississippi State
University. Plants were treated for boll weevils with Guthion,
and normal fertilizer, herbicide, and other cultural practices
were applied. The field design was normally a randomized
complete block with 4 replications. Plant material (terminals,

[1] Mention of a trademark, proprietary product or vendor does
not constitute a guarantee or warranty of the product by the U.S.
Department of Agriculture and does not imply its approval to the
exclusion of other products or vendors that may also be suitable.

larger leaves, squares, bracts, flowers, and bolls) was collected at intervals throughout the season, freeze dehydrated, ground, and stored at -10°C until used.

First instar tobacco budworm larvae were restrained in 15 cm long dialysis casings that were slipped over the terminals, and collected and weighed after five days. Adequate numbers and replications were employed for statistical evaluation.

Laboratory bioassays were performed by placing 1st instar tobacco budworm larvae on a commercial medium (742-A) prepared by Bioserv Inc., Frenchtown, NJ. Organic solvent-soluble test compounds were diluted in either hexane or ethanol, added to casein and evaporated to dryness in a rotary evaporator. Water soluble compounds were dissolved in water and added in place of the prescribed water content. The casein was then incorporated with the remainder of the ingredients and poured into 10 cm petri plates to gel. Diet cylinders of 10 mm diameter, 5 mm height were cut with a cork borer and transferred into 15 x 45 (1 dram) shell vials. A neonate larva was added and the vials were incubated at 26°C, 60%RH, 12L:12D for 5 days at which time the larvae were weighed. A randomized complete block design with 8 replicates of 5 larvae each was used. Compounds were tested at 5-8 concentrations ranging from 0.006 to 0.6% of the diet on a dry weight basis.

Chemical analyses and fractionations. Freeze-dehydrated cotton tissue was analyzed for condensed tannin (heated n-BuOH-HCl), gossypol (phloroglucinol-HCl), and anthocyanins-anthocyanidins (alc. HCl at 540 nm). Other analyses performed but not reported here were for catechin, total phenols, E_{11} (tannin), and aniline reactive terpenes (gossypol).

Freeze dehydrated plant tissue was extracted by Soxhlet first with cyclohexane/ethyl acetate/acetic acid:500/500/1 (CHEA) and then with acetone/water:7/3. The CHEA extract was chromatographed on a 40 cm silicic acid column with hexane and solvents of increasing polarity to yield gossypol and several other components, each of which was formulated for laboratory bioassay testing. The aqueous acetone fraction was chromatographed on a 1 m Sephadex LH-20 column with 70% aqueous methanol, methanol and dimethyl formamide/methanol:10/90 and 25/75 to yield the flavonoids, anthocyanin, and condensed tannin. Similarly, each was formulated for laboratory bioassay testing. The isolated components were rechromatographed on LH-20, polyamide, and cellulose columns as required to achieve purity for biological evaluation and identification work.

Identification of cotton plant compounds. The identity of gossypol was confirmed by comparison of the spectral (NMR, MS) and chromatographic properties with an authentic sample. The identity of the cotton leaf anthocyanin was confirmed by comparison of the chromatographic and spectral properties of the chrysanthemin isolated from cotton flowers, by ^{1}H NMR, and by procedures that we described previously (26). The condensed tannins (polymeric proanthocyanidins) were characterized with regard to their stereochemistry, structural units, and molecular weight by the procedures of Czochanska et al. (27). The condensed tannins were found to consist of a mixture of related polymers, the molecular weight ranging from 1500–6000, the prodelphinidin: procyanidin ratio from 1.8–3.7, and the stereochemistry of the monomer units primarily cis (81–95%). Figure 1 is a 13C MNR spectrum of a highly purified cultivar BJA-592 condensed tannin. The spectra were obtained with an acquisition time of 0.2 sec, a pulse width of 13 μ sec, and a 45° flip angle. The average molecular weight was deduced from the ratio of the signals at 72/67 σ to be 4221.

Histological Examinations. For the histological-histochemical work, fresh samples were frozen at -20C in a cryostat (Int. Equip. Co., Model CTI) , mounted on specimen holders, trimmed, and sliced with wedge shaped knives at -20C to produce 20-30 μm tissue slices. For tannins, small pieces of leaf (or other tissue) were fixed for 78h in $FeSO_4$-5% formol saline, sliced at 30 μm and picked up on warm (room temp.) slides. Ferric chloride (1%) or 10% KOH were frequently used as stain intensifiers. Tannin granules (cells) stained blackish in unmodified fix, bluish with $FeCl_3$ modified fix, and brown to red-brown in KOH modified fix. In unfixed tissues, gossypol stained red with phloroglucinol-HCl, and lignified elements stained light violet. To visualize anthocyanins and gossypol, 30 μm fresh frozen sections were fixed over formaldehyde vapors for 24h, and treated with 5% HCl or NaOH to produce red or green anthocyanin reactions respectively. Gossypol remained yellow with acid or base treatment, in contrast to the anthocyanins.

Flower petals were examined under 10-70X magnification and photographed in situ after treatment with 5% KOH which stained anthocyanins green. Phloroglucinol-HCl treated gossypol glands were stained red, and flavonoids yellow.

Results and Discussion

Toxicity of cotton plant compounds in diets to TBW larvae. Table I gives ED_{50} values for a number of cotton constituents tested as inhibitors of tobacco budworm larval growth. For comparison, values obtained independently by Chan et al. (12) and Stipanovic (unpublished data) using essentially identical procedues are included. The ED_{50} values were quite reproducible by each laboratory. Further separation of the tannin by Sephadex LH-20 chromatography gave fractions with average molecular weights ranging from 1500 to 6000. When bioassayed, the ED_{50} values varied no more than from 0.05 to 0.10%, so it is deduced that within limits, the size of the molecule does not

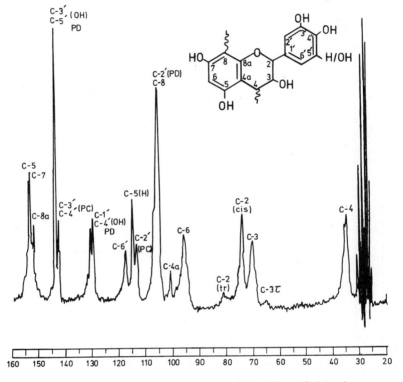

Figure 1. ^{13}C-NMR spectrum of cultivar BJA-592 purified tannin.

TABLE I

Inhibition of Tobacco Budworm Larval Growth by Cotton Constituents, ED_{50} as Percent of Diet

Constituent	Chan et al.	Stipanovic	Miss. State	Regression equation
Gossypol	0.12	0.05	0.113	$y=104.66-504.30x+596.12$ $r=0.68$ P>F=0.01
Hemigossypolone	0.03	0.29	—	
Heliocide H_1	0.12	0.10	—	
Methyl sterculate	0.41	—	N.T.	
(+)-Catechin	0.13	—	0.052	$y=7.06x^{-.562}$ $r^2=0.90$ P>F=0.001
Condensed tannin	0.15	—	0.063	$y=2.07x$ $r^2=0.87$ P>F=0.00
Quercetin	0.05	—	0.042	$y=3.29x^{-.705}$ $r^2=0.90$ P>F=.001
Isoquercitrin	0.10	—	0.060	$y=4.49x^{-.888}$ $r^2=0.85$ P>F=.001
Cyanidin	—	—	0.166	$y=105.66-332.89x$ $r^2=0.71$ P>F=.001
Delphinidin	—	—	0.138	$y=124.43-540.36x$ $r^2=0.89$ P>F=.001
Chrysanthemin	—	—	0.070	$y=7.91x^{-.707}$ $r^2=0.81$ P>F=.001

appreciably affect toxicity. In recent years as the chemical
bases for resistance were explored, there was the expectation
that some individual compound or group of related compounds would
be found that could account for the resistance of cotton to this
insect. This has been shown not to be the case based on dietary
incorporation bioassays, because the toxicities are essentially
equivalent at the ED_{50} values. However, the growth response
becomes curvilinear at higher levels. While it is possible to
reduce growth to less than 10% of the control with gossypol and
chrysanthemin, it was impossible to reduce growth to less than 20%
with some others.

Effect of gossypol, tannin, and chrysanthemin in terminals on
TBW larval growth. Figures 2, 3 and 4 give percent concentrations of gossypol, chrysanthemin, and tannins in terminal
leaves, and also give tobacco budworm larval weights for the 20
cotton lines, 15 glanded and 5 non-glanded. The weights are for
larvae feeding on intact plants in the field. The percent contents of gossypol and chrysanthemin were negatively correlated
with larval weights ($r = -.38$ and $-.40$) while the tannins were
weakly positively correlated ($r = +.16$); in fact, glandless lines
that produced large larvae were as high in tannin as most of the
lines that produced small larvae. Thus, these studies corroborate the work of Hanny et al. (23,24) and suggest that the
absolute concentration of the tannins in the total tissue does
not explain the expected toxic effects of the insect feeding on
intact tissue. Female tobacco budworm moths oviposit mainly on
the terminal leaves, although in mid- and late-season they oviposit significant numbers on the square bracts. Consequently, the
initial feeding site is most often the terminal leaves. Our studies (25) have shown that the larvae migrate down the plant,
feeding on bracts, meristematic (older) tissue buds, flower
petals, and bolls in the process where the content of anthocyanins, gossypol, and tannins is mostly similar.

Table II is a summary of the averages of tannins, gossypol,
and chrysanthemin in 18 cultivars harvested as 3 replicates in
August 1981. The values (somewhat lower than in 1980) show that
the 3 constituents are present in comparable quantities in all
tissues analyzed except lower in medium bolls. It is to the
medium bolls that the larvae eventually migrate, perhaps to avoid
high levels of allelochemics. As the season progresses, tannin
increases in all tissues sharply while gossypol and chrysanthemin
gradually decrease (Figure 5). During the same period, the larval weight gains remained similar, however.

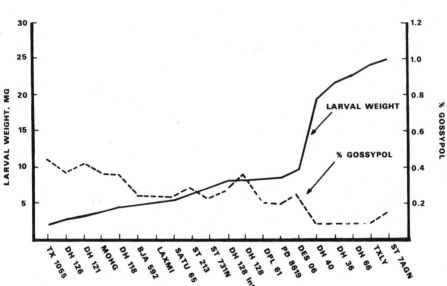

Figure 2.   Percent gossypol in cotton terminal leaves and ranked tobacco budworm larval weights for 20 strains (r = 0.3771).

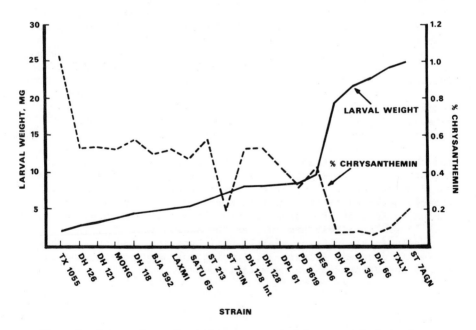

Figure 3.   Percent chrysanthemin in cotton terminal leaves and ranked tobacco budworm larval weights for 20 strains (r = 0.3958).

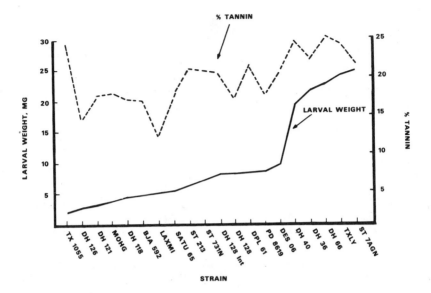

Figure 4. Percent tannins in cotton terminal leaves and ranked tobacco budworm larval weights for 20 strains (r = 0.1613).

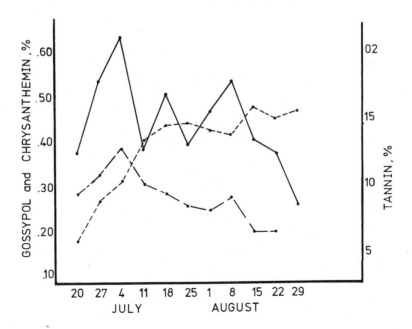

Figure 5. Seasonal trends in chrysanthemin (——), gossypol (— — —), and tannin (– – –) content during 1981.

TABLE II
Content of Tannins, Gossypol, and Chrysanthemin in Cotton
Plant Tissues; Seasonal Averages for 3 August 1981 Replicates,
18 Cultivars

Tissue	Tannins, %	Gossypol, %	Chrysanthemin,%
Terminals	6.02	0.21	0.14
Leaves	8.10	0.23	0.18
Squares	7.92	0.50[a]	0.10
Square bracts	6.02	0.21	0.14
Small bolls	11.71	0.29	0.11
Medium bolls	9.36	0.04	0.05
Medium boll carpels	17.07	0.18	0.10

[a] Somewhat high; average of 0.28 in 1980, 0.23 for 16 other
lines in 1981, but 0.47 by Elliger et al. (15).

Effect of gossypol, tannin, and chrysanthemin in flower petals
on TBW larval growth and survival. Our recent observations
demonstrated that in the laboratory, larvae fed more successfully
on white, first day petals (intact) than on red, second day
petals, and that they fed more successfully on non-glanded petals
than on glanded petals. Table III presents tannin, gossypol, and
chrysanthemin concentrations of white and red petals of 2 glanded
and 2 non-glanded (glandless) lines. Values for tobacco budworm 5
day larval weights and percent survival are also included. The
tannin contents of the non-glanded white petals were higher than
those of the glanded strains, but there was little difference in
the red petal tannin contents. The gossypol contents of the
petals of the glanded lines were much higher than those of the
non-glanded lines as could be expected. The chrysanthemin contents of all red petals were also much higher than those of white
petals as was expected. The weights and survival of larvae
feeding on red petals were considerably less than those feeding on
white petals. Both percents tannin and chyrsanthemin were higher
in red petals than in white petals (Table III). Larvae weights
and survival were reduced on red petals when compared with white
petals. Based collectively on data in Table III and Figures 3 and
4, we conclude that chrysanthemin is more important than tannin
for the reduced larval size and survival on glandless
(gossypol-low) red petals. Gossypol and chrysanthemin contribute
to the toxicity of glanded red petals, and gossypol to that of
glanded white petals. We now have preliminary data that larvae
fed leaves and bracts of red cottons gained 20% less (statistically significant) than those fed leaves and bracts from com-

TABLE III

Relative Effects of Cotton Flower Petal Constituents on Tobacco Budworm Growth and Survival[a]

Cultivar	Petal color	Tannins, %	Gossypol, %	Chrysanthemin, %	Larval Wt, mg[d]	Larval survival, %
ST-7AGN (NG)[b]	W[c]	5.79	0.10	0.07	4.72 a	39.5
	R	8.68	0.11	0.67	0.46	7.0
DH 66 (NG)	W	5.40	0.17	0.07	4.18 a	31.0
	R	8.55	0.13	0.73	0.56	16.5
ST-213 (G)	W	3.49	0.52	0.13	1.52 b	28.0
	R	8.75	0.79	0.59	0.41	8.5
DH 126 (G)	W	3.16	1.72	0.18	--[e]	--
	R	6.25	2.46	0.65	--	--

a/ % of dry weight.
b/ NG = Nonglanded, G = glanded.
c/ W = white; first day flower color, R = red; second day flower color. Means of larvae fed on white petals not significantly different at .05 level if followed by the same letter.
d/ Average tobacco budworm weight after feeding 5 days on petals.
e/ Not fed.

parable green strains. Thus, red coloration now appears to be a factor of considerable importance in insect feeding of both petals and leaves. There is still a residual mortality of insects feeding on white glandless petals (Table III). This can be attributed at least in part to the flavonoids, some of which we have previously identified (28) and demonstrated to be toxic to this insect (Table I).

Histochemical studies on localization of tannins, gossypol, and anthocyannins in plant tissues. We were able to observe by magnification that tobacco budworm larvae avoided gossypol glands during feeding (Figure 6). Waiss et al. (29) had made similar observations with H. zea. To determine where the tannins, gossypol, and anthocyanin are localized in the plant, some histological studies were carried out. Figure 7 shows magnifications of tissue slices of a glanded line, DH-126, fixed in $FeSo_4$-5% formal saline and stained with phloroglucinol-HCl to visualize tannins. The midrib is prominent in the center of Figure 7, and the tannins appear to be concentrated near the surface in granular form. In glandless lines, tannin is more diffuse. Figure 8 is a 5 μm paraplast section through a cotton leaf at the gossypol gland site which shows the outer anthocyanin-containing envelope (halo) surrounding the gossypol gland. In fresh tissue sections, the outer halo stains bright red in acid. The halo when subsequently neutralized with KOH is converted to green, verifying that it is anthocyanin. It is tempting to speculate from this that the biosyntheses of leaf gossypol and anthocyanin are pleiotropically related because the content of each is much higher in glanded than in non-glanded leaves (Figures 2 and 3). However, the anthocyanin content can be high and gossypol low as in flower petals (Table III). Gossypol and related aldehydes are biosynthesized via the acetogenic pathway and anthocyanins and other flavonoids by the mixed acetate/shikimic pathways. Thus each should be able to increase independently in the plant as we have demonstrated in glanded and glandless red petals. This halo effect may also alter the interpretation of Figure 6 that shows the apparent avoidance of gossypol glands by tobacco budworm larvae. The presumed feeding deterrence of gossypol may be in fact caused by the anthocyanin halo. Although both the anthocyanin and gossypol are toxic when ingested, an antifeedant mechanism may be expressed in this instance.

Figure 9 shows a Pima (G. barbadense L.) flower petal treated with HCl to visualize red anthocyanin "granules". Treatment with alkali converted the granule color to green. The halo surrounding the gossypol gland inside the outer carpel wall of DH-126 buds (Figure 10) is also anthocyanin. The anthocyanin (chrysanthemin) in very young Stonville-213 bolls on the day of anthesis was found by dissecting, extraction, and TLC analysis to be located primarily in the outer carpel wall.

Figure 6. Avoidance of gossypol glands by tobacco budworm while feeding.

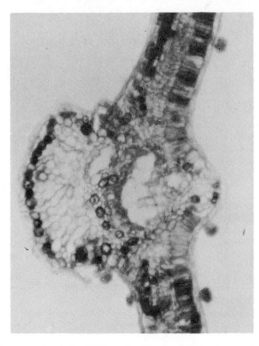

Figure 7. Cross-section of midrib and adjacent blade tissue of DH-126 leaves showing tannin cells.

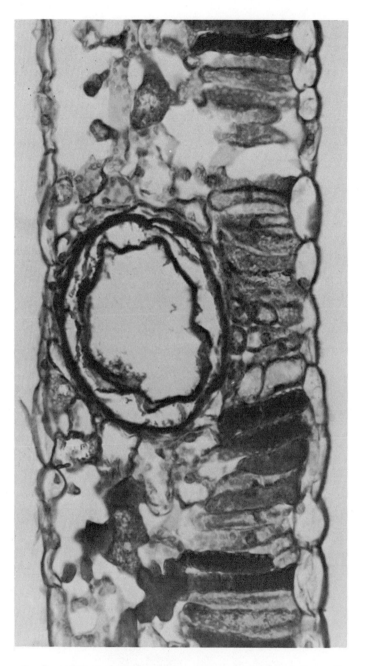

Figure 8. Paraplast section (5 m) of a gossypol gland and the anthocyanin envelope.

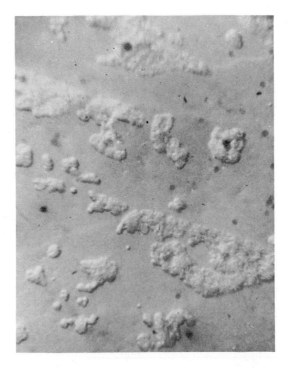

Figure 9. Acid treatment of Pima flower petal to visualize anthocyanin.

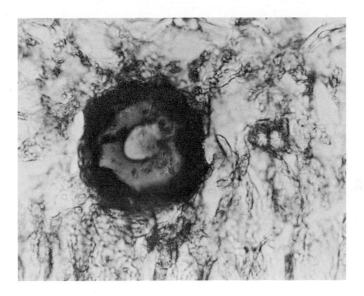

Figure 10. Acid visualization of anthocyanin halo in Stoneville 213 bud.

In summary, the contents of chrysanthemin and gossypol were shown to be negatively correlated with tobacco budworm larval growth in the field while the tannins were slightly positively correlated. The recognition of chrysanthemin as a resistance factor provides a basis for developing resistant cotton cultivars that are low or devoid of gossypol, a long sought objective. It also suggests that anthocyanins in general may be a basis for selecting for resistance in other crops to Heliothis.

Literature Cited

1. McMichael, S. C. Agr. J. 1959. 51, 30.
2. Bottger, G. T., Sheehan, E. T. Lukefahr, M. J. J. Econ. Entomol. 1964, 57, 283.
3. Jenkins, J. N., Maxwell, F. G., Lafever, H. N. J. Econ. Entomol. 1966. 59, 352.
4. Lukefahr, M. J., Nobel, L. W., Houghtaling, J. E. J. Econ. Entomol. 1966, 59, 817.
5. Lukefahr, M. J., Martin, D. F. J. Econ. Entomol. 1966, 59, 176.
6. Lukefahr, M. J., Shaver, T. N., Cruhm, D. E., Houghtaling, J. E. Proc. Beltwide Cotton Prod. Res. Conf., 1974, Memphis, TN, p. 93.
7. Bell, A. A., Stipanovic, R. D., Howell, C. R., Mace, M. E. Proc. Beltwide Cotton Prod. Res. Conf. 1974, Jan. 7-9, Dallas, TX, 40.
8. Gray, J. R., Mabry, R. J., Bell, A. A., Stipanovic, R. D., Lukefahr, M. J. J. C. S. Chem. Comm. 1976. 109.
9. Bell, A. A., Stipanovic, R. D. Howell, C. R., Fryxell, P. A. Phytochemistry. 1975, 14
10. Bell, A. A., Stipanovic, R. D. Proc. Beltwide Cotton Production Res. Conf. 1977, Jan. 10-12, Atlanta, GA, 244.
11. Stipanovic, R. D., Bell, A. A., O'Brien, D. H., Lukefahr, M. J. Tetrahedron Letters. 1977. 6, 567.
12. Howell, C. R., Bell, A. A., Stipanovic, R. D. Physiol. Plant Pathol. 1976. 8, 181.
13. Chan, B. G., Waiss, A. C., Lukefahr, M. J. J. Insect Physiol. 1978. 24, 113.
14. Chan, B. C., Waiss, A. C., Binder, R. G., Elliger, C. A. Ent. Exp. & Appl. 1978. 24, 94.
15. Elliger, C. A., Chan, B. G., Waiss, A. C. J. Econ. Entomol. 1978. 71, 161.
16. Stipanovic, R. D.; personal communication.
17. Shaver, T. W., Parrott, W. L. J. Econ. Entomol. 1970. 63, 1802.
18. Waiss, A. C.; personal communication.
19. Shaver, T. N., Garcia, J. A., Dilday, R. H. Environ. Entomol. 1977. 6, 82.
20. Shaver, T. N., Dilday, R. H., Wilson, F. D. Crop Sci. 1980. 20, 545.

21. Schuster, M. F., Maxwell, F. G., Jenkins, J. N., Parrott, W. L. J. Econ. Entomol. 1972. 65, 1104.
22. Schuster, M. F., Lane, H. C. Proc. Beltwide Cotton Prod. Res. Conf. 1980. Jan. 6-10, St. Louis, MO, p. 83.
23. Hanny, B. W., Bailey, J. C., Meredith, W. R. Environ. Entomol. 1979. 8, 706.
24. Hanny, B. W. J. Agr. Food Chem. 1980. 28, 504.
25. Hedin, P. A., Collum, D. H., White, W. H., Parrott, W. L., Lane, H. C., Jenkins, J. N. Proc. Int. Conf. on Regulation of Insect Development and Behavior, Wroclaw Tech. Univ. Press, Wroclaw, Poland. 1980, 1071.
26. Hedin, P. A., Minyard, J. P., Thompson, A. C., Struck, R. E., Frye, J. Phytochemistry. 1967. 6, 1165.
27. Czochanska, Z., Foo, L. P., Newman, R. H., Porter, L. J. J.C.S. Perkin I. 1980, 2278.
28. Hedin, P. A., Miles, L. R., Thompson, A. C., Minyard, J. P. J. Agr. Food Chem. 1968. 16, 505.
29. Waiss, A. C., Chan, B. G., Elliger, C. A., and Binder, R. G. Proc. Beltwide Cotton Proc. Res. Conf. 1981. Jan. 4-8. New Orleans, La. p. 61.

RECEIVED September 28, 1982

INDEX

Production by Robin Giroux
Indexing by Deborah Corson

Elements typeset by Service Composition Co., Baltimore, MD
Printed and bound by The Maple Press Co., York, PA